Mammalian Thermogenesis

Mammalian Thermogenesis

EDITED BY

Lucien Girardier

Professor of Physiology
Faculty of Medicine
University of Geneva
Switzerland

and

Michael J. Stock

Reader in Physiology
St. George's Hospital Medical School
University of London
UK

London   New York

CHAPMAN AND HALL

First published 1983 by
Chapman and Hall Ltd
11 New Fetter Lane, London EC4P 4EE
Published in the USA by
Chapman and Hall
733 Third Avenue, New York NY10017
© *1983 Chapman and Hall Ltd*

Printed in Great Britain
at the University Press, Cambridge

ISBN 0 412 23550 1

British Library Cataloguing in Publication Data

Mammalian thermogenesis.
  1. Body temperature—Regulation
  I. Girardier, Lucien   II. Stock, Michael J.
  591.19'12   QP135

  ISBN 0–412–23550–1

Library of Congress Cataloging in Publication Data

Main entry under title:

Mammalian Thermogenesis.

  Bibliography: p.
  Includes index.
  1. Animal heat.   2. Energy metabolism.
3. Brown adipose tissue.   4. Mammals—Physiology.
I. Girardier, Lucien.   II. Stock, Michael, J.
[DNLM: 1. Adaptation, Physiological.   2. Body
temperature regulation. QT 165 M265]
QP135.M35 1983   599'.01912   83–1929
ISBN 0-412-23550-1

Contents

List of Contributors

L. Howard Aulick US Army Institute of Surgical Research, Brooke Army Medical Center, Fort Sam Houston, Texas 78234, USA.

Steven M. Eiger Department of Physiology, The University of Michigan Medical School, Ann Arbor, Michigan 48109, USA.

Lucien Girardier Department of Physiology, Centre Medical Universitaire, University of Geneva, Switzerland.

Jean Hims-Hagen Department of Biochemistry, University of Ottawa, 451 Smyth Road, Ottawa, Ontario K1H 8M5, Canada.

W.P.T. James Dunn Nutrition Laboratory, University of Cambridge and Medical Research Council, Milton Road, Cambridge CB4 1XJ, UK.

Matthew J. Kluger Department of Physiology, The University of Michigan Medical School, Ann Arbor, Michigan 48109, USA.

Lewis Landsberg Beth Israel Hospital, Harvard Medical School, Boston, Massachusetts 02215, USA.

Rebecca Locke Neurochemistry Laboratory, Department of Psychiatry, Ninewells Medical School, University of Dundee, Dundee DD1 9SY, UK.

David Nicholls Neurochemistry Laboratory, Department of Psychiatry, Ninewells Medical School, University of Dundee, Dundee DD1 9SY, UK.

Nancy J. Rothwell Department of Physiology, St. George's Hospital Medical School, University of London, London SW17, UK.

Michael J. Stock Department of Physiology, St. George's Hospital Medical School, University of London, London SW17, UK.

Donald Stribling, ICI Pharmaceuticals Division, Alderley Park, Macclesfield, Cheshire SK10 4TH, UK.

P. Trayhurn Dunn Nutritional Laboratory, University of Cambridge and Medical Research Council, Milton Road, Cambridge CB4 1XJ, UK.

A.J.F. Webster Department of Animal Husbandry, University of Bristol, Langford, Bristol BS1 7DU, UK.

Douglas W. Wilmore US Army Institute of Surgical Research, Brooke Army Medical Center, Fort Sam Houston, Texas 78234, USA.

James B. Young Department of Medicine, Beth Israel Hospital, Harvard Medical School, Boston, Massachusetts 02215, USA.

Chapter One

Mammalian Thermogenesis: An Introduction

Lucien Girardier and Michael J. Stock

The task of editing a book should never be taken lightly because, as most editors soon realise, there is much more to it than simply acting as a postbox for the contributing authors. Initially, our problem was to determine whether there was a requirement for a book devoted to current knowledge of the effectors of mammalian thermogenesis. It soon appeared that, even by limiting the field to non-shivering thermogenesis, i.e. to mechanisms which liberate chemical energy by processes not involving muscular contraction, there was ample room for a new monograph. Firstly, the most recent comprehensive treatment was the book on *Brown Adipose Tissue* (ed. O. Lindberg, Elsevier, New York) which was published in 1970 and, although the proceedings of conferences (e.g. Effectors of Thermogenesis, eds L. Girardier and J. Seydoux, Birkhäuser, Basel, 1978) have since appeared, sufficient time has now elapsed to justify the appearance of an updated treatment of the topic.

Apart from the need to collate the information that is bound to accumulate over a decade, there are additional and more important reasons for producing this book. For example, there have been some significant developments in brown adipose tissue research and we now know that in spite of its relatively small mass this tissue can, in small rodents, make a major contribution to total heat production in certain situations (see Chapter 3). Furthermore, considerable advances have been made in our biochemical understanding of this tissue's remarkable capacity for heat production, since it now appears to possess a thermogenic pathway that is unique to brown fat (see Chapter 2). There have also been important developments in our understanding of the autonomic and endocrine control of thermogenesis (see Chapter 4 and Chapter 5) which, in their own right, would justify the appearance of a new book on brown-fat thermogenesis. However, there has also been a quite remarkable, almost coincidental, fusion of research interests between those working on brown fat and cold-induced thermogenesis on the one hand, and those studying diet-induced thermogenesis and obesity on the other. Thus, we had to add to our list of contributors those whose interests are more concerned with the problems of

1

of nutrition, energy balance regulation and obesity than with thermo-regulation. Hence the inclusion of chapters on livestock production (Chapter 6), diet-induced thermogenesis (Chapter 7), obesity (Chapter 8) clinical hypermetabolic conditions (Chapters 9 and 10) and the pharmacological manipulation and control of heat production (Chapter 11). In providing this more general view of thermogenesis we hope to demonstrate how variations in heat production play a key role in the body's response to a variety of physiological, nutritional and pathological conditions.

The attention given to thermogenesis and brown adipose tissue metabolism by investigators from many diverse research areas is represented to some extent in the two editors. Although we are both physiologists, one has been principally concerned with cellular bioenergetics and the other with problems in whole body energy balance regulation. When compiling this book, we realized that, in spite of our familiarity with each others work and ideas, the terminology was often different and at times confusing. This problem was not restricted to the editors, since at least five of the chapters we received devoted some space to a definition and description of terms. In

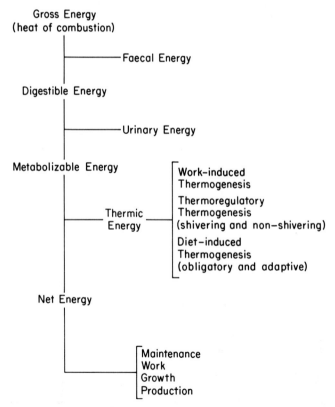

Fig. 1.1 The fate of ingested food energy.

an attempt to rationalize these terminological and conceptual difficulties we have considered the origins of mammalian heat production from two different points of view.

The scheme depicted in Fig. 1.1 illustrates the fate of energy in the body as seen by the nutritionist. After allowing for losses of energy in faeces and urine, the metabolizable energy obtained from food is utilized for maintaining and increasing body energy content (maintenance, external work, growth and production). The transformation of metabolizable energy into these forms of net energy also involves inevitable energy losses in the form of heat – thermic energy. Similarly, maintaining homeothermy in cold environments involves shivering and non-shivering thermogenesis (NST) and the energy costs of assimilating nutrients and retaining net energy results in obligatory heat losses due to diet-induced thermogenesis (DIT). This obligatory DIT is mainly due to the energy cost of protein and fat synthesis but, in addition to this, there is an adaptive component of DIT that helps maintain body energy content (i.e. body weight) by dissipating the metabolizable energy consumed in excess of the requirements for maintenance, growth and production.

In Fig. 1.2, we have converted this nutritionist's scheme (A) into one that

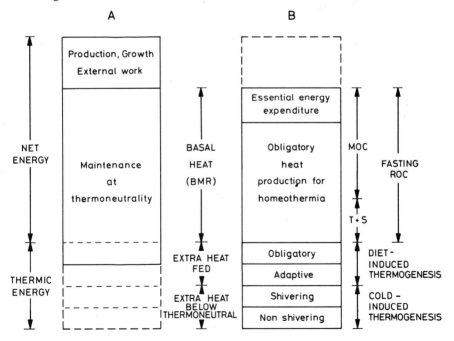

Fig. 1.2 The utilization of metabolizable energy represented in terms of energy balance (Scheme A) and thermal balance (Scheme B). MOC = minimal oxygen consumption; ROC = resting oxygen consumption; T+S = resting muscle tone + sympathetic activity.

can be compared with the thermal physiologist's view (B) of energy dissipation. The differences are more profound than appear at first sight. The nutritionist focuses on net energy and is interested in the efficiency of energy utilization, which could be defined as the ratio of net energy over total energy intake. For the nutritionist, the thermic energy is of secondary interest in normal conditions and only becomes of major importance in unusual or extreme situations. In the cold, for instance, when the heat production of the subject follows the thermostatic heat requirement and increases in proportion to the decrease in environmental temperature, net energy will decrease reciprocally with the increased heat production. Another example is in animals fed an unbalanced diet, particularly a very low protein diet in which, instead of producing body fat in proportion to the excess of non-protein food energy, the animal increases its heat production, and thus the efficiency of food utilization will decrease. In many normal situations, however, heat is dissipated at minimal energy cost, and mainly in proportion to the environmental temperature.

Someone once said that, depending on the point of view, the 'noise' for some is the 'signal' for others and this seems to be particularly true for our present, somewhat manichaen, analysis. The 'lumped' thermic energy of the nutritionist has been carefully dissected by the thermophysiologist, who works in a time scale in which the net energy of production is negligible, and the maintenance heat of the nutritionist has been divided into two parts, essential energy expenditure and the obligatory heat production required for homeothermy. Essential expenditure is the low level of resting metabolism of reptiles and other non-avian and non-mammalian animals, and is 8–10 times less than that of a bird or mammal of the same body weight and at the same temperature. It represents the chemical 'friction' for synthesis, transport and convection. This essential energy expenditure does not produce sufficient heat to maintain the body temperature above that of the environment for any except the largest of the homeotherms. For smaller homeotherms, including man, another form of continuous heat production is necessary, i.e. the obligatory heat for homeothermy. Else and Hulbert (1981) have shown, comparing a lizard and a mouse of the same weight and reared at the same temperature, that the mammal had relatively larger internal organs, the organs had a greater proportion of mitochondria and these mitochondria had a greater relative membrane surface. They estimated the capacity for energy expenditure to be three to six times higher in the mouse compared to the lizard, which is less than the eightfold difference in their basal metabolic rate. A point of major interest, therefore, is to explain how the dissipation of this extra heat is controlled – we can be certain it does not depend on a single factor. A minimal oxygen consumption (MOC) has been operationally defined by Denckla (1973) as the oxygen consumption in an anaesthetized animal immersed in water at a neutral temperature. It represents 40–80% of the resting oxygen consumption

(ROC) and is affected in a given subject only by its thyroid status. The remaining fraction of ROC depends on, among other things, muscle tone and the activity of the adrenergic nervous system. When the intact animal is fed, and/or placed below thermoneutrality, heat-production increases further still and the control and origins of this extra heat is the subject of several chapters in this book.

Independently of these differences in their theoretical approach, nutritionists and thermophysiologists have, for quite obvious reasons, chosen experimental conditions in which the variables of interest are easiest to evaluate. Most of the studies on the regulation of heat production have been carried out on rodents, which have developed a powerful thermo-effector system and many animal nutritionists are working with large domesticated species, particularly selected for their high food efficiency. Thus, there are bound to be disputes concerning the quantitative importance of regulatory heat production. Nevertheless, there is general agreement about what is meant by work induced and cold-induced thermogenesis, and the division of the latter into shivering and non-shivering thermogenesis. There is, however, considerable confusion and debate about the concepts of obligatory and adaptive DIT, and this is partly due to an excess of synonyms. For example, obligatory DIT is described by various workers as Specific Dynamic Action (SDA), the Heat Increment of Feeding (HIF) or the Thermic Effect (TE). Adaptive DIT can also be Regulatory DIT, Facultative DIT or, to give its original name, Luxus-konsumption. Further confusion arises when some of the so-called 'obligatory' forms appear to contain an adaptive element. The acute Thermic Effect of a single meal, for example, is assumed by many workers to represent the obligatory energy cost of assimilating the nutrients supplied. However, this post-prandial rise in heat production varies considerably, depending on the animal's capacity for adaptive thermogenesis, and often contains a large β-adrenergic component.

This terminological mess certainly needs cleaning up, but we have resisted the temptation to enforce a strict code of our own design. Thus, each chapter that follows will reflect the author's own approach and definition of terms. Nevertheless, we would like to see a move towards conformity and suggest than the terms 'obligatory' and 'adaptive' could be used to distinguish between different forms of DIT. There is some practical basis for this distinction that resembles that used by thermal physiologists to assess the relative contributions of shivering and non-shivering thermogenesis. Originally, the inhibition of shivering by curare was used, but this has now been replaced by measurements of either thermogenic responses to β-agonists (to assess maximum capacity for NST) or the inhibition of thermogenesis by β-adrenergic antagonists. The latter technique can also be used to distinguish between obligatory and adaptive DIT, if it is assumed that the adaptive form is mediated by sympathetic activation of β-receptors (see Chapter 6). This is still a tentative suggestion, since the practical advantages of a pharmacological distinction such as this have to be weighed against various other assumptions and, of course, one has to eliminate variations due to NST. The assessment of both NST and adaptive DIT by this technique might be an overestimation because β-adrenergic blockade could affect metabolic rate indirectly, via effects on tissue blood flow (due to cardiovascular changes), central neural and neuroendocrine activity (due to the non-specific, anaesthetic actions of drugs like propranolol) as well as effects on substrate mobilization. The latter effect can prove particularly difficult to dissociate from direct sympathetic activation of thermogenesis. An example would be the inhibition of free fatty acid release from white adipose tissue in response to circulating adrenaline. These free fatty acids can be oxidized in liver and muscle and could make a contribution to total heat production that was independent of, say, noradrenergic activation of NST or DIT.

Throughout this book there is a strong emphasis on effector mechanisms in thermogenesis and, although many other aspects are discussed, it will become evident that one of the least studied and understood areas concerns the central activation and control of thermogenesis. This is particularly true for adaptive dietary changes, where evidence is only just beginning to suggest that hypothalamic areas involved in the control of feeding behaviour may also have a role in controlling DIT. An important area for future research will be studying the interactions between hypothalamic control of feeding, NST and DIT, as well as trying to identify the afferent pathways and signals responsible for activating DIT. These aspects have been covered by several of our contributors but there is no doubt that the potential and scope for making rapid advances is now enormous, particularly as the list of putative neurotransmitters and modulators is growing almost daily. We would predict a plethora of observations and theories relating many of these to the control of thermogenesis, and considerable caution and restraint will have to be exercised if the subject is not to be submerged in a sea of conflicting mechanisms.

Whichever way future research will take the subject, we can be sure that mammalian thermogenesis will continue to provide numerous challenges to investigators from many different disciplines. We are indebted to our colleagues who have responded to our invitation to describe the recent developments in thermogenesis, and thereby help direct our thoughts to future research problems. At the same time, we have to apologize to those who could have made their own additional and unique contribution but, for reasons of space and organization, could not be accommodated.

REFERENCES

Denckla, W.D. (1973) Minimal O_2 consumption as an index of thyroid status: standardization of method. *Endocrinology*, **93**, 61–73.

Else, P.L. and Hulbert, A. (1981) Comparison of the 'mammalian machine' and the 'reptile machine': energy production. *Am. J. Physiol.*, **240**, R3–R9.

Chapter Two

Cellular Mechanisms of Heat Dissipation

David Nicholls and Rebecca Locke

ABBREVIATIONS

$\Delta\psi_p$, the membrane potential across the plasma membrane

$\Delta\psi_m$, the membrane potential across the mitochondrial inner membrane

$\Delta\mu_H+$, the proton electrochemical potential across the mitochondrial inner membrane

C_mH^+, the effective proton conductance of the mitochondrial inner membrane

FCCP, carbonylcyanide-*p*-trifluoromethoxyphenylhydrazone.

2.1 INTRODUCTION

The past five years have seen a steadily accelerating advance in our understanding of the mechanisms and scope of mammalian thermogenesis. At the physiological level the recognition that brown fat is the major thermogenic organ not only in hibernators (Smith and Horwitz, 1969) but also in non-hibernators such as the rat (Foster and Frydman, 1978, 1979), together with the implication of brown fat in diet-induced thermogenesis (Rothwell and Stock, 1979) and the evidence that grown-fat thermogenesis is defective in certain examples of genetic obesity (Himms-Hagen and Desautels, 1978) have all centred interest upon this tissue. At the sub-cellular level the demonstrations that the mitochondria from brown fat possess a unique regulatable proton short circuit (Nicholls, 1974*a*) enabling ATP synthesis to be bypassed, and that this function is performed by a specialized inner membrane protein (Heaton *et al.*, 1978) whose amount varies in synchrony with the thermogenic capacity of the tissue, have provided a biochemical mechanism to account for the physiological findings.

In this chapter we shall review the current state of knowledge of these biochemical mechanisms. Although we shall concentrate upon brown-fat thermogenesis, we shall also consider the thermodynamic constraints common to the generalized cell.

2.2 THE THERMODYNAMICS OF HEAT DISSIPATION

The heat produced by any biochemical process is equal to the difference in enthalpy content between the reactants and products. The enthalpy change itself (ΔH) is related to the change in Gibbs free energy (ΔG) and the change in entropy (ΔS) by the equation:

$$\Delta H = \Delta G + T\Delta S \qquad (2.1)$$

In most biochemical reactions which have a high enthalpy change the entropy term is relatively minor (Prusiner and Poe, 1970) and it is therefore a good approximation to equate the heat production with the Gibbs free energy change. The Gibbs free energy change for a reaction is dependent only upon the nature and concentration of the initial reactants and final products and is completely independent of the pathway by which the reactants are converted into products. For example, the direct chemical combustion of one mole of palmitate to CO_2 and H_2O will generate precisely the same heat as when the fatty acid is oxidized inside a cell, even though the latter might involve a sequence of thirty reactions. It is important to emphasize that there is no difference between the efficiency of thermogenic processes which involve 'coupled' mitochondria (i.e. those capable of stoichiometric ATP synthesis) and 'uncoupled' mitochondria (in which respiration can occur without obligatory ATP turnover). This is so because any ATP synthesized in the former process is immediately recycled and does not appear in the final equation of reactants and products (Fig. 2.1). In fact any thermogenic process which occurs by whatever pathway has an efficiency of *zero,* as long as there is no net accumulation of intermediates, accumulation of ions across a membrane, or useful mechanical work.

It follows that the rate of heat production for a given overall reaction is merely a question of kinetics, and will be controlled by the activity of the rate-limiting step in any reaction sequence. It is therefore not necessary to dissect out individual reaction steps for analysis by either equilibrium- (Prusiner and Poe, 1970) or network-thermodynamics (Horowitz and Plant, 1978).

The mitochondrial respiratory chain is central to all hypotheses of cellular thermogenesis, and since the rate of electron transfer to oxygen is carefully regulated in normal cells, it is necessary at this stage to discuss the nature of resiratory control. In the conventional chemiosmotic proton circuit (Fig. 2.2) there is an obligatory coupling between electron transfer down the respiratory chain and proton translocation across the membrane, and also between proton re-entry through the ATP synthetase and net ATP synthesis (for reviews see Greville, 1969; Mitchell, 1976; Nicholls, 1982). As a result of this coupling the cell exhibits an automatic respiratory control, a decreased cellular ATP demand causing a feed-back inhibition upon

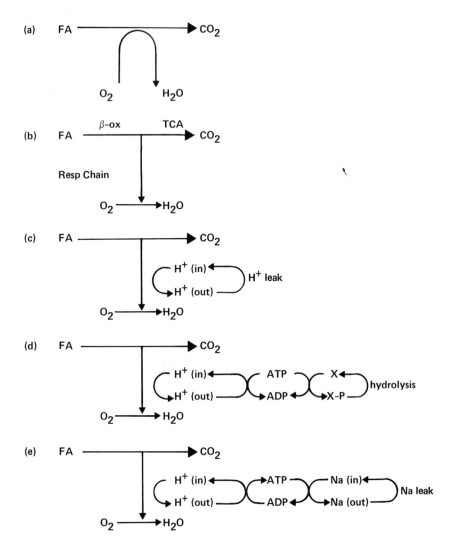

Fig. 2.1 The free energy change of a given overall reaction does not depend on the number or nature of cyclical reaction steps. In (a) fatty acid is directly combusted to CO_2; in (b) fatty acid is oxidized to CO_2 by a series of intracellular reactions, while electrons are transferred down a non-proton translocating respiratory chain to oxygen; in (c) the respiratory chain is proton translocating but the proton circuit is completed by a dissipative proton leak; in (d) the proton circuit drives a conventional ATP synthetase, but ATP is re-hydrolysed by a dissipative metabolic cycle, and in (e) the ATP is re-hydrolysed by a dissipative ionic cycle. In each of these examples the oxidation of one mole of fatty acid is associated with the same standard free energy change. β-ox, fatty acid β-oxidation; TCA, tricarboxylic acid cycle.

respiration through the intermediacy of decreased mitochondrial ATP synthesis, decreased proton re-entry through the ATP synthetase, enhanced proton electrochemical potential, decreased proton extrusion, decreased electron flow in the respiratory chain, decreased respiration and hence a decreased rate of heat production. In order to utilize mitochondria for deliberate thermogenesis it is necessary either to exploit or circumvent this control, allowing respiration to proceed more rapidly than would be possible within the constraints of normal cellular ATP turnover.

The proton circuit suggests three ways in which thermogenesis might be enhanced (Fig. 2.2). Firstly conventional respiratory control may be maintained by the mitochondrion, but a means of rapidly hydrolysing ATP

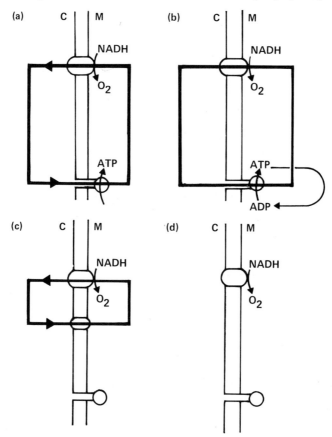

Fig. 2.2 The chemiosmotic proton circuit. (a) Conventional circuit, respiration controlled by cellular ATP demand; (b) conventional circuit, extra-mitochondrial ATPase allows respiration to increase (shivering); (c) proton short-circuit allows respiration to be uncoupled from stoichiometric ATP synthesis (brown-fat mitochondria); (d) alternative respiratory chain allows electron transfer to occur without stoichiometric proton translocation (Arum lily mitochondria).

subsequent to its synthesis could be induced, either in the matrix or the cytoplasm. Although no physiological intra-mitochondrial ATP hydrolysing mechanisms have been described, the 'uncoupling' action of arsenate occurs by an analogous mechanism. Arsenate can substitute for phosphate at the ATP synthetase with the resultant synthesis of $ADP-AsO_4$, which is unstable and hydrolyses spontaneously to regenerate ADP. The effect upon respiration is thus equivalent to an intra-mitochondrial ATPase. Diagnostic of such a mechanism in a cell would be a sensitivity to oligomycin (which inhibits the ATP synthetase) but no inhibition by atractylate (which inhibits exchange of adenine nucleotides across the mitochondrial inner membrane).

Several extra-mitochondrial ATP-hydrolysing mechanisms have been described. The best established example is of course shivering, in which small random contractions of the skeletal muscle actomyosin complex serve to hydrolyse ATP without doing useful mechanical work (Hemingway, 1964). Within the context of non-shivering thermogenesis a number of dissipative substrate cycles or ionic cycles have been proposed, which will be discussed below. One would predict that thermogenesis due to an extra-mitochondrial ATPase would be immediately inhibited by oligomycin, by atractylate, or by inhibiting the ATP-hydrolytic pathway itself.

The second and third classes of thermogenic mechanisms both involve specific lesions in mitochondrial energy transduction prior to the ATP synthetase. One of these would be a modification to the respiratory chain such that electron transfer could be 'uncoupled' from proton translocation (Fig. 2.2). In this way substrate oxidation could be divorced completely from ATP turnover. Since it would be desirable for any dissipative mechanism to be capable of regulation, a variable effective proton stoichiometry might be produced by dividing the electron flow from the substrate between parallel proton-translocating and non-translocating pathways.

The best established example of a non-translocating respiratory chain occurs in the plant kingdom, with the Arum lily, whose mitochondria can induce a very active alternative electron transfer pathway allowing 'uncoupled' respiration to occur at a rate sufficient to raise the temperature of the plant's spadix several degrees above ambient (Moore and Rich, 1980). In mammalian thermogenesis, a proposal for a non-conserving electron transfer pathway in mitochondria from the liver of cold stressed rats has been made by Skulachev and colleagues (Skulachev *et al.*, 1963; Mokhova, Skulachev and Zhigacheva, 1977). These authors observed that the cytochrome *c* of liver mitochondria from cold-stressed rats was more mobile than under control conditions, allowing electrons from cytosolic NADH to be transferred from the outer membrane–cytochrome *c* reductase to the inner membrane cytochrome *c* oxidase. However, since the great majority of NADH during thermogenesis is generated in the matrix, it is not clear what the significance of such a pathway would be *in vivo,* particularly since the liver is not usually considered to be a thermogenic organ (Himms-Hagen, 1976).

The third class of modification to the chemiosmotic proton circuit requires specialized mitochondria which possess conventional, proton-translocating respiratory chains and ATP synthetases, but which are able to present the protons with alternate pathways of re-entry into the matrix – either conventionally via the ATP synthetase, or through a short-circuit pathway of proton re-entry which is purely dissipative (Fig. 2.2). By controlling the division of proton flux between these two pathways the overall rate of substrate oxidation can be increased with no change in the rate of extra-mitochondrial ATP turnover. Such a pathway may conduct protons directly (or the equivalent OH^- in the opposite direction), or may be a component of a more complex series of cyclic ion movements, such as that catalysed *in vitro* by the simultaneous addition of valinomycin and nigericin or the physiological cycling of Ca between independent uptake and efflux pathways. However Ca cycling across the inner mitochondrial membrane *in vivo* is slow and performs a regulatory rather than dissipative role (for reviews see Nicholls and Crompton, 1980; Åkerman and Nicholls, 1982).

It is possible in theory to conceive of a pathway for the oxidation of fatty acids which by-passes the mitochondrion completely, relying on a cytoplasmic 'fatty acid oxidase'. At first sight the peroxisomes appear to perform such a role; in brown fat these organelles possess a fatty acid β-oxidation complex generating H_2O_2 which is broken down by catalase (Ahlabo and Barnard, 1971; Pavelka *et al.,* 1976; Kramer *et al.,* 1978; Nedergaard, Alexson and Cannon, 1980). However although there can be a ten-fold increase in total peroxisomal palmitoyl CoA oxidation during cold-adaptation, the maximal oxygen consumption would be only about 1–2% of the tissue's respiration (Nedergaard *et al.,* 1980).

Since there are no significant alternatives to the mitochondrial respiratory chain it would be expected that a close proportionality would exist between respiration and heat production (measured by microcalorimetry). This correlation has been confirmed in studies of isolated brown-fat mitochondria (Ricquier, Gaillard and Turc, 1979*a*), isolated brown adipocytes (Nedergaard *et al.* 1977) or whole tissue (Girardier *et al.,* 1976; Chinet *et al.,* 1977).

The various strategies for de-restricting electron flow down the respiratory chain will only result in enhanced thermogenesis if the supply of substrate remains adequate. In the case of glycolysis the Pasteur effect ensures that any decrease in cytosolic ATP, for example upon induction of shivering, leads to an immediate increase in the rate of pyruvate production (for review see Krebs, 1972). When the predominant substrate for thermogenesis is fatty acid, as in the case of brown fat, an adequate substrate supply is ensured by noradrenaline-induced β-adrenergic activation of lipolysis via a cyclic AMP-dependent protein kinase (for reviews see Nicholls, 1977; Skala, Hahn and Knight, 1978). Indeed a central problem in the study of brown-fat thermogenesis is the way in which lipolysis and the dissipative mechanisms are activated and inhibited synchronously for the acute regulation of heat production. Failure to achieve synchrony could result in a dissipative pathway remaining active after substrate supply had decreased to basal rates, with a consequent complete collapse of the cell's energy level. As we shall discuss, brown fat appears to avoid this problem by a precise auto-regulation of substrate supply and demand during thermogenesis.

2.3 EXTRA-MITOCHONDRIAL ATP-HYDROLYSING MECHANISMS: GENERAL CONSIDERATIONS

Extra-mitochondrial ATP hydrolysis is the most complex of the mechanisms discussed in the previous Section, since it involves the most extensive chain of cyclical events. For such a mechanism to be feasible the following conditions must be satisfied:

(a) There must exist, in the cytoplasm or at the plasma membrane, a means or hydrolysing ATP at a rate which is adequate to allow conventionally 'coupled' mitochondria to respire at the rate required by the thermogenic tissue.

(b) The mitochondrial ATP synthetase must be of adequate activity to allow for the stoichiometric synthesis of ATP by the mitochondria at the respiration rates observed during thermogenesis.

(c) One of the final steps involved in the actual hydrolysis of ATP must be capable of regulation in order to allow acute induction and termination of thermogenesis in an individual cell. The regulated step must be at this stage in order to allow the cell to maintain a high cytosolic ATP level under non-thermogenic conditions.

(d) Inhibition of the ATP hydrolytic mechanism, the mitochondrial ATP synthetase or the adenine nucleotide translocator must result in an instantaneous inhibition of the accelerated respiration during thermogenesis.

Three types of ATP-hydrolytic pathways have been discussed in the thermogenic context. The most clearly established mechanism is shivering, in which ATP is utilized both to drive the random contractions of the actomyosin complex, and also for the repetitive accumulation of Ca within the sacroplasmic reticulum. Shivering satisfies each of the four criteria discussed above (Hemingway, 1963).

Substrate cycling (for reviews see Newsholme and Crabtree, 1976) occurs when two intermediates can be interconverted by two independent pathways, one of which requires ATP which is not salvaged in the reverse reaction. A classic example is the interconversion of fructose-6-phosphate and fructose-1,6-biphosphate in the glycolytic sequence (Fig. 2.3). The majority of the substrate cycles which have been described perform a regulatory rather than dissipative function, even though a measure of ATP utilization must accompany any cycle (Newsholme and Crabtree, 1976). Dissipation is usually minimized by reciprocal control of the forward and reverse reactions, enabling a very precise regulation to be achieved of the net flux into and out of the cycle. If however the forward and reverse reactions were to be activated simultaneously, a potent cytosolic ATPase activity could be created.

There is good evidence for substrate cycling as a thermogenic mechanism in the flight muscle of the *Bombus* species of bumble-bee (see Newsholme and Crabtree, 1976). The glycolytic cycle discussed above appears modified in that the fructose-1,6-bisphosphatase lacks the inhibition by AMP which is characteristic of the enzyme from other species, allowing both enzymes in the substrate cycle to be active simultaneously. However, although the two enzymes have a high activity there is doubt as to whether this mechanism alone can account for the heat required to maintain the thorax temperature (Newsholme and Crabtree, 1976). The significance of substrate cycling in mammalian thermogenesis is, we feel, less convincing (Section 2.4).

The final type of ATP-hydrolytic mechanism which has been proposed relies upon enhanced Na-cycling at the plasma membrane, which has been invoked to account for both thyroid-induced thermogenesis (see Smith and

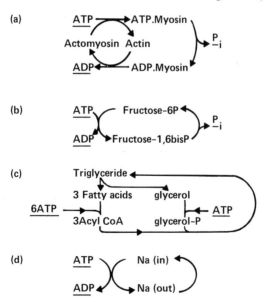

Fig. 2.3 Proposed ATP-hydrolysing cycles. (a) Actomyosin contraction–relaxation cycles (shivering); (b) phosphofructokinase plus fructose-6-phosphatase; (c) triglyceride lipolysis and re-esterification; (d) Na-cycling at the plasma membrane.

Edelman, 1979) and noradrenaline-induced thermogenesis in brown fat (see Horwitz, 1979). The activity of the Na^+K^+-ATPase of the plasma membrane is restricted in most tissues by the limited rate at which Na can re-enter the cell. An increase in the Na-permeability of the plasma membrane would thus increase the rate of Na-cycling and consequently increase the rate of ATP hydrolysis by the Na^+K^+-ATPase. However while a potential pathway for ATP dissipation does exist, it is necessary to demonstrate for each individual case that the activity of every component of the pathway is adequate for the respiratory rates observed during thermogenesis. As we shall discuss in more detail below, we feel that particularly in the case of brown fat this criterion has not been met.

2.4 THE STATUS OF ATP-HYDROLYSING MECHANISMS IN BROWN FAT

In Table 2.1 we have assembled representative determinations of respiratory capacity and enzyme activities from the brown-fat literature. The values are not intended to be definitive, since the type and status of the animals used differs and since assay conditions are not standardized. Nevertheless it is possible to obtain order-of-magnitude comparisons of the kinetic competence of a number of proposed thermogenic mechanisms in the tissue.

It is apparent from Table 2.1 that the rate of noradrenaline-stimulated lipolysis is comfortably in excess of the requirements of the mitochondrion. Indeed in incubations of isolated brown adipocytes a substantial export of free fatty acids occurs during the period when respiration is stimulated by noradrenaline (Bieber, Pettersson and Lindberg, 1975; Nedergaard and Lindberg, 1979). The respiration of the intact tissue, calculated from microsphere distributions and arterial–venous oxygen differences (Foster and Frydman, 1978, 1979; Thurlby and Trayhurn, 1980; Rothwell and Stock, 1981) correlates rather closely with that obtained with isolated adipocytes (Prusiner *et al.*, 1968*a,b*; Williamson, 1970; Bieber *et al.*, 1975; Pettersson and Vallin, 1976; Nedergaard and Lindberg, 1979) whereas rates with superfused tissue are substantially lower, perhaps due to limitations of oxygenation (see Girardier, Seydoux and Clausen, 1968). The rate at which the uncontrolled mitochondria can oxidize palmitoyl-L-carnitine in the presence of malate is sufficiently similar to the rates seen with the intact cells to suggest that this is the rate limiting step during thermogenesis, and that the full uncontrolled respiration is required by the thermogenic cell (Hittelman, Lindberg and Cannon, 1969; Nicholls, Grav and Lindberg, 1972; Cannon, Nicholls and Lindberg, 1973). In contrast the rate of ADP-stimulated respiration is substantially lower (Cannon *et al.*, 1973; Lindberg, Bieber and Houstek, 1976; Nicholls and Bernson, 1977) due to the uniquely low activity (Lindberg *et al.*, 1967; Guillory and Racker, 1968; Bulychev *et* *al.,* 1972) and amount (Cannon and Vogel, 1977; Houstek and Drahota, 1977) of ATP synthetase in brown-fat mitochondria, a deficiency which is found in mitochondria from the brown fat of rats (Skaane, Christiansen and Grav, 1972), hamster (Cannon *et al.,* 1973; Cannon and Vogel, 1977; Houstel and Drahota, 1977) guinea-pig (Nicholls and Bernson, 1977), rabbit (Lindberg *et al.,* 1967) or hedgehog (Skaane *et al.,* 1972). Typically the ATP synthetase subunits are present in one-tenth of the molar ratio to the respiratory chain of that found in 'normal' mitochondria such as those from heart (Cannon and Vogel, 1977; Houstek and Drahota, 1977). For such a specialized tissue as brown fat to have evolved an elaborate thermogenic mechanism based upon ATP turnover, and then to render the mechanism largely inoperative due to a uniquely deficient ATP synthetase appears highly unreasonable, and provides one of the many lines of evidence that brown fat thermogenesis utilizes a dissipative mechanism which by-passes the enzyme (Section 2.6).

The next difficulty faced by the ATP-hydrolysis group of mechanisms is the search for an actual means of hydrolysing ATP at up to 1.5 μmol ATP min^{-1} per 10^6 cells. No comparable ATPase activity for brown fat homogenates has ever been reported although it could be argued that the dissipative mechanism became inhibited as a result of cell disruption. A proposal for a dissipative substrate cycle based upon lipolysis and re-esterification (Fig. 2.3) was made by Dawkins and Hull (1964). However the activity of such a cycle, which would hydrolyse the equivalent of 7 ATP per triglyceride hydrolysed and re-esterified, was shown to be two orders of magnitude too low (Lindberg *et al.,* 1967; Pursiner *et al.,* 1968c).

The concept that the activity of the Na$^+$+K$^+$-ATPase is responsible for a major fraction of the basal ATP turnover in the generalized cell has provided the basis for hypotheses concerning both thyroid hormone-induced thermogenesis (for review see Smith and Edelman, 1979) and brown fat thermogenesis (for review see Horwitz, 1979). Electrophysiological studies of brown fat either *in situ* or in excised tissue show that binding of noradrenaline to the plasma membrane is followed by a rapid decrease in plasma membrane potential ($\Delta\psi_p$) from a resting value of about 50 mV to 25–30 mV (Girardier *et al.,* 1968; Horwitz *et al.,* 1969; Krishna *et al.,* 1970; Williams and Matthews, 1974*a,b*; Fink and Williams, 1976; Flaim *et al.,* 1977). When the tissue is stimulated electrically depolarization can be detected within 1 s and can be maximal within 2 s, while respiration begins to increase by 3 s (Seydoux and Girardier, 1978). This drop in potential is consistent with an 8-fold increase in Na-permeability (Girardier *et al.,* 1968) and independent qualitative evidence of an increase in plasma membrane conductance has been presented (Horowitz, Horwitz and Smith, 1971). An apparent two-fold activation of the Na$^+$K$^+$-ATPase by noradrenaline has also been reported (Herd, Horwitz and Smith, 1970; Horwitz and Eaton, 1975) although since the assays were performed on the microsomal fraction

Table 2.1 Typical activities of proposed intermediate reactions in brown-fat thermogenesis.

Parameter	Rate required*	Rate observed	Conditions	Reference
A. Noradrenaline-stimulated respiration				
(i) Tissue *in situ*	(nmol O min^{-1} per 10^6 cells)			
	750	750	c/a rat IBAT 37°C	Foster and Frydman (1978)
(ii) Superfused tissue	750	40	w/a rat IBAT 37°C	Chinet *et al.* (1978)
(iii) Isolated brown adipocyte	750	700	w/a hamster 37°C	Nedergaard (1981)
(iv) Isolated mitochondria (palmitoyl-L-carnitine plus malate)	750	12 (controlled 32 (synthesizing ATP) 200 (uncontrolled)	c/a hamster 20°C	Lindberg *et al.* (1976)
B. Noradrenaline-stimulated lipolysis				
(v) Isolated brown adipocyte	(nmol fatty acid min^{-1} per 10^6 cells) 16	40	w/a hamster 37°C	Nedergaard (1981)
C. Mitochondrial proton conductance				
(vi) Isolated mitochondria	(nmol H$^+$ min^{-1} mV^{-1} per 10^6 cells) 20	14	c/a guinea-pig 23°C	Nicholls and Bernson (1977)
D. Mitochondrial ATP synthesis				
(vii) Isolated mitochondria	(nmol ATP min^{-1} per 10^6 cells) 2200	150	c/a hamster 30°C	Bulchev *et al.* (1972)
E. Plasma membrane Na$^+$+K$^+$-ATPase				
(viii) 1400g supernatant of homogenate	(nmol ATP min^{-1} per 10^6 cells) 2200	12	c/a rat 37°C	Horwitz and Eaton (1975)
F. Triglyceride lipolysis/re-esterification				
(ix) Isolated brown adipocyte	(nmol fatty acid min^{-1} per 10^6 cells) 940	<40	w/a hamster 37°C	Nedergaard (1981)

*Rate required to support respiration at 750 nmol O min^{-1} per 10^6 cells if all energy flow is channelled through the given pathway.
c/a, cold-adapted: w/a, warm-adapted. IBAT, interscapular brown adipose tissue. Conversion factors: 10^6 cells are equivalent to 10 mg wet weight, 1.3 mg mitochondrial protein or 0.09 mg 14000g supernatant nitrogen.

from brown fat, the possibility that the increased ATP hydrolysis was secondary to an enhanced Na-permeability in plasma membrane vesicles cannot be excluded.

The Na-cycling hypothesis runs into difficulties when the activities reported by the proponents of the hypothesis (Herd *et al.*, 1970; Horwitz and Eaton, 1975) are examined. Although the heterogeneous use of units makes rate comparisons difficult, we calculate a rate of ATP hydrolysis of 5 μmol P_i h^{-1} per mg of microsomal nitrogen to approximate to 12 nmol P_i min^{-1} per 10^6 cells (Table 2.1), or more than two orders of magnitude slower than required.

An independent approach is to look for an instantaneous effect of oligomycin, ouabain or atractylate upon the rate of noradrenaline respiration. It is necessary for an effect to be apparent within a few seconds since otherwise a change in respiration could be due to secondary changes in cellular ATP levels or cytosolic ions. The problems in interpretation of results with ouabain have been reviewed by Himms-Hagen (1976). Ouabain slowly depolarizes the plasma membrane of non-stimulated brown adipocytes (Girardier *et al.*, 1968; Williams and Matthews, 1974*a*) and causes a parallel inhibition of the lipolysis evoked by a subsequent addition of noradrenaline (Herd, Hammond and Mamolsky, 1973; Fain *et al.*, 1973; Williams and Matthews, 1974*b*). The observation that ouabain requires preincubation in order to decrease noradrenaline-stimulated respiration (Horwitz, 1973; Herd *et al.*, 1973; Chinet *et al.*, 1978) suggests therefore that other factors such as substrate supply or cytosolic ATP concentration might be contributing to the observed drop in respiration.

An ATP-hydrolytic mechanism would predict an immediate decrease in noradrenaline-stimulated respiration upon addition of either oligomycin, to inhibit the mitochondrial ATP synthetase, or atractylate, to inhibit the subsequent export of ATP to the cytosol, and would also predict that this inhbition would be reversible by the subsequent addition of a proton translocator such as FCCP. Experiments with hamster brown adipocytes show that oligomycin does inhibit, but that the inhibition develops only over 1–2 min, is not releasable by FCCP and occurs even in cells to which the proton translocator had already been added (Prusiner *et al.*, 1968*b*). The effect of atractylate is also ambiguous; Williamson (1970) found no inhibition of of noradrenaline-stimulated respiration even at very high inhibitor concentrations, whereas Nedergaard and Lindberg (1979) observed an immediate 40% inhibition by lower atractylate concentrations, but only at sub-optimal noradrenaline concentrations. Although the behaviour of isolated brown adipocytes presents a number of ambiguities, the effects of these inhibitors are clearly more complex than the ATP-hydrolytic mechanisms would predict.

Although the evidence is against a significant direct dissipative role of plasma membrane Na-cycling, the noradrenaline-induced plasma membrane depolarization does remain of interest as a possible signal in the acute thermogenic response. Williams and Matthews (1974*a,b*) studying brown fat from young warm-adapted rats showed that the depolarization was not analogous to stimulation-secretion coupling (Douglas, 1974) since enhanced lipolysis still accompanied depolarization in the absence of external Ca or in the presence of the Ca-channel inhibitor D600. All agents which increased lipolysis (measured by glycerol release) also caused depolarization. An association between depolarization and lowered cellular ATP might be suggested by the slow depolarization induced by the glycolytic inhibitor iodoacetate (Williams and Matthews, 1974*a*). Although the mitochondrial respiratory chain inhibitors CN^- and antimycin A did not depolarize in Williams' and Matthews' study, Girardier *et al.* (1968) found a slow depolarization when the oxygen content of the medium was lowered from 95% to 20%; however proton translocators depolarized (Williams and Matthews, 1974*b*). It should be noted that proton translocators deplete ATP more effectively than respiratory chain inhibitors since the mitochondrial ATP synthetase becomes free to reverse and hydrolyse cytosolic ATP. It is possible that these effects and the rapid depolarization induced by noradrenaline may be superimposed in some studies.

More recent studies have thrown doubt upon the relation of depolarization to lipolysis. The brown adipocyte has been found to possess both α- and β-adrenergic receptors (Fink and Williams, 1976; Flaim, Horwitz and Horowitz, 1977) activation of which both lead to depolarization; lipolysis however is β-specific (Hamilton and Horwitz, 1979).

2.5 MITOCHONDRIAL DISSIPATORY MECHANISMS: GENERAL CONSIDERATIONS

In contrast to the search for extra-mitochondrial ATP-hydrolysing mechanisms, which must be directed towards the cytoplasm or plasma membrane, a direct energy-dissipating pathway operating prior to the synthesis of ATP should be detectable in isolated mitochondria. The presence of such a pathway would be indicated by the ability of the mitochondrion to oxidize substrates at a high rate without the need for a stoichiometric ATP turnover in the incubation.

The chemiosmotic theory has shown that the link between the respiratory chain and the ATP synthetase in mitochondria is a proton electrochemical potential whose maintenance depends on the proton impermeability of the inner membrane (for reviews see Greville, 1969; Mitchell, 1976; Nicholls, 1982). This proton impermeability may readily be overcome in any mitochondrion by the addition of synthetic proton translocators such as FCCP or 2,4-dinitrophenol, by allowing free fatty acids to accumulate (Pressman and Lardy, 1956; Wojtchak and Wojtchak, 1960), by allowing the mitochondria to accumulate Ca and P_i under inappropriate conditions (see Nicholls and Crompton, 1980), or by simple mistreatment of the mitochondria.

It follows from this fragile coupling between the respiratory chain and the ATP synthetase that the simple failure to see respiratory control or ATP synthesis in a particular preparation of mitochondria from a thermogenic tissue does not provide adequate evidence that the tissue possesses a physiological mechanism to uncouple the mitochondria. For example, while the first preparations of brown fat mitochondria from thermogenic animals showed no evidence of energy conservation (Lepkovsky *et al.,* 1959; Smith, Roberts and Hittelman, 1966; Lindberg *et al.,* 1967) later studies showed that by altering the incubation conditions mitochondria could be prepared which showed respiratory control, ATP synthesis and Ca accumulation (for review see Nicholls, 1979). A much more detailed analysis is therefore required in order to support the hypothesis that these mitochondria possess a physiologically relevant dissipative mechanism (Section 2.6).

The dissipation of the proton electrochemical potential ($\Delta\mu_H^+$) which would be required for a mitochondrial thermogenic pathway could be achieved by inducing a direct proton conductance in the membrane, an experimentally equivalent OH^- conductance, or a more indirect proton permeability resulting from the cycling of a cation across the membrane between a uniport and a cation/proton antiport. A physiologically relevant example of this last possibility would be the cycling of Ca across the inner membrane, which under normal conditions is considered to be a slow, regulatory process with little energy utilization (for reviews see Nicholls and Crompton, 1980; Åkerman and Nicholls, 1982). The possibility that an enhanced Ca cycling at the brown fat mitochondrial inner membrane, following a noradrenaline-induced increase in cytosolic Na, might contribute significantly to the cell's energy dissipation has been eliminated (Nedergaard, 1981).

The kinetic competence of a hypothetical mitochondrial dissipative mechanism is easier to assess than in the case of an ATP-hydrolysis pathway, since it is only necessary to show that a pathway is adequate to allow fatty acid oxidation to occur at uncontrolled rates in isolated mitochondria, Table 2.1. It is of course still necessary to show that the mechanism can operate at the same activity in the intact cell.

2.6 THE PROTON SHORT-CIRCUIT OF BROWN-FAT MITOCHONDRIA

By 1968 it was possible to prepare brown-fat mitochondria from thermogenic animals which could phosphorylate ADP to ATP with conventional stoichiometry provided that albumin was present in the medium to remove endogenous fatty acids (Joel, Neaves and Rabb, 1968; Guillory and Racker, 1968; Aldridge and Street, 1968). The conclusion from these studies was that fatty acid removal was a prerequisite for oxidative phosphorylation, and this was strengthened when Hittelman *et al.* (1969) showed that brown-fat mitochondria from cold-adapted rats could synthesize ATP and show conventional respiratory control as long as the mitochondria were pre-incubated with CoA, carnitine and ATP to allow them to oxidize their endogenous fatty acids. Since the cytosolic fatty acid concentration would be predicted to increase on the induction of lipolysis and decrease at its termination, a simple and elegant theory of physiological fatty acid uncoupling of brown-fat mitochondria became current (Drahota, Honova and Hahn, 1968; Hittelman and Lindberg, 1970; Williamson, 1970; Reed and Fain, 1970; Bulychev *et al.*, 1972). As we shall now discuss, the discovery of a specific, nucleotide-sensitive dissipatory pathway in brown fat led to a general abandonment of the 'fatty acid uncoupling' hypothesis (see Nicholls, 1979; Lindberg, Nedergaard and Cannon, 1981), although we shall later argue for a re-evaluation of the regulatory role of free fatty acids.

The first indication that fatty acid removal was not the only requirement for the recoupling of brown-fat mitochondria came with the failure to observe a respiratory stimulation when low concentrations of ADP were added to albumin-containing incubations of brown-fat mitochondria from cold-adapted or new-born rats, new-born guinea pigs or hibernating hedge-hogs (Drahota *et al.*, 1968; Hohorst and Rafael, 1968; Pedersen, Christiansen and Grav, 1968; Rafael *et al.*, 1968, 1969; Hittelman *et al.*, 1969; Christiansen, Pederson and Grav, 1969; Pedersen 1970; Grav *et al.*, 1970; Skaane *et al.*, 1972). The observation which revolutionized these investigations was made by Rafael and colleagues (Hohorst and Rafael, 1968; Rafael *et al.*, 1968, 1969). They found that the presence of 2 mM ATP in an albumin-containing incubation of cold-adapted guinea-pig brown-fat mitochondria allowed good respiratory control to be seen upon subsequent addition of ADP. While an apparent protective effect of ATP had some precedent in that the nucleotide was known to prevent mitochondrial damage during Ca uptake (for review see Nicholls and Crompton, 1980; Åkerman and Nicholls, 1982), the finding that GTP was even more effective with the guinea-pig mitochondria was unexpected and pointed to a novel site of action. Further studies (Pedersen, 1970; Heaton and Nicholls, 1977) have shown that the nucleotide requirement has a broad specificity. Thus GDP, GTP, ATP, ADP and to a lesser extent IDP and ITP all increase ATP synthesis and respiratory control in new-born or cold-adapted animals, while nucleotide monophosphates and pyrimidine nucleotides are in-effective.

In order to understand this phenomenon and its relation to non-shivering thermogenesis it was necessary to reduce the complexity of the system. In particular there was a need to avoid incubation conditions in which purine nucleotides were performing multiple roles. For example, in many early experiments ADP was added to test for respiratory control in mitochondrial suspensions oxidizing 2-oxoglutarate. In this system ADP would be acting as substrate for oxidative phosphorylation, as substrate for substrate-level phosphorylation by succinate thiokinase, in the brown fat-specific 're-coupling' of oxidative phosphorylation and possibly in protecting the mito-chondria against any Ca-induced uncoupling. It was therefore not surprising that many respiratory traces proved difficult to interpret (for discussion see Nicholls, 1979).

The first investigations in which mitochondrial energy conservation was studied in the complete absence of exogenous purine nucleotides were made by Christiansen (1971) who used the transient increase in respiration induced by Ca uptake as a means of testing for respiratory control. The addition of Ca to new-born guinea-pig brown fat mitochondria in albumin medium gave no respiratory increase and little Ca was accumulated. The prior addition of ATP, ADP, GTP or GDP in the presence of the ATP synthetase inhibitor oligomycin resulted in slowed controlled respiration, an increase on adding Ca, and a greatly increased Ca uptake, all of which were indicative of a nucleotide-induced 'recoupling' of the mitochondria (Christiansen, 1971).

The apparent contradiction between the requirement for nucleotide when fatty acids were removed by albumin and the immediate respiratory control when fatty acids were oxidized endogenously (Hittelman *et al.,* 1969) was resolved when it became clear that the ATP added to activate the fatty acids was also 'recoupling' the mitochondria (Cannon *et al.,* 1973). The two effects could be separated by their concentration dependencies, since 10^{-5}–10^{-4}M ATP provides adequate substrate for fatty acid activation while 10^{-3}M ATP is required (in the presence of Mg) for optimal respiratory control (Cannon *et al.,* 1973). There is a similar explanation for the ability of investigators to use 100 μM ADP to test for respiratory control without saturating the 'recoupling' mechanism since this nucleotide concentration appears again to be inadequate for optimal recoupling of the brown-fat mitochondrial inner membrane.

These results gave a clear indication that the brown-fat mitochondria prepared from animals adapted for thermogenesis possessed a unique lesion in their energy conservation, however a molecular understanding of this phenomenon required the application of chemiosmotic theory.

The oligomycin insensitivity of the nucleotide 'recoupling' (Christiansen, 1971) eliminated the possibility of an intra-mitochondrial ATP-hydrolysing system as the means of allowing rapid respiration in the absence of exo-genous phosphate acceptor. The two remaining mechanisms of Fig. 2.2 therefore remained as possibilities. A defective coupling of respiratory chain electron transfer to proton extrusion was eliminated by the obser-vation (Nicholls, 1974a) that a pulse of oxygen added to an anaerobic incubation of hamster brown-fat mitochondria caused protons to be extruded into the medium even in the absence of nucleotide. In this experiment, however, the decay of the proton gradient after exhaustion of the oxygen pulse was more rapid in the absence of nucleotide, suggesting an enhanced proton permeability of the membrane.

Confirmation of this abnormally high proton permeability came when it was found that rapid osmotic swelling of non-respiring hamster brown-fat mitochondria occurred in albumin-containing potassium acetate medium in the presence of the K^+-ionophore valinomycin (Nicholls, 1974a) even though such swelling would be dependent upon a proton permeability of the membrane (Fig. 2.4). The addition of purine nucleotide greatly inhibited swelling in this medium without affecting the permeability of acetic acid or the action of valinomycin (Nicholls and Lindberg, 1973; Nicholls, 1974a). This indicated that the nucleotide was able to inhibit an unusually high proton permeability while the ability to observe this effect in 'de-energized'

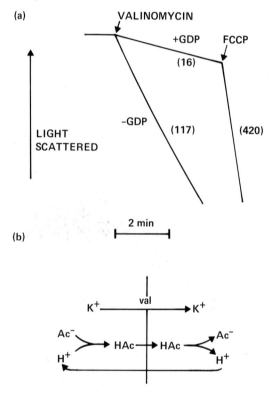

Fig. 2.4 Nucleotide-sensitive proton permeability in brown-fat mitochondria detected by swelling in potassium acetate. Hamster brown-fat mitochondria were suspended in 100 mM potassium acetate medium in the presence of albumin (to remove fatty acids) and valinomycin (to include a potassium uniport permeability). (a) Swelling was determined from the rate of light-scattering decrease, additions 1 mM GDP, 5μM FCCP. (b) Ion movements, note that swelling is dependent upon a pathway for proton permeation across the membrane. For details see Nicholls 1974a.

non-respiring mitochondria clearly dissociated this phenomenon from any effect at the level of the respiratory chain or ATP synthetase.

The study of ion permeation in non-respiring mitochondria revealed an unexpectedly high permeability also to Cl^- and Br^-, which could be inhibited by the same nucleotides, and at the same concentrations, as the proton permeation (Nicholls and Lindberg, 1973; Nicholls, 1974b). The observation of competition between Cl^- and H^+ movements (Nicholls, 1974b; Nicholls and Heaton, 1978) could be explained most readily if what was thought to be a pathway allowing protons to enter the matrix was in fact an anion-specific channel capable of transporting OH^- (or Cl^-) out of the matrix, since it is not possible to distinguish experimentally between these options. We shall however continue to refer to a proton permeability in this review to avoid confusion. The halide permeation itself would not be predicted to affect energy conservation since once Cl^- achieves electrochemical equilibrium with the membrane potential net flux will cease and no steady-state dissipation due to the Cl^- permeability would be expected.

A more quantitative analysis of the ease of proton permeation may be made by determining the effective proton conductance of the inner membrane (Mitchell and Moyle, 1967; Nicholls, 1974a), the units for $C_m H^+$ being nmol of H^+ min^{-1} (mg protein)$^{-1}$ per mV of proton electrochemical potential. This conductance, obtained by dividing the steady-state proton current flowing across the membrane by $\Delta\mu_H^+$, may be readily calculated for isolated mitochondria. $\Delta\mu_H^+$ itself can be estimated (for reviews see Rottenberg, 1975; Nicholls, 1982) by determining membrane potential ($\Delta\psi$) from the Nernst equilibrium of $^{86}Rb^+$ in the presence of valinomycin, or of a lipophilic cation such as tetraphenylphosphonium, and determining ΔpH from the equilibrium distributiuon of weak acids or bases. Since the transfer of two electrons down the respiratory chain to oxygen is always associated with the extrusion of a proportionate number of protons, which must then re-enter the matrix, the proton current may be calculated from the steady-state respiration multiplied by the $H^+/2e^-$ stoichiometry for the substrate in question. A typical calculation is shown in Table 2.2. showing that $C_m H^+$ for the inner membrane of fatty acid depleted brown-fat mitochondria is at least an order of magnitude greater (in the absence of nucleotide) than in the case of a 'normal' liver mitochondrion (Nicholls, 1979). Two factors compound to give this enhanced conductance: firstly $\Delta\mu_H^+$ under the conditions of this particular experiment is only 80 mV instead of 220 mV, and secondly respiration, being uncontrolled, is much greater than for the liver mitochondrion. Addition of exogenous GDP, as a representative purine nucleotide, reduces $C_m H^+$ to a low value comparable to that of rat liver mitochondria and hence raises $\Delta\mu_H^+$ to 220 mV and inhibits respiration (Nicholls, 1974a). Titration of $C_m H^+$ with varying nucleotide concentrations enables inhibition constants K_i to be obtained (Nicholls, 1974a, Nicholls and Bernson, 1977). It is notable that whereas GDP at pH 7.0 in the absence of

Table 2.2 The effect of fatty acid removal and nucleotide addition on the proton conductance of mitochondria from brown fat. For experimental conditions see Nicholls (1974a)

Mitochondria from:	Additions	Respiration (nmol O min^{-1} mg^{-1})	Proton current (nmol H$^+$ min^{-1} mg^{-1})	$\Delta\mu_{H^+}$ (mV)	Proton conductance (nmol H$^+$ min^{-1} mV^{-1} mg^{-1})
Rat liver	—	12	72	226	0.21
Hamster brown fat	—	61	366	<10	>35
Hamster brown fat	5 mg per ml albumin	140	840	80	7
Hamster brown fat	1 mM GDP	65	390	140	1.9
Hamster brown fat	albumin+GDP	48	288	220	0.88

Mg has a K_i close to $1\,\mu\text{M}$, the more physiologically relevant nucleotide ATP, in the presence of millimolar Mg, has a much lower affinity. Also apparent is the highly pH dependent nature of the conductance inhibition, the nucleotides being considerably less effective at elevated pH (Nicholls, 1974*a*).

Despite the totally different approaches involved in estimating proton permeability from swelling rates in potassium acetate plus valinomycin and in calculating $C_m\text{H}^+$ as just described, a remarkably good qualitative agreement is obtained between the methods (Nicholls, 1974*a*).

2.6.1 Proton conductance and respiratory control

Having established that the locus of the purine nucleotide action lies in a modulation of the 'leak' proton conductance of the inner membrane it is now necessary to discuss how this influences the rate of controlled respiration and the division of proton flux between ATP synthesis and the leak. In the high conductance state, $\Delta\mu_\text{H}^+$ is low and exerts little back pressure upon the respiratory chain. As a consequence the rate of electron flow down the respiratory chain is not limited by $\Delta\mu_\text{H}^+$ but rather by kinetic factors or substrate supply: this is the condition of uncontrolled respiration. As $C_m\text{H}^+$ is decreased steadily the potential required to drive protons back into the matrix increases in inverse proportion to $C_m\text{H}^+$. Initially (Fig. 2.5) no effect is seen upon respiration; however when $\Delta\mu_\text{H}^+$ increases above a threshold which depending on the substrate is typically 150–175 mV, the back-pressure exerted by $\Delta\mu_\text{H}^+$ upon the respiratory chain becomes sufficient to limit electron transfer and hence induce respiratory control. The rate of this controlled respiration then continues to slow down as $C_m\text{H}^+$ is further decreased and $\Delta\mu_\text{H}^+$ continues to rise (Fig. 2.5).

It is clear from this that to achieve uncontrolled respiration upon induction of thermogenesis in the cell it is only necessary to increase $C_m\text{H}^+$ sufficiently to lower $\Delta\mu_\text{H}^+$ to the 150–175 mV threshold at which respiration is no longer restricted by the back-pressure of the proton gradient (Nicholls, 1974*a*; Nicholls and Bernson, 1977). This has two important consequences: firstly this lowered potential is still thermodynamically competent for ATP synthesis, it is therefore possible to achieve maximal rates of thermogenesis while still retaining some capacity for the mitochondrion to synthesize ATP for cellular energy requiring events (Nicholls and Bernson, 1977). Secondly, this relatively high $\Delta\mu_\text{H}^+$ would not be predicted greatly to disturb the ability of the mitochondria to regulate cytosolic Ca or to transport metabolites.

In summary, therefore, the chemiosmotic theory would predict two states for the brown-fat mitochondrion, a non-thermogenic state when the leak pathway is closed, $\Delta\mu_\text{H}^+$ is some 220 mV, and respiration is stoichiometric with ATP synthesis (as in normal mitochondria); and a thermogenic state in which the leak conductance is increased above 5 nmol $\text{H}^+\text{min}^{-1}\,\text{mg}^{-1}\,\text{mv}^{-1}$,

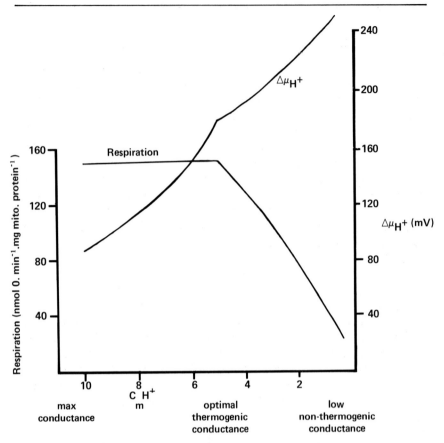

Fig. 2.5 The effect of decreasing membrane proton conductance upon proton electrochemical potential and respiration of brown-fat mitochondria. Low (non-thermogenic) conductance state C_mH^+ about 0.5 units; optimal conductance for uncontrolled respiration and high $\Delta\mu_H + 5$ units; maximal observed conductance (in presence of albumin) 10 units. Data recalculated from Nicholls and Bernson (1977).

$\Delta\mu_H^+$ drops in consequence to 150–175 mV and the fatty acids liberated by β-adrenergic lipolysis are oxidized at maximal rates. However before this model can be accepted a number of questions must be asked. Firstly, what is the component responsible for the leak pathway and how do nucleotides interact with it? Secondly, can a plausible *in vitro* model be constructed which can explain both the chronic adaptive regulation of thermogenic capacity during cold- and diet-induced thermogenesis and the acute regulation of the leak on initiation and termination of thermogenic episodes? Thirdly, is there direct evidence for the operation of the leak in intact cells or tissues?

2.6.2 The nature of the proton conductance pathway

The matrix of isolated brown-fat mitochondria retain millimolar concentrations of adenine nucleotides (Pedersen and Grav, 1972). Since the mitochondria remain in a high conductance state until micromolar exogenous nucleotides are added, it follows that the matrix nucleotides do not have access to the regulatory site, i.e. that this is located on the outer face of the inner membrane. This is supported by the contrast between the broad specificity for nucleotide of the regulatory site (Heaton and Nicholls, 1977) and the specificity for ADP and ATP of the adenine nucleotide translocator spanning the inner membrane (Nicholls, 1976). The nucleotides do not bind covalently and are not modified chemically during their interaction with the regulatory site (Nicholls, 1976). It is however possible to detect a binding site on the outer face of the membrane whose affinity, specificity and pH dependency for nucleotide correlate very closely with the conditions for nucleotide-induced inhibition of the conductance (Nicholls, 1976).

Depending on the adaptive status of the animal between 0.1 and 0.7 nmol of nucleotide per mg, protein can be bound in intact mitochondria (Section 2.7). In order further to identify the regulatory site it was necessary to achieve a covalent binding of nucleotide. This was accomplished by utilizing an azido derivative of ATP (Heaton *et al.*, 1978) which was photosensitive, forming a highly reactive nitrene free radical upon UV irradiation, with a resultant covalent binding to groups in the immediate vicinity. The nucleotide, which was labelled with $^{32}P_i$, inhibited C_mH^+ in the dark, showing that it interacted with the regulatory site. After UV illumination the inner membrane proteins were separated by polyacrylamide electrophoresis, and radioautography was performed in order to locate the labelled components. Two major bands were seen; one at 30000 molecular weight was not observed if the mitochondria had previously been incubated with the adenine nucleotide translocase inhibitor atractylate, indicating that this was the translocase. The remaining band, at 32000 molecular weight was abolished by including GDP prior to illumination; this displacement enabled this band to be identified as the regulatory site of proton conductance (Heaton *et al.*, 1978).

The 32000 protein has been partially (Ricquier *et al.*, 1979b) or more extensively (Lin and Klingenberg, 1980) purified from hamster brown-fat mitochondria and shown to retain its nucleotide binding after solubilization. Hydrodynamic evidence suggests that it may function in the membrane as a dimer of molecular weight 64000 (Lin *et al.*, 1980). The nucleotide-binding 32000 protein appears to be brown-fat specific (Nicholls, 1976; Rafael and Heldt, 1976; Lin and Klingenberg, 1980).

2.7 THE MOLECULAR BASIS FOR THE ADAPTIVE COLD-AND DIET-INDUCED INCREASE IN BROWN-FAT THERMOGENIC CAPACITY

In the preceding section the discussion was limited specifically to mitochondria prepared from the brown fat of hibernators such as the hamster, or new-born and cold-adapted homeotherms such as the rat or guinea pig. In these cases a nucleotide-sensitive conductance pathway is found with sufficient activity to allow uncontrolled respiration (Table 2.1). However when brown-fat mitochondria are made from foetal or warm-adapted adult non-hibernators the properties of the isolated mitochondria reflect the decreased thermogenic capacity of the animal. The temperature required to 'warm-adapt' an animal is dependent upon the species. Skaane *et al.* (1972) found that temperatures in excess of 30°C were needed for the rat, therefore the '23°C' rat used in many laboratories as the control may still be significantly cold-adapted.

Although a lower yield of mitochondria is obtained from warm-adapted brown fat (Rafael *et al.*, 1968) the highly distinctive pattern of inner membrane proteins seen with sodium dodecyl sulphate polyacrylamide electrophoresis of the cold-adapted mitochondria is conserved with one exception, as will be discussed below (Ricquier and Kader, 1976; Heaton *et al.*, 1978; Desautels *et al.*, 1978). The bioenergetic properties of the mitochondria are however significantly modified: some measure of respiratory control and synthesis of endogenous ATP can be observed in albumin-containing media in the absence of exogenous purine nucleotides (Christiansen, 1971; Nicholls *et al.*, 1974; Rafael *et al.*, 1974), while low concentrations of ADP, which fail to demonstrate respiratory control in the mitochondria from thermogenically active animals, can usually produce a conventional respiratory increase in the warm-adapted preparation (Rafael *et al.*, 1968, 1969; Christiansen *et al.*, 1969; Grav *et al.*, 1970; Pedersen, 1970).

These respiratory experiments imply that the maximal capacity of the proton short-circuit is reduced (per mitochondrion). There are two lines of evidence that this decreased capacity is due to a repressed synthesis of the 32 000 protein, rather than an inhibition of the protein. The first is that the capacity of purine nucleotide binding to the outer face of the inner membrane varies in synchrony with the proton conductance deduced from respiratory traces. With guinea-pig brown-fat mitochondria the amount of nucleotide bound per mg protein increases six-fold at birth, and then declines as the animal is reared at thermo-neutral temperatures (Rafael and Heldt, 1976). This correlates precisely with the respiratory response of the mitochondria to low ADP concentrations (Rafael *et al.*, 1968). A very similar pattern is observed in the developing rat (Sundin and Cannon, 1980). Cold adaptation of the adult rat leads to a ten-fold increase in GDP-binding per mg protein within 3–7 days (Desautels *et al.*, 1978). Surprisingly for an adaptive process, a significant increase could be detected after only 1 h (Desautels *et al.*, 1978). The increased binding appears to represent an increased number of binding sites rather than an increased binding affinity, since Scatchard plots indicate that binding affinity does not change on cold adaptation (Sundin and Cannon, 1980). An increased incorporation of labelled phenylalanine into the 25000–35000 molecular weight group of inner membrane proteins during cold-acclimation (Himms-Hagen *et al.*, 1980) is consistent with induction of synthesis of the 32000 protein. Although an inability to survive cold-adaptation in the presence of cyclo-hexamide prevented definitive conclusions as to the necessity of nuclear-coded protein synthesis, an insensitivity to oxytetracycline suggested that synthesis of the 32000 protein did not involve mitochondrial DNA (Desautels and Himms-Hagen, 1980).

Adaptive changes in nucleotide binding capacity have been seen also in the context of diet-induced thermogenesis (see Chapter 7). The adaptation of rats to a high calorie diet during 'cafeteria feeding' (Rothwell and Stock, 1979, 1981) results in a doubling of the GDP-binding per mg mitochondrial protein and a threefold increase in the total GDP bound per depot (Brooks *et al.*, 1980), although the changes were somewhat less dramatic than during cold-adaptation when a 14-fold increase in total nucleotide binding per depot was seen (Brooks *et al.*, 1980).

The possibility of a metabolic defect in brown-fat thermogenesis under-lying certain cases of genetic obesity (see Chapter 8) has been advanced as a result of studies on the genetically obese ob/ob mouse (Himms-Hagen and Desautels, 1978; Hogan and Himms-Hagen, 1980; Thurlby and Trayhurn, 1980). The ob/ob mouse does not lack brown fat (Himms-Hagen and Desautels, 1978) but the total response to noradrenaline *in vivo* is only 40% of that in the lean control (Thurlby and Trayhurn, 1980). The purine nucleotide binding capacity of brown-fat mitochondria from the warm-adapted ob/ob mouse is unusually low (Himms-Hagen and Desautels, 1978, Hogan and Himms-Hagen, 1980). The ob/ob mouse however does appear to adapt to mild cold (i.e. 14°C) with a decreased weight gain, a doubling of brown-fat protein and a threefold increase in GDP-binding per mg mito-chondrial protein (Hogan and Himms-Hagen, 1980).

Purine nucleotide binding therefore appears to give a reasonable corre-lation with thermogenic capacity in a wide range of animals and experi-mental conditions, although the acute increase in nucleotide binding observed by Desautels *et al.* (1978) remains to be clarified.

An independent approach to the relation between thermogenic state and mitochondrial function is to look for changes in the amount of protein in the 32000 band on polyacrylamide gels. The first report of an increase in this band in response to cold-adaptation (Ricquier and Kader, 1976) preceded the study which identified the short-circuit protein (Heaton *et al.*, 1978). In guinea-pigs the 32 000 band is prominent in new-born and cold-adapted animals but not in warm-adapted, three-week-old animals (Heaton *et al.*, 1978; Ricquier *et al.*, 1979*c*). An increase in the 32 000 band has also been reported for the rat and mouse upon cold-adaptation (Desautels *et al.*, 1978; Ricquier *et al.*, 1979*c*; Hogan and Himms-Hagen, 1980). As would be predicted from both respiratory and binding studies, the hamster retains a prominent 32 000 band even in the warm (Ricquier *et al.*, 1979*c*). It should however always be borne in mind that the 32 000 band is a heterogeneous mixture of different proteins, so that a large change in the short-circuit protein could be partially masked by the presence of the other components.

2.8 THE ACUTE REGULATION OF THE PROTON SHORT-CIRCUIT IN BROWN-FAT MITOCHONDRIA

The respiration of brown adipose tissue *in situ* can increase 40-fold on noradrenaline stimulation (Foster and Frydman, 1978, 1979; Thurlby and Trayhurn, 1980; Rothwell and Stock, 1981). An almost equivalent increase can be achieved with isolated brown adipocytes (Prusiner *et al.*, 1968*a,b*; Williamson, 1970). The stimulation in respiration is accompanied by a shift towards a more oxidized state of both cytochrome *b* and NADH (Prusiner *et al.*, 1968*c*; Williamson, 1970), indicative of a decreased $\Delta\mu_H^+$ across the mitochondrial inner membrane. The respiratory increase and spectral changes are both optimal by 2 min (Prusiner *et al.*, 1968*a,b,c*) and a similar rate of reversal to basal conditions is seen upon addition of the β-adrenergic antagonist propranolol (Pettersson and Vallin, 1976).

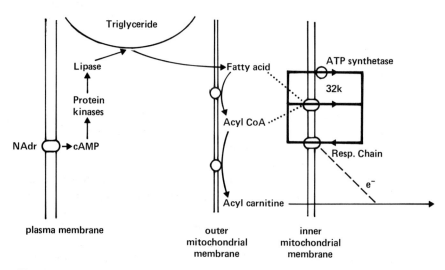

Fig. 2.6 Metabolic events during the acute induction of thermogenesis in brown fat; the possible roles of fatty acid or acyl-CoA in the regulation of the 32 000 protein.

The proton conductance pathway is, presumably, closed in the non-thermogenic state. If this were not so the combination of a slow substrate supply and a high conductance state of the mitochondrial membrane would result in a virtual collapse of $\Delta\mu_H^+$, with the result that the mitochondria would be unable to contribute to cellular ATP production or to regulate cellular Ca. It is therefore reasonable to search for a messenger in the cytosol which can increase the mitochondrial proton conductance at least 40-fold within 2 min of noradrenaline binding to the cell membrane, and decrease the conductance with equal rapidity upon the termination of an acute thermogenic episode.

Starting from first principles, the 'uncoupling messenger' could be a normal component of the reaction sequence from the β-receptor to the mitochondrion, such as cyclic AMP, a cyclic-AMP activated protein kinase, fatty acids, fatty acyl-CoA or fatty acylcarnitine. These would provide the simplest mechanisms since the levels of the messenger would be automatically synchronized with the induction and termination of lipolysis. In addition a number of indirect messengers can be considered, diverging from the β-adrenergic lipolytic sequence at various points, or even being under independent hormonal control (Fig. 2.6). Based upon isolated mitochondrial experiments, putative messengers might include cytosolic purine nucleotides or cytosolic pH. With an indirect messenger it is also necessary to consider the mechanism by which its concentration changes on inducing thermogenesis.

Since the lipolytic and respiratory responses of isolated adipocytes to noradrenaline are both β-specific (Hamilton and Horwitz, 1979; Mohell, Nedergaard and Cannon, 1980) the possibility of an independent hormonal control of the 32 000 protein seems remote. We shall therefore confine our discussion to putative messengers originating from the binding of noradrenaline to the cell.

2.8.1 Decreased cytosolic purine nucleotides

The noradrenaline-stimulation of respiration in isolated brown adipocytes is accompanied by a decrease in total cellular ATP, both as a proportion of the adenine nucleotide pool and in assayable amount (Prusiner *et al.*, 1968*b*; Williamson, 1970; Pettersson and Vallin, 1976). However the decrease in ATP is not commensurate with a 40-fold increase in proton conductance, particularly since the fall in ATP is partially compensated for by an increase in ADP which is also an inhibitory nucleotide although somewhat less effective than ATP (Heaton and Nicholls, 1977; Nicholls and Bernson, 1977). The apparent contradiction between the micromolar concentrations of nucleotide which inhibit the proton conductance of isolated mitochondria (Nicholls, 1974*a*; Nicholls and Bernson, 1977) and the millimolar concentrations present in the cytosol of brown adipocytes (Grav *et al.*, 1970) is not as apparent as it seems at first sight, since most *in vitro* experiments have been performed in the absence of Mg and using the highly inhibitory GDP (e.g. Nicholls, 1974*a*), whereas *in vivo* the predominant nucleotide would be the less potent ATP (Nicholls and Bernson, 1977). With ATP in the presence of Mg, the concentration of nucleotide required to give optimal inhibition of respiration is close to 1 mM (Cannon *et al.*, 1973; Locke and Nicholls, 1981).

A problem common to any 'indirect' messenger is the means by which its own concentration would be regulated. The existing nucleotide analyses with isolated brown adipocytes do not enable a distinction to be made between a decreased cytosolic ATP preceding and initiating increased respiration, and a decrease which is secondary to the decreased ability of the mitochondria to synthesize ATP (Prusiner *et al.*, 1968*b*, Pettersson and Vallin, 1976). We therefore consider the case for a regulation of proton conductance by altered nucleotide concentrations to be unproven.

2.8.2 Increased cytosolic pH

The ability of purine nucleotides to bind to the 32 000 protein (Nicholls, 1976) and inhibit the proton conductance (Nicholls, 1974*a*) decreases very rapidly as the pH of a mitochondrial incubation is increased over the range 6.8–7.5. In theory therefore an increase in cytosolic pH, although unprecedented, could activate the proton conductance of the mitochondrion *in situ*. This proposal has been made by Chinet *et al.* (1978). These authors observed that the slow inhibition of noradrenaline-stimulated respiration by ouabain of superfused tissue (see Section 2.4) only occurred under conditions of metabolic acidosis, and suggested that a primary role of the plasma membrane Na^+K^+-ATPase was to counteract intracellular acidification which would otherwise prevent the conductance increase. However the authors did not present any evidence that noradrenaline action was associated with an increased cytosolic pH, nor that an artifactual increase in pH could cause an increase in basal respiration in intact tissue.

The respiratory response of isolated hamster brown adipocytes to noradrenaline is enhanced if a Krebs-phosphate medium is bubbled with CO_2 even though the extra-cellular pH falls from 7.3 to 6.9 (Pettersson, 1977). Although the effects of CO_2 appear mainly related to the provision of substrate for pyruvate carboxylase (Cannon and Nedergaard, 1979) the results do indicate that a substantial acidification can occur with no adverse effects on thermogenesis. Conversely 10 mM NH_4Cl, which would presumably cause an alkalinization of the cytoplasm due to protonation of NH_3 within the cell, *inhibits* noradrenaline-stimulated respiration (Cannon and Nedergaard, 1979). These lines of evidence are not immediately compatible with the messenger role of an increased cytosolic pH.

2.8.3 Fatty acids and fatty acyl-CoA

Rapid respiration in isolated brown adipocytes may be induced not only by noradrenaline but also by dibutyryl cyclic AMP and/or theophylline (Reed and Fain, 1968) and by externally added fatty acids (Prusiner *et al.*, 1968*a*; Williamson, 1970). If it is accepted that the nature of the respiratory response is the same in each case, this implies that the messenger signalling an enhanced proton conductance must either be fatty acid itself, a metabolite such as acyl CoA or acyl carnitine, or must be generated in response to an elevation in one of these metabolites. Although the amount of fatty acid which must be added to brown adipocyte incubations in order to obtain a respiratory stimulation appears large (typically 400 μM, see Prusiner *et al.*, 1968*a*) the use of high albumin concentrations in the medium means that the molar ratio of added fatty acid to albumin is only about 1 : 1. Since albumin possesses eight binding sites for fatty acids (Spector, Fletcher and Ashbrook, 1971) the unbound free fatty acid in the incubation would be much lower.

Acyl carnitine can reasonably be eliminated as a possible uncoupling messenger since isolated mitochondria retain respiratory control (e.g. Cannon *et al.*, 1973) and show no increase in proton conductance (Locke and Nicholls, 1981) in the presence of 30–50 μM palmitoyl-L-carnitine. We shall therefore now consider the respective merits of fatty acids and acyl-CoA as putative physiological uncouplers.

'Uncoupling' by fatty acids is a pre-chemiosmotic concept of the 1950s (for review see Heaton and Nicholls, 1976) and is to some extent intellectually unsatisfying. Consequently in reviews from our own (Nicholls, 1979) and other laboratories (Lindberg *et al.*, 1981) there has been a tendency to reject the concept as over-simplistic, particularly since fatty acids will uncouple any mitochondrion when added at sufficient concentration (e.g. Pressman and Lardy, 1956; Wojtchak and Wojtchak, 1960). We shall however argue in this section that there are compelling reasons for resurrecting this hypothesis, and to argue that free fatty acids generated by lipolysis can specifically reverse the low conductance state of the 32 000 protein maintained by cytosolic purine nucleotides (Nicholls and Locke, 1981; Locke and Nicholls, 1981). In so doing we are updating an early suggestion by Rafael *et al.* (1969).

The endogenous fatty acid content of brown-fat mitochondria prepared in the absence of albumin is unsually high, typically 10–60 nmol palmitate equivalent per mg (Pedersen and Grav, 1972; Heaton and Nicholls, 1976). The even higher estimates of earlier papers (Hittelman *et al.*, 1969; Bulychev *et al.*, 1972) have been ascribed to errors in the assay (Heaton and Nicholls, 1976). In the presence of these fatty acids the effectiveness of added purine nucleotides in inhibiting the proton conductance is considerably reduced such that respiratory control is not usually observed except at unphysiologically acidic pH (Nicholls, 1974a).

The ability of fatty acids to increase C_mH^+ can be quantitated by measuring the proton conductance (by parallel determinations of respiration and $\Delta\mu_H{}^+$, as in Table 2.2) as a function of the amount of fatty acid bound to the mitochondrion. A comparison of liver and brown-fat mitochondria (Heaton and Nicholls, 1976) showed that the latter were no less than 30 times more susceptible to fatty acids, even in the presence of GDP, than were liver mitochondria. Conversion into acyl-CoA in these experiments was precluded by the absence of ATP.

The case for acyl-CoA as a specific antagonist of purine nucleotide action at the 32000 protein (Cannon *et al.*, 1977, 1980; Lindberg *et al.*, 1981) is largely based upon the ability of added palmitoyl-CoA to induce swelling of brown-fat mitochondria suspended in a KCl medium in the presence of valinomycin and GDP. Since swelling in this medium is limited by Cl⁻ permeability, and since Cl⁻ can permeate through the 32000 protein by a nucleotide-sensitive mechanism (Nicholls and Lindberg, 1973; Nicholls, 1974b) this was advanced as evidence that added palmitoyl-CoA could antagonize the action of GDP at the 32000 protein (Cannon *et al.*, 1977). Unfortunately very low mitochondrial protein concentrations were employed by these authors so that the acyl-CoA per mitochondrion was

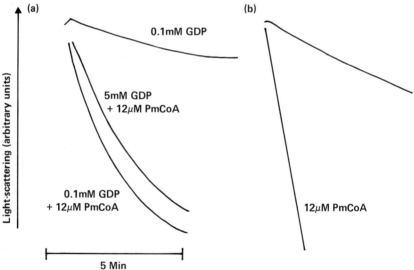

Fig. 2.7 The action of palmitoyl-CoA on the Cl⁻ permeability of brown-fat and liver mitochondria. (a) Hamster brown-fat mitochondria (0.5 mg protein ml⁻¹) incubated in 100 mM KCl, 10 mM K-Tes, pH 7.2, 5 μM rotenone and 0.5 μM valinomycin, with further additions of GDP or palmitoyl-CoA (PmCoA) as indicated. (b) Rat liver mitochondria incubated in the same medium. note that palmitoyl-CoA causes a Cl⁻ permeability which is not restricted to brown-fat mitochondria.

unphysiologically high, varying from 20 to 100 nmol (mg protein)$^{-1}$. While we have been able to confirm the original effect of palmitoyl-CoA (Fig. 2.7), we have been unsuccessful in reversing the swelling by high GDP concentrations (contrast Cannon *et al.*, 1977). Since we find that the same conditions induce rapid swelling in liver mitochondria (Fig. 2.7) which lack the 32 000 protein, we conclude the acyl-CoA action cannot be directed towards the specific thermogenic short-circuit. It may in any case be an over-simplification to assume that Cl$^-$ and proton conductances of the 32 000 protein always change in parallel since, in contrast to palmitoyl CoA, fatty acids increase the proton conductance in the presence of high nucleotide concentrations without affecting the Cl$^-$ permeability (Nicholls and Lindberg, 1973).

In a recent study (Locke and Nicholls, 1981, see also Nicholls and Locke, 1981) we have re-investigated the relative roles of fatty acids and acyl derivatives in the acute regulation of the 32 000 protein by attempting to mimic physiological conditions as closely as possible. We have studied the effects of a controlled transition from a glycolytic substrate (pyruvate) to a 'thermogenic' substrate by slowly infusing fatty acids (or derivatives) while monitoring continuously respiration and membrane potential in order to detect changes in proton conductance (Fig. 2.8). Infusion of substrate not only mimics the steady supply of fatty acids during lipolysis, but also enables the threshold for conductance changes or detergent effects to be accurately determined.

The major conclusion from this study is that the infusion of palmitate, under conditions where it can be activated and oxidized, results in an immediate membrane depolarization. There is no perceptible threshold concentration of palmitate which has to be exceeded before the increase in proton conductance can be detected. During the infusion a steady state is achieved when the fatty acid is being oxidized at the same rate as it is being supplied. As soon as the infusion stops, the membrane potential rises showing that the conductance is decreasing, and respiration returns to basal levels. Notice the similarity in time-course and relative rates with the response of intact cells to noradrenaline and propranolol (Fig. 2.9). It is interesting that Bieber *et al.* (1975) reported that 5 μM palmitate (10 nmol (mg protein)$^{-1}$) was oxidized at uncontrolled rates by hamster brown-fat mitochondria previously depleted of fatty acids in the presence of ATP.

We therefore believe that fatty acid is a sufficient messenger to reverse the inhibition of proton conductance induced by millimolar ATP, although the detailed kinetics of the regulation remain to be investigated. There would thus appear to be no need to search for an independent 'uncoupling' messenger to modulate the induction and termination of a high conductance state required for non-shivering thermogenesis.

The possibility that the proton conductance is due to acyl CoA rather than fatty acid itself may be readily eliminated. Firstly infused palmitoyl CoA

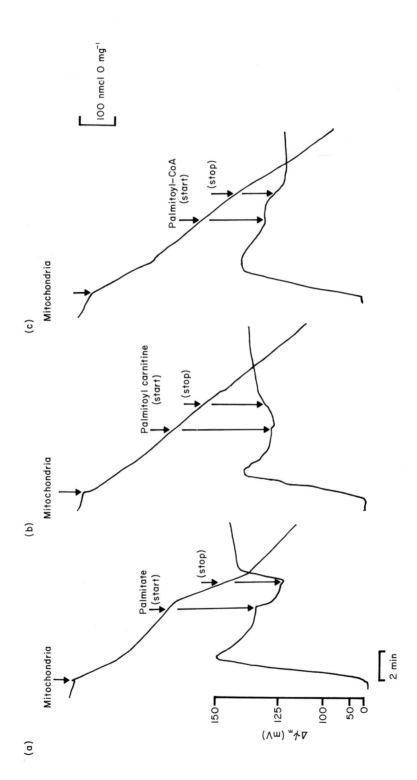

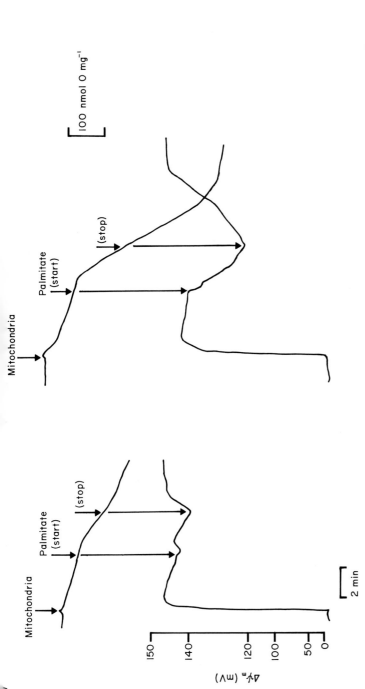

Fig. 2.8 Increased membrane proton conductance during palmitate infusion. Hamster brown-fat mitochondria (0.5 mg ml^{-1}) were incubated in 50 mM KCl, 10 mM Na-Tes, 1 mM P$_i$, 2 mM MgCl$_2$, 1 µg per ml oligomycin, 2 mM DL-carnitine and 3 mM ATP. In (a–c) 5 mM pyruvate and 30 µM CoA were initially present. After oxidation of endogenous fatty acids was complete palmitate (a), palmitoyl-L-carnitine (b) or palmitoyl-CoA (c) were infused for 3 min at 17 nmol min^{-1} (mg protein^{-1}). Respiration and membrane potential were monitored continuously. Note that palmitate alone induces an immediate reversible depolarization and respiratory increase in this preparation of mitochondria. Traces (d) and (e) were performed with a preparation of mitochondria which apparently showed a lower proton conductance during palmitate infusion. However limiting the rate of conversion of palmitate into palmitoyl-CoA by reducing the CoA from 30 µM (d) to 0.3 mM (e) *enhanced* the depolarization and respiratory increase, showing that palmitate rather than palmitoyl-CoA or palmitoyl carnitine is the uncoupling intermediate.

only induces a membrane depolarization after 17 nmol of acyl-CoA per mg protein have been infused (Fig. 2.8). Secondly the potential fails to return at the end of the infusion, suggesting detergent-induced membrane damage. Thirdly the conductance increase is *potentiated* by inhibiting the conversion of fatty acids into acyl-CoA, either by lowering the concentration of CoA or by substituting GDP for ATP (i.e. depriving the fatty acyl-CoA ligase of the necessary substrate ATP).

It is interesting that a decrease of only about 10 mV in membrane potential occurs during the rapid oxidation of the fatty acids. This agrees with earlier observations (Nicholls, 1974a; Nicholls and Bernson, 1977) of the relation between respiratory rate and proton electrochemical potential, and implies that even during a thermogenic episode, the mitochondria might well retain a sufficiently high proton electrochemical potential to allow oxidative phosphorylation to continue.

Although the extreme sensitivity of brown-fat mitochondria to fatty acid 'uncoupling' coincides with the presence of the 32 000 protein, direct evidence is needed that fatty acids act on the protein, rather than at an independent locus. Current evidence is incomplete: the presence or absence of fatty acids appears to have no effect upon the Cl⁻ conductance of the protein (Nicholls and Lindberg, 1973; Nicholls, 1974b) or on the binding of purine nucleotides (Nicholls, 1976). On the other hand the overall proton conductance of the membrane is not consistent with 'fatty acid channels' independent of the 32 000 protein, since the total proton conductance of the freshly prepared mitochondria is far greater than the sum of the conductance supposedly due to fatty acids (purine nucleotide present) plus the conductance due to the 32 000 protein (albumin present) (Nicholls, 1974a, and Table 2.2).

A further area where more clarification is needed is the relationship between experiments with isolated mitochondria and those with isolated brown adipocytes. In particular, reports of very high free fatty acid levels even in non-stimulated brown adipocytes (Bieber *et al.*, 1975) are difficult to reconcile with the extreme sensitivity to fatty acids of the isolated mitochondria, or with the ability of propranolol to inhibit noradrenaline-stimulated respiration within 2 min (Pettersson and Vallin, 1976 and Fig. 2.9). In our opinion, it is not possible to eliminate the possibility that a substantial overestimate of fatty acid concentrations is being made in the adipocyte studies.

2.9 CONCLUSIONS

The final picture of thermogenesis in the brown adipocyte is one of great simplicity and one which, apart from translation into chemiosmotic terms and provision of a molecular mechanism, resembles the 'fatty acid uncoupling' hypothesis of twelve years ago (Drahota *et al.*, 1968; Hittelman

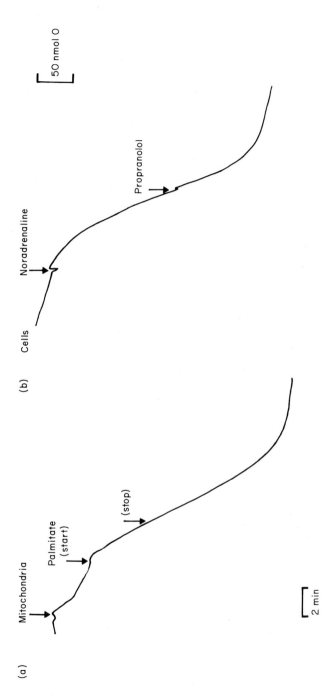

Fig. 2.9 A comparison of the respiratory responses of brown adipocytes to the induction and termination of lipolysis by noradrenaline and propranolol respectively, with the response of isolated brown-fat mitochondria to the initiation and termination of fatty acid infusion. (a) Hamster brown-fat mitochondria were infused with 17 nmol palmitate min^{-1} (mg protein)$^{-1}$ under the conditions of Fig. 2.8(e). (b) Brown adipocytes from cold-adapted guinea pigs were incubated at a concentration of 140000 cells ml^{-1} in an incubation containing 110 mM NaCl, 20 mM Na$_2$HPO$_4$, 5.5 mM KCl, 1.5 mM KH$_2$PO$_4$, 1.4 mM MgSO$_4$, 1.5 mM CaCl$_2$, 5 mM NaHCO$_3$, 10 mM glucose, 10 mM fructose, 40 mg per ml albumin, pH 7.4. Where indicated 5 μM noradrenaline and 10 μM propranolol were added.

and Lindberg, 1970; Williamson, 1970; Reed and Fain, 1970). What remains is to identify the way in which fatty acids modulate the conductance of the 32 000 protein, to show unequivocally the operation of this mechanism in the intact cell, and to unravel the regulatory mechanisms which control the synthesis and breakdown of the 32 000 protein during non-shivering- and diet-induced-thermogenesis. These tasks should prove sufficient to maintain interest in brown fat for a number of years to come.

ACKNOWLEDGEMENTS

Our investigations have been supported by grants from the British Science Research Council and Medical Research Council.

REFERENCES

Ahlabo, I. and Barnard, T. (1971) Observations on peroxisomes in brown adipose tissue of the rat. *J. Histochem. Cytochem.*, **19**, 670–5.

Åkerman, K.E.O. and Nicholls, D.G. (1982) Physiological and bioenergetic aspects of mitochondrial calcium transport. *Rev. Physiol. Biochem. Pharmacol.*, **95**, 149–20.

Aldridge, W.N and Street, B.W. (1968) Mitochondria from brown adipose tissue. *Biochem. J.*, **107**, 315–7.

Bieber, L.L., Pettersson, B. and Lindberg, O. (1975) Studies on norepinephrine-induced efflux of free fatty acids from hamster brown-adipose-tissue cells. *Eur. J. Biochem.*, **58**, 375–81.

Brooks, S.L., Rothwell, N.J., Stock, M.J., Goodbody, A.E. and Trayhurn, P. (1980) Increased proton conductance pathway in brown adipose tissue mitochondria of rats exhibiting diet-induced thermogenesis. *Nature (Lond.)*, **286**, 274–6.

Bulychev, A., Kramer, R., Drahota, Z. and Lindberg, O. (1972) Role of specific endogenous fatty acid fraction in the coupling–uncoupling mechanism of oxidative phosphorylation of brown adipose tissue. *Exp. Cell Res.*, **72**, 169–87.

Cannon, B. and Nedergaard, J. (1979) The physiological role of pyruvate carboxylation in hamster brown adipose tissue. *Eur. J. Biochem.*, **94**, 419–26.

Cannon, B. and Vogel, G. (1977) The mitochondrial ATPase of brown adipose tissue. Purification and comparison with the mitochondrial ATPase from beef heart. *FEBS Lett.*, **76**, 284–9.

Cannon, B., Nicholls, D.G. and Lindberg, O. (1973) Purine nucleotides and fatty acids in energy coupling of mitochondria from brown adipose tissue. In *Mechanisms in Bioenergetics* (eds G.F. Azzone, L. Ernster, S. Papa, E. Quagliariello and N. Siliprandi) Academic Press, New York, pp. 357–63.

Cannon, B., Sundin, U. and Romert, L. (1977) Palmitoyl coenzyme A: a possible physiological regulator of nucleotide binding to brown adipose tissue mitochondria. *FEBS Lett.*, **74**, 43–6.

Cannon, B., Nedergaard, J. and Sundin, U. (1980) Physiological uncoupling in brown fat mitochondria. *Proc. Int. Symp. Thermal Physiol.*, Pecs Hungary, pp. 479–81.

Chinet, A., Friedli, C. and Girardier, L. (1977) Indirect contribution of active Na-K transport in brown fat thermogenesis. *Experientia*, **33**, 778.

Chinet, A., Friedli, C., Seydoux, J. and Girardier, L. (1978) Does cytoplasmic alkanization trigger mitochondria energy dissipation in the brown adipocyte? in *Effectors of Thermogenesis* (eds L. Girardier and J. Seydoux) *Experientia Suppl.*, **32**, 25–32.

Christiansen, E.N. (1971) Calcium uptake and its effect on respiration and phosphorylation in mitochondria from brown adipose tissue. *Eur. J. Biochem.*, **19**, 276–82.

Christiansen, E.N., Pederson, J.I. and Grav, H.J. (1969) Uncoupling and recoupling of oxidative phosphorylation in brown adipose tissue mitochondria. *Nature*, **222**, 857–60.

Dawkins, M.J.R. and Hull, D. (1964) Brown adipose tissue and the response of new-born rabbits to cold. *J. Physiol. (Lond.)*, **172**, 216–38.

Desautels, M. and Himms-Hagen, J. (1980) Roles of noradrenaline and protein synthesis in the cold-induced increase in purine nucleotide binding by rat brown adipose tissue mitochondria. *Can. J. Biochem.*, **57**, 868–976.

Desautels, M., Zaror-Behrens, J. and Himms-Hagen, J. (1978) Increased purine nucleotide binding, altered polypeptide composition, and thermogenesis in brown adipose tissue mitochondria of cold-acclimated rats. *Can. J. Biochem.*, **56**, 378–83.

Douglas, W.W. (1974) Involvement of calcium in exocytosis and in the exocytosis vesiculation sequence. *Biochem. Soc. Symp.*, **39**, 1–28.

Drahota, Z., Honova, E. and Hahn, P. (1968) The effect of ATP and carnitine on the endogenous respiration of mitochondria from brown adipose tissue. *Experentia*, **24**, 431–2.

Fain, J.N., Jacobs, M.D. and Clement-Cormier, Y.C. (1973) Interrelationship of cyclic AMP, lipolysis and respiration in brown fat cells. *Am. J. Physiol.*, **224**, 346–51.

Fink, S.A. and Williams, J.A. (1976) Adrenergic receptors mediating depolarisation in brown adipose tissue. *Am. J. Physiol.*, **231**, 700–6.

Flaim, K.E., Horwitz, B.A. and Horowitz, J.M. (1977) Coupling of signals to brown fat: α- and β-adrenergic responses in intact rats. *Am. J. Physiol.*, **232**, R101–9.

Foster, D.O. and Frydman, M. (1978) Nonshivering thermogenesis in the rat. II Measurements of blood flow with microspheres point to brown adipose tissue as the dominant site of the calorigenesis induced by noradrenaline. *Can. J. Physiol. Pharmacol.*, **56**, 110–22.

Foster, D.O. and Frydman, M.L. (1979) Tissue distribution of cold-induced thermogenesis in conscious warm- or cold-adapted rats: the dominant role of brown adipose tissue. *Can. J. Physiol. Pharmacol.*, **57**, 257–70.

Girardier, L., Seydoux, J. and Clausen, T. (1968) Membrane potential of brown adipose tissue. *J. Gen. Physiol.*, **52**, 925–40.

Girardier, L., Seydoux, J., Giacobino, J.P. and Chinet, A. (1976) Catacholamine binding and modulation of the thermogenic response in brown adipose tissue of the rat. in *Regulation of Depressed Metabolism and Thermogenesis* (eds L. Jansky and X.J. Musacchia) Thomas, Springfield, Ill, USA, pp. 196–212.

Grav, H.J.., Pederson, J.I. and Christiansen, E.N. (1970) Conditions *in vitro* which effect respiratory control and capacity for respiration-linked phosphorylation in brown adipose tissue mitochondria. *Eur. J. Biochem.*, **12**, 11–23.

Greville, G.D. (1969) A scrutiny of Mitchell's chemiosmotic hypothesis of respiratory chain and photosynthetic phosphorylation. in *Current Topics in Bioenergetics* (ed. D.R. Sanadi) Academic Press, New York, **3**, 1–78.

Guillory, R.J. and Racker, E. (1968) Oxidative phosphorylation in brown adipose mitochondria. *Biochim. Biophys. Acta*, **153**, 490–3.

Hamilton, J. and Horwitz, B. (1979) Adrenergic and cyclic nucleotide-induced glycerol release from brown adipocytes. *Eur. J. Pharmacol.*, **56**, 1–5.

Heaton, G.M. and Nicholls, D.G. (1976) Hamster brown adipose tissue mitochondria. The role of fatty acids in the control of the proton conductance of the inner membrane. *Eur. J. Biochem.*, **67**, 511–7.

Heaton, G.M. and Nicholls, D.G. (1977) The structural specificity of the nucleotide binding site and the reversible nature of the inhibition of proton conductance induced by bound nucleotides in brown adipose tissue mitochondria. *Biochem. Soc. Trans.*, **5**, 210–2.

Heaton, G.M., Wagenvoord, R.J., Kemp, A. and Nicholls, D.G. (1978) Brown adipose tissue mitochondria. Photoaffinity labelling of the regulatory site for energy dissipation. *Eur. J. Biochem.*, **82**, 515–21.

Hemingway, A. (1963) Shivering. *Physiol. Rev.*, **43**, 397–422.

Herd, P.A., Horwitz, B.A. and Smith, R.E. (1970) Norepinephrine-sensitive $Na^+ + K^+$/ATPase activity in brown adipose tissue. *Experientia*, **26**, 825–6.

Herd, P.A., Hammond, R.P. and Hamolsky, M.W. (1973) Sodium pump activity during norepinephrine-stimulated respiration in brown adipocytes. *Am. J. Physiol.*, **224**, 1300–4.

Himms-Hagen, J. (1976) Cellular thermogenesis. *Ann. Rev. Physiol.*, **38**, 315–51.

Himms-Hagen, J. and Desautels, M. (1978) A mitochondrial defect in brown adipose tissue of the obese (ob/ob) mouse: reduced binding of purine nucleotides and a failure to respond to cold by an increase in binding. *Biochem. Biophys. Res. Commun.*, **83**, 628–34.

Himms-Hagen, J., Dittmar, E. and Zaror-Behrens, G. (1980) Polypeptide turnover in brown adipose tissue mitochondria during acclimation of rats to cold. *Can. J. Biochem.*, **58**, 336–44.

Hittelman, K.J. and Lindberg, O. (1970) Fatty acid uncoupling in brown fat mitochondria. in *Brown Adipose Tissue*, (ed. O. Lindberg) Elsevier, New York, pp. 245–62.

Hittelman, K.J., Lindberg, O. and Cannon, B. (1969) Oxidative phosphorylation and compartmentation of fatty acid metabolism in brown fat mitochondria. *Eur. J. Biochem.*, **11**, 183–92.

Hogan, S. and Himms-Hagen, J. (1980) Abnormal brown adipose tissue in obese (ob/ob) mice: response to acclimation to cold. *Am. J. Physiol.*, **239**, E301–9.

Hohorst, H.-J. and Rafael, J. (1968) Oxydative Phosphorylierung durch Mitochondrien aus braunem Fettgewebe. *Hoppe-Seyler's Z. Physiol. Chem.*, **349**, 268–70.

Horowitz, J.M. and Plant, R.E. (1978) Controlled cellular energy conversion in brown adipose tissue thermogenesis. *Am. J. Physiol.*, **235**, R121–9.

Horowitz, J.M., Horwitz, B.A. and Smith, R.Em. (1971) Effect *in vivo* of norepinephrine on the membrane resistance of brown fat cells. *Experientia*, **27**, 1419–21.

Horwitz, B.A. (1973) Ouabain-sensitive component of brown fat thermogenesis. *Am. J. Physiol.*, **224**, 352–5.

Horwitz, B.A. (1979) Cellular events underlying catecholamine-induced thermogenesis: cation transport in brown adipocytes. *Fed. Proc.*, **38**, 2170–6.

Horwitz, B.A. and Eaton, M. (1975) The effect of adrenergic agonists and cyclic AMP on the Na^+/K^+ ATPase activity of brown adipose tissue. *Eur. J. Pharmacol.*, **34**, 241–5.

Horwitz, B.A. Horowitz, J.M. and Smith, R.E. (1969) Norepinephrine-induced depolarisation of brown fat cells. *Proc. Nat. Acad. Sci. USA*, **64**, 113–20.

Houstek, J. and Drahota, Z. (1977) Purification and properties of mitochondrial ATPase of hamster brown adipose tissue. *Biochim. Biophys. Acta*, **484**, 127–39.

Joel, C.D., Neaves, W.B. and Rabb, J.M. (1968) Mitochondria of brown fat: oxidative phosphorylation sensitive to 2,4-dinitrophenol. *Biochem. Biophys. Res. Commun.*, **29**, 490–5.

Kramer, R., Hüttinger, M., Gmeiner, B. and Goldenberg, H. (1978) β-Oxidation in peroxisomes of brown adipose tissue. *Biochim. Biophys. Acta*, **531**, 353–6.

Krebs, H.A. (1972) The Pasteur effect and the relations between respiration and fermentation. *Essays Biochem.*, **8**, 1–34.

Krishna, G., Moskowitz, J., Dempsey, P. and Brodie, B.B. (1970) The effect of norepinephrine and insulin on brown fat cell membrane potentials. *Life Sci.*, **9**, 1353–61.

Lepkovsky, S., Wang, W., Loike, T. and Dimick, M.L. (1959) The oxygen uptake and phosphorus uptake by particulate suspensions (mitochondrial) from liver, brown and white fatty tissues. *Fed. Proc.*, **18**, 272.

Lin, C.S. and Klingenberg, M. (1980) Isolation of the uncoupling protein from brown adipose tissue mitochondria. *FEBS Lett.*, **113**, 299–303.

Lin, C.S., Hackenberg, H. and Klingberg, M. (1980) The uncoupling protein from brown adipose tissue mitochondria is a dimer. *FEBS Lett.*, **113**, 304–6.

Lindberg, O., Bieber, L.L. and Houstek, J. (1976) Brown adipose tissue metabolism: an attempt to apply results from *in vitro* experiments on tissue *in vivo*. in *Regulation of Depressed Metabolism and Thermogenesis* (eds L. Jansky and X.J. Musacchia) Thomas, Springfield, Ill, USA, pp. 117–36.

Lindberg, O., de Pierre, J., Rylander, E. and Afzelius, B.A. (1967) Studies of the mitochondrial energy-transfer system of brown adipose tissue. *J. Cell Biol.*, **34**, 293–310.

Lindberg, O., Nedergaard, J. and Cannon, B. (1981) Thermogenic mitochondria. in *Mitochondria and Microsomes* (eds C.P. Lee, G. Schatz and G. Dallner) Addison-Wesley, pp. 93–119.

Locke, R.M. and Nicholls, D.G. (1981) A re-evaluation of the role of fatty acids in the physiological regulation of the proton conductance of brown adipose tissue mitochondria. *FEBS Lett.*, **135**, 249–52.

Mitchell, P. (1976) Vectorial chemistry and the molecular mechanisms of chemiosmotic coupling. *Biochem. Soc. Trans.*, **4**, 399–430.

Mitchell, P. and Moyle, J. (1967) Acid–base titration across the membrane system of rat liver mitochondria. *Biochem. J.*, **104**, 588–600.

Mohell, N., Nedergaard, J. and Cannon, B. (1980) An attempt to differentiate between α- and β-adrenergic respiratory responses in hamster brown fat cells. *Proc. Int. Symp. Thermal Physiol.* Pecs, Hungary, pp. 495–7.

Mokhova, E.N., Skulachev, V.P. and Zhigacheva, I.V. (1977) Activation of the external pathway of NADH oxidation in liver mitochondria of cold-adapted rats. *Biochim. Biophys. Acta*, **501**, 415–23.

Moore, A.L. and Rich, P.R. (1980) Plant mitochondria. *Trends Biochem. Sci.*, **5**, 284–8.

Nedergaard, J. (1981) Effects of cations on brown adipose tissue in relation to possible metabolic consequences of membrane depolarisation. *Eur. J. Biochem.*, **114**, 159–67.

Nedergaard, J. and Lindberg, O. (1979) Norepinephrine-stimulated fatty-acid release and oxygen consumption in isolated hamster brown-fat cells. Influence of buffers, albumin, insulin and mitochondrial inhibitors. *Eur. J. Biochem.*, **95**, 139–45.

Nedergaard, J., Cannon, B. and Lindberg, O. (1977) Microcalorimetry of isolated mammalian cells. *Nature*, **267**, 518–20.

Nedergaard, J., Alexson, S. and Cannon, B. (1980) Cold-adaptation in the rat: increased brown fat peroxisomal β-oxidation relative to maximal mitochondrial oxidative capacity. *Am. J. Physiol.*, **239**, C208–16.

Newsholme, E.A. and Crabtree, B. (1976) Substrate cycles in metabolic regulation and in heat production. *Biochem. Soc. Symp.*, **41**, 61–109.

Nicholls, D.G. (1974a) Hamster brown adipose tissue mitochondria: The control of respiration and the proton electrochemical potential gradient by possible physiological effectors of the proton conductance of the inner membrane. *Eur. J. Biochem.*, **49**, 573–83.

Nicholls, D.G. (1974b) Hamster brown adipose tissue mitochondria. The chloride permeability of the inner membrane under respiring conditions. The influence of purine nucleotides. *Eur. J. Biochem.*, **49**, 585–93.

Nicholls, D.G. (1976) Hamster brown adipose tissue mitochondria. Purine nucleotide control of the ion conductance of the inner membrane. The nature of the nucleotide binding site. *Eur. J. Biochem.*, **62**, 223–8.

Nicholls, D.G. (1977) Hormonal control of brown adipose tissue metabolism. *Biochem. Soc. Trans.*, **5**, 908–12.

Nicholls, D.G. (1979) Brown adipose tissue mitochondria. *Biochim. Biophys. Acta*, **549**, 1–29.

Nicholls, D.G. (1982) *Bioenergetics: an introduction to the chemiosmotic theory*. Academic Press, London.

Nicholls, D.G. and Bernson, V.S.M. (1977) Inter-relationships between proton electrochemical potential, adenine-nucleotide phosphorylation potential and respiration, during substrate-level and oxidative phosphorylation by mitochondria from brown adipose tissue of cold-adapted guinea-pigs. *Eur. J. Biochem.*, **75**, 601–12.

Nicholls, D.G. and Crompton, M. (1980) Mitochondrial calcium transport. *FEBS Lett.*, **111**, 261–8.

Nicholls, D.G. and Heaton, G.M. (1978) Anion uniport across the inner membrane of brown adipose tissue mitochondria. in *The Proton and Calcium Pumps* (eds G.F. Azzone *et al.*) Elsevier, Amsterdam, pp. 309–18.

Nicholls, D.G. and Lindberg, O. (1973) Brown adipose tissue mitochondria. The influence of albumin and nucleotide on passive ion permeabilities. *Eur. J. Biochem.*, **37**, 523–30.

Nicholls, D.G. and Locke, R.M. (1981) Heat generation by mitochondria. in *The Proton Cycle* (eds P.C. Hinkle and V.P. Skulachev) Addison-Wesley, pp. 567–576.

Nicholls, D.G., Grav, H.J. and Lindberg, O. (1972) Mitochondria from hamster brown adipose tissue. Regulation of respiration *in vitro* by variations in volume of the matrix compartment. *Eur. J. Biochem.*, **31**, 526–33.

Nicholls, D.G., Cannon, B., Grav, H.J and Lindberg, O. (1974) Energy dissipation in non-shivering thermogenesis. in *Dymamics of Energy Transducing Membranes* (eds L. Ernster *et al.*) Elsevier, Amsterdam, pp. 529–37.

Pavelka, M., Goldenberg, H., Hüttinger, M. and Kramer, R. (1976) Enzymic and morphological studies on catalase positive particles from brown fat of cold adapted rats. *Histochem.*, **50**, 47–55.

Pedersen, J.I. (1970) Coupled endogenous respiration in brown adipose tissue mitochondria. *Eur. J. Biochem.*, **16**, 12–8.

Pedersen, J.L., Christiansen, E.N. and Grav, H.J. (1968) Respiration-linked phosphorylation in mitochondria of guinea-pig brown fat. *Biochem. Biophys. Res. Commun.*, **32**, 492–500.

Pedersen, J.I. and Grav, H.J. (1972) Physiologically-induced loose coupling of brown-adipose-tissue mitochondria correlated to endogenous fatty acids and adenosine phosphates. *Eur. J. Biochem.*, **25**, 75–83.

Pettersson, B. (1977) CO_2-mediated control of fatty acid metabolism in isolated hamster brown-fat cells during norepinephrine stimulation. *Eur. J. Biochem.*, **72**, 235–40.

Pettersson, B. and Vallin, I. (1976) Norepinephrine-induced shift in levels of adenosine 3′:5′-monophosphate and ATP parallel to increased respiratory rate and lipolysis in isolated hamster brown-fat cells. *Eur. J. Biochem.*, **62**, 383–90.

Pressman, B.C. and Lardy, H.A. (1956) Effects of surface-active agents on latent ATPase of mitochondria. *Biochim. Biophys. Acta*, **21**, 458–66.

Prusiner, S. and Poe, M. (1970) Thermodynamic considerations of mammalian heat production in *Brown Adipose Tissue* (ed. O. Lindberg) Elsevier, pp. 263–82.

Prusiner, S.B., Cannon, B. and Lindberg, O. (1968a) Oxidative metabolism in cells isolated from brown adipose tissue. I. Catecholamine and fatty acid stimulation of respiration. *Eur. J. Biochem.*, **6**, 15–22.

Prusiner, S., Cannon, B., Ching, T.M. and Lindberg, O. (1968b) Oxidative metabolism in cells isolated from brown adipose tissue. 2. Catecholamine regulated respiratory control. *Eur. J. Biochem.*, **7**, 51–7.

Prusiner, S., Williamson, J.R., Chance, B. and Paddle, B.M. (1968c) Pyridine nucleotide changes during thermogenesis in brown fat in tissue *in vivo*. *Arch. Biochem. Biophys.*, **123**, 368–77.

Rafael, J. and Heldt, H.W. (1976) Binding of guanine nucleotides to the outer surface of the inner membrane of guinea-pig brown fat mitochondria in correlation with the thermogenic capacity of the tissue. *FEBS Lett.*, **63**, 304–8.

Rafael, J., Klaas, D. and Hohorst, H.-J. (1968) Mitochondria aus braunem Fettgewebe: Enzyme und Atmungskettenphosphorylierung während der prä und postnatalen Entwichlung des interscapularen Fettkörpers des Meerschweinches. *Hoppe Seyler's Z. Physiol. Chem.*, **349**, 1711–24.

Rafael, J., Ludolph, H.-J. and Hohorst, H.-J. (1969) Mitochondrien aus braunem

Fettgewebe: Entkopplung der Atmungsketttenphosphorylierung durch langkettige Fettsauren und Rekopplung durch Guanosintriphosphat. *Hoppe-Seyler's Z. Physiol. Chem.*, **250**, 1121–31.

Rafael, J., Wiemer, G. and Hohorst, H.-J. (1974) Mitochondria from brown adipose tissue: Influence of albumin, guanine nucleotides and of substrate level phosphorylation on the internal adenine nucleotide pattern. *Physiol. Hoppe Seyler's Z. Physiol. Chem.*, **355**, 341–52.

Reed, N. and Fain, J.N. (1968) Stimulation of respiration in brown fat cells by epinephrine, dibutyryl-3′,5′-adenosine monophosphate, and *m*-chloro (carbonyl cyanide) phenylhydrazone. *J. Biol. Chem.* **243**, 2843–8.

Reed, N. and Fain, J.N. (1970) Hormonal control of the metabolism of free brown fat cells. in *Brown Adipose Tissue* (ed. O. Lindberg) Elsevier, pp. 207–24.

Ricquier, D. and Kader, J.C. (1976) Mitochondrial protein alteration in active brown fat: a SDS: polyacrylamide gel electrophoretic study. *Biochem. Biophys. Res. Commun.*, **73**, 577–83.

Ricquier, D., Gaillard, L. and Turc, J.M. (1979*a*) Microcalorimetry of isolated mitochondria from brown adipose tissue. Effect of GDP. *FEBS Lett.*, **99**, 203–6.

Ricquier, D. Gervais, C., Kader, J.C. and Hemon, P. (1979*b*) Partial purification by GDP-agarose affinity chromatography of the 32,000 m.w. polypeptide from mitochondria of brown adipose tissue. *FEBS Lett.*, **101**, 35–8.

Ricquier, D., Mory, G. and Hemon, P. (1979*c*) Changes induced by cold adaptation in the brown adipose tissue from several species of rodents, with special reference to the mitochondrial components. *Can. J. Biochem.*, **57**, 1262–6.

Rothwell, N.J. and Stock, M.J. (1979) A role for brown adipose tissue in diet-induced thermogenesis. *Nature (Lond.)*, **281**, 31–5.

Rothwell, N.J. and Stock, M.J. (1981) Influence of noradrenaline on blood flow to brown adipose tissue in rats exhibiting diet-induced thermogenesis. *Pflügers Arch*, **389**, 237–42.

Rottenberg, H. (1975) The measurement of transmembrane electrochemical proton gradients. *Bioenergetics*, **7**, 61–74.

Seydoux J. and Girardier, L. (1978) Control of brown fat thermogenesis by the sympathetic nervous system. *Experientia Suppl.*, **32**, 153–67.

Skaane, O., Christiansen, E.N. and Grav, H.J. (1972) Oxidative properties of brown adipose tissue mitochondria from rats, guinea-pigs and hedgehogs. *Comp. Biochem. Physiol.*, **42B**, 91–107.

Skala, J.P., Hahn, P. and Knight, B.L. (1978) The 'second messenger' system in brown adipose tissue of developing rats. Its molecular composition and mechanism of function. *Experientia Suppl.*, **32**, 69–74.

Skulachev, V.P., Maslov, S.P., Siukova, V.G., Kalinichenko, L.P. and Maslova, G.M. (1963) Uncoupling of oxidation from phosphorylation in muscles of cold adapted white mice. *Biokhimiya*, **28**, 70–9.

Smith, R.E. and Horwitz, B.A. (1969) Brown fat and thermogenesis. *Physiol. Rev.*, **49**, 330–425.

Smith, R.E., Roberts, J.C. and Hittelman, K.J. (1966) Non-phosphorylating respiration of mitochondria from brown adipose tissue of rats. *Science*, **154**, 653–4.

Smith, T.J. and Edelman, I.S. (1979) The role of sodium transport in thyroid thermogenesis. *Fed. Proc.*, **38**, 2150–3.

Spector, A.A., Fletcher, J.E. and Ashbrook, J.D. (1971) Analysis of long chain free fatty acid binding to bovine serum albumin. *Biochemistry*, **10**, 3229–32.

Sundin, U. and Cannon, B. (1980) GDP-binding to the brown fat mitochondria of developing and cold-adapted rats. *Comp. Biochem. Physiol.*, **65B**, 463–71.

Thurlby, P.L. and Trayhurn, P. (1980) Regional blood flow in genetically obese (ob/ob) mice. The importance of brown adipose tissue to the reduced energy expenditure on non-shivering thermogenesis. *Pflügers Arch.*, **385**, 193–201.

Williams, J.A. and Matthews, E.K. (1974a) Effects of ions and metabolic inhibitors on membrane potential of brown adipose tissue. *Am. J. Physiol.*, **227**, 981–6.

Williams, J.A. and Matthews, E.K. (1974b) Membrane depolarisation, cyclic AMP, and glycerol release by brown adipose tissue. *Am. J. Physiol.*, **227**, 987–92.

Williamson, J.R. (1970) Control of energy metabolism in hamster brown adipose tissue. *J. Biol. Chem.*, **245**, 2043–50.

Wojtchak, L. and Wojtchak, A.B. (1960) Uncoupling of oxidative phosphorylation and inhibition of ATP-Pi exchange by a substance from insect mitochondria. *Biochem. Biophys. Acta*, **39**, 277–86.

Chapter Three

Brown Fat:
An Energy Dissipating Tissue

Lucien Girardier

3.1 INTRODUCTION

This chapter is not intended to be a review on brown adipose tissue. Throughout the different chapters of this book, the mitochondriologist, the nutritionist, the endocrinologist and the pharmacologist have described facets of this tissue and this adequately covers the metabolic potential of brown fat. In this chapter, I will be concerned with the problem of how and under what conditions this potential can be called upon by the organism. Brown fat mitochondria are able to develop tremendous respiratory rates, and this requires equally high substrate supply, oxygen delivery and heat transfer capabilities. These flows must be matched in some way, both in the steady state and dynamically, otherwise an acute shortage of either substrate or oxygen could develop. A shortage of substrate at high respiratory rates leads to a failure of cellular ATP production and cellular ionic regulation. A shortage of oxygen supply on the other hand, besides limiting the heat production, would also lead to a collapse of cellular ATP synthesis, unless the brown adipocyte exhibited a Pasteur effect, which is not the case – at least in tissue from adult animals. The dynamic matching of supply to utilization therefore requires a concerted regulation of blood supply and cellular activation, which is probably mainly neural. When trying to integrate the biochemical, physiological and behavioural information relating to brown fat, one soon realizes that this is available mostly for one species, namely, the laboratory rat. Thus, one has to focus on the rat as a frame of reference.

3.2 THE SURVIVAL VALUE OF AN ENERGY DISSIPATOR

What evolutionary pressure can be evoked for the development of an energy dissipating tissue? The energy released by respiration, and other processes of energy metabolism, becomes available to the cell mainly through the build-up of energy-rich compounds. The best examples of these compounds are the nucleotides of adenylic acid, which have the considerable advantage of being relatively resistant to non-enzymatic hydrolysis. Among these, ATP holds a leading place as a universal energy transfer agent. Now, the central dogma of metabolic regulation is the so-called *respiratory control,* which states that the respiratory rate is controlled by the concentration of the reactants of ATP synthesis, i.e. ADP and inorganic phosphate. Thus, the rate of respiration is mastered by the energy require-ment of the cell. The evolutionary pressure for more and more efficient pathways for ATP synthesis is easily conceivable and the mitochondrial respiration is its realization. It is functionally a compact, integrated system, oxidizing substrates at a rate much higher than would be possible with a redox system in solution. This system is unique, and the respiratory chain structure is remarkably similar in widely different cells; hence the inhibitors, uncoupling substances and the pH effect on mitochondrial respiration in these cells are similar.

The evolutionary pressure which led to the creation of a dissipating mechanism, short-circuiting the energy conservation system, is much more difficult to conceive. What can the survival value of an energy dissipator be? A clear case is that of hibernators, in which brown fat allows adaptation of the animal to seasonal variations in the quality and/or quantity of food supply. Endothermia is, when compared to ectothermia, an extremely 'expensive' system for homeothermia, especially for small animals. The basal metabolism of a small endotherm is 8–10 times higher than that of an ectotherm of the same body weight and temperature. An efficient way to ensure survival of a small mammal during seasonal food shortage is to turn off the endothermic regulatory system and reduce the energy requirement to the essential energy expenditure. In this way, energy re-serves may last 8–10 times longer. However, a warming device is neces-sary to switch back to endothermia and a non-hibernator has no means of increasing body temperature once it has become hypothermic. It must rely entirely upon extraneous heat sources. Hibernators, on the other hand, are able, owing to strategic brown fat deposits, to reactivate their endo-thermy. In response to sympathetic stimulation, this tissue will export heat and substrate, primarily to the animal's brain and thoracic organs, then to the splanchnic organs, allowing a return to the high metabolic rates char-acteristic of endotherms. As Bernson and Nicholls (1974) have shown, brown fat mitochondria are perfectly adapted for their role of rewarming source. They must be able to function at 4°C, but the citric acid cycle, however, can scarcely function at this low temperature. Thus, acetyl-CoA from β-oxidation tends to accumulate and in the absence of an alternative pathway, thermogenesis would therefore be limited. However, such a path-way does exist in brown fat mitochondria, whereby acetyl-CoA is hydro-lysed to acetate, and fatty acid oxidation to acetate still occurs at 4°C. Thus brown fat can spark the rise of body temperature required for arousal from hibernation.

It is interesting to note that qualitative changes in food supply can induce hibernation. Ambid (1980) has shown that in the garden dormouse, the mere suppression of alimentary proteins induces hypothermia even if the caloric supply remains abundant. Fed with such a diet, hypothermia develops in two to seven days, and this at any season or ambient temperature, and the hibernation is always preceded by a decrease in activity of the adrenal medulla. It is reasonable to suppose that the reduction in circulating adrenaline facilitates the decrease of the metabolic rate. Quite the reverse is observed in the rat. It has been reported that, in this animal, a low protein diet induces a hypertrophy of brown fat (Teague *et al.*, 1981) and even a single low protein, high carbohydrate meal increases the weight of the interscapular brown fat and its rate of respiration (Glick *et al.*, 1981). It is known that the latter is controlled by the sympathetic innervation of the tissue, and this would appear to be stimulated by the low protein diet. This suggests the dangerously attractive hypothesis that brown fat may serve as an effector in a homeostatic mechanism by which a diet unbalanced in its macronutrients can become more balanced by oxidation of the macronutrients in excess. This would explain why, when mixed meals are offered, the diet-induced thermogenesis is considerably smaller than would be expected by the sum of the 'specific dynamic effect' produced when each of the macronutrients is provided separately. The problem of the coupling of sympathetic activity with diet composition and caloric content is a fast growing field which is treated in Chapter 4. It is fascinating to see that, to solve the same problem of diet unbalance and using the same effector, Nature has found different solutions in closely related species: a temporary shut off in garden dormouse, a hibernator, and a stimulation in the rat, a permanent homeotherm.

Most of the studies focusing on brown adipose tissue are concerned with its role in cold adaptation in adult animals. When adult homeotherms are exposed to a cold environment, the emergency extra heat is produced by two regulatory processes: non-shivering and shivering thermogenesis. The latter interferes with body movement. Its progressive suppression by the adaptive increase of the capacity for non-shivering thermogenesis is certainly of survival value, and brown adipose tissue plays a major role in this adaptation (Bruck, 1970). However, this adaptative reaction appears as a remanifestation of a mechanism that is prevalent during the early stage of life. Indeed, the neonates of many species are able to produce a relatively high rate of non-shivering thermogenesis even though they had certainly not experienced cold exposure during intrauterine life. One might think that brown fat provided the neonate with a special heat dissipator whereby its thermoregulatory system is adjusted to the smaller body size. An extreme example of the importance of brown fat in the survival of a neonate is that of harp seal pups as described by Grav *et al.* (1974). They are born each year in February and March on the open ice-floes of the North Atlantic Ocean. The recorded air temperature can be as low as $-30°C$. This implies that the pup is exposed to a thermic gradient approaching 70°C at birth. Moreover the newborn pup has to contend with a low insulation since the thick layer of subcutaneous blubber is a post-natal development and the infantile fur of the harp seal insulates efficiently when dry, but poorly when wet. The cold stress at birth is probably exacerbated by the arctic wind. Grav *et al.* (1974) report that very young pups have not been observed to shiver, even when wet, and so compensation for heat loss has to come from non-shivering thermogenesis. Indeed, these authors have found that the carcass of the pup is covered in a yellowish-brown adipose tissue forming a continuous layer which varies in thickness from 2 to 8 mm and which they histologically identified as brown adipose tissue. The authors conclude that 'it is tempting to propose that the pups meet the thermal challenge offered at birth by stimulating oxidation of the lipid stores contained in the brown adipose tissue'. Seal pups represent an extreme. First, the subcutaneous location of brown fat is not typical. As a rule, brown adipose tissue does not belong to the panniculus adiposus. It is essentially a deep tissue, on or close to blood vessels. Second, the pups of harp seals have a body weight of 10–14 kg. According to the data of Davydov and Makarova (1964) their resting metabolism in an environmental temperature of $-11°C$ is 5.7 W kg^{-1} which is three times the basal metabolism of other homeotherms of this weight. Immersed in ice-cold water, their resting metabolism rises to 13.4 W kg^{-1}.

For smaller pups, the 5–6 g newborn rat for instance, the cost of homeothermia through adaptive increases in heat dissipation of this magnitude would be unbearably high for the cardiovascular and respiratory system. Conklin and Heggeness (1971) have shown that, despite the fact that newborn rats have a high capacity for heat production per kilogram of body weight, they have so little control over heat loss, that utilization of adaptive heat production for body temperature regulation is limited to a very narrow ambient temperature range. Survival during the first days of life is probably aided by the young rat's ability to sustain a high metabolic rate despite appreciable body cooling, which is a striking characteristic of the pups. In fact, Farkas *et al.* (1972) are inclined to think that the failure of the newborns of a number of mammalian species to maintain deep body temperature at ambient temperatures actually below their thermoneutral zone represents a regulated response mediated by a central controller. They have shown that newborn rabbits and guinea pigs, submitted to a moderately cold environment (20°C) respond by a marked fall in deep body temperature despite a sharp rise in heat production. Exposure to hypoxia reduced body temperature further. However, after termination of hypoxia, heat production increased significantly, above the level observed under the same ambient conditions prior to hypoxia, indicating that the full capacity for heat production was not exploited. Furthermore, they showed that no correlation exists between body weight and the cooling rate observed during exposure to mild cold. They reason that, since the fall in deep temperature of these newborns cannot be attributed either to an inability to increase heat production sufficiently, nor to the mass:surface ratio, it must depend on some central proportional controller. It appears that infant huddling, maternal behaviour and a nesting structure, adjusted to provide a micro environment placing nearly minimal heat-producing requirement upon the infant, is of overwhelming importance. This, together with the characteristic of the newborn rat to be able to sustain high metabolic rate despite appreciable body cooling (Conklin and Heggeness, 1971), their very low lethal temperatures (down to about 7°C, Fairfield, 1948) and the fact that, in contrast with the adults, their heat production and body temperature are resistant to hypercapnia (Varnai *et al.*, 1970), are powerful means to aid newborn survival. With this in mind, the selective pressure for brown fat development as thermic effector in the newborn does not appear to be so obvious, although the heat it produces might have been originally brought to bear upon a relatively small size heat sink of high thermic sensitivity.

Smith (1964) has insisted on the fact that the topology of brown fat is such that, in the rat, the heat it produces is applied to the vascular supply of the thoracocervical spinal cord, the heart and related thoracic structures. Furthermore, brown fat overlays the dorsomedial course of the thoracic sympathetic chain, thus, direct conductive heating could be available to these neuronal elements. This, as well as the convective heating of the cervical and thoracic segments of the spinal cord, are the structures from which all of the major neuronal controls of the thoracic organs emerge. The activity of brown fat could considerably reduce the extent of temperature fluctuations of these vital regions during incidental periods of cold exposure. With these regions being preserved, the temperature of the periphery can be allowed to drift towards the ambient temperature, thus decreasing the heat gradient, and consequently the rate of heat loss. A cold rectum is no vital threat *per se,* whereas cold has a pronounced effect on the integrative function of the nervous system. In many organisms, a rapid drop in temperature will delay reflex responses and interfere with learned behaviour and so the selective pressure for temperature regulation of structures controlling behaviour, essential for survival, is easily conceived.

Along these lines, it is interesting to note that some fish possess brain heaters, and a recent report by Carey (1982) gives a good description of the heater in the swordfish. This large pelagic fish spends the night near the surface, but goes as deep as 600 m during the day. In these vertical excursions, water temperature may change as much as 19°C. The temperature of the core of the fish is tightly coupled to water temperature by circulation of the blood, but the temperature in the cranial cavity is 10–14°C warmer than the water. Associated with the eye muscle is a mass of brown tissue applied to the ventral side of the brain case. This tissue is described as made of cuboidal cells with dense brown cytoplasm packed with mitochondria and containing numerous vacuoles. Its cytochrome *c* concentration is of the same magnitude as that reported for the mammalian brown fat. This mass of thermogenic tissue is supplied with blood through a vascular heat exchanger preventing the convective dissipation and confining the heat to the vicinity of the brain. This heater is probably important in allowing the fish to hunt effectively during the large temperature changes which are part of its daily routine.

The subcutaneous brown fat of the seal pups, the perivascular brown adipose tissue of the rat and the brown tissue of the swordfish are focal and extreme expressions of an 'unthrifty' gene. One wonders if the risk to develop such an 'uncoupler' is commensurate with the survival value of cold adaptation and if the 'original purpose' might not be deeper. In this perspective the work of Stucki (1980) is of interest. This author developed, using the formalism of non-equilibrium thermodynamics, a phenomenological theory allowing the calculation of optimal efficiency of oxidative phosphorylation. He showed that this optimum is a function of the degree of coupling. For instance, a maximal net rate of oxidative phosphorylation occurs at a degree of coupling of 0.78, but maximal power output of oxidative phosphorylation (net rate times established phosphate potential) results at a degree of coupling of 0.91. It could be speculated that the variable coupling of brown fat evolved from a mechanism of optimalization of the degree of coupling to output requirement of the different cell types.

3.3 ANATOMY

3.3.1 Macroscopy

Macroscopic anatomical studies are expected to provide two important parameters for the physiology of brown adipose tissue: its topology and its total mass for a given individual. The topology gives clues on tissue function, and the total mass, together with the thermogenic capacity per unit weight of tissue, yield a value of the maximal heat production that can be attributed to this source. The prerequisite of these determinations is that brown fat can be macroscopically distinguished from white adipose and connective tissue. This is the case in hibernators, bats and true rodents. In this group of mammals the tissue is distinctly brown and retains the same appearance throughout the life of the animal; but in rabbit, cat, man, and presumably the majority of the mammals, it is only distinct in the newborn, and in later life comes to resemble white adipose tissue. This evidently sets a serious problem in evaluating the putative role of brown fat in diet-induced thermogenesis in adult man, for instance. In the first group and in the newborn of the second group, brown fat occurs in organized patches or pads at several loci where they can be identified by their brownish colour. An important part of the colour is undoubtedly due to their blood content. This can be verified by perfusing, through an afferent artery, the tissue's vascular bed with an haemoglobin-free, balanced salt solution. At the onset of perfusion, as blood is washed away, the tissue colour fades away to a light tan. The faint brown–yellow colour of saline-perfused brown fat is thought to be mainly due to the high flavin and cytochrome contents of these cells which are exceptionally rich in mitochondria. Remarkably, the brown fat deposits have a comparable distribution in the various species, although there are wide variations in the relative size of the deposits (see Afzelius, 1970). The larger deposits can be categorized in the following way: a bilobar mass situated in a depression of the muscles in the mid-dorsal line between the scapulae and two symmetrical lobes in the cervical region make up the superficial components; the great vessels entering the thorax are covered with brown fat; deposits are lying in contact with the blood vessels of the neck, of the axillae and, in the mediastinum, deposits are found between the oesophagus and trachea as well as along the descending aorta; thin sheets are found along the course of the intercostal vessels. In the abdominal region, the most conspicuous deposits are the suprailiac and perirenal, with smaller deposits enveloping the adrenals, aorta and autonomic ganglia.

This description is not exhaustive and scattered brown fat is likely to be found between and among various muscle groups. Feyrter (1973) noted that the tissue never belongs to the *panniculus adiposus*. Merklin (1974) pointed out that 'Viewed in its totality, one is struck by the vest-like arrangement of brown fat'. Smith (1964) drew attention to the close relationship between brown fat and the vascular bed, a feature which allows the metabolic heat of the tissue to be applied directly to the blood stream as it returns to the thorax, and the vital structures are thus protected from undue cooling. Moreover, Smith (1964) noted that countercurrent heat exchange, i.e. when arterial supply and venous drainage of a deposit are juxtaposed, can result in a positive feedback on heat production. When the thermogenesis of the tissue is stimulated, the venous blood will be warmer than arterial blood and heat will be transferred to the arteries with the resulting conservation and even amplification of the heat produced by virtue of the direct effect of temperature on reaction rate (Arrhenius effect) added to the local metabolic heat. According to Smith and Robert's (1964) anatomical studies, such an effect could occur at the level of the interscapular and superior cervical pads, which are supplied bilaterally by arterial channels closely juxtaposed with their corresponding venous return. Hence, countercurrent heat exchange may readily occur along these routes. Such a system would be unstable without some form of damping, but this is provided by an alternative venous drainage, unattended by an artery, which feeds directly into the inner vertebral sinus of the cervical and thoracic segments and hence, via the unpaired vertebral and thoracic veins, into the great veins entering the heart. The important point is that vasomotor control of the route of blood flow might, not only permit very rapid adjustment of the on-off thermogenic responses, but also favour the capacity for transient injections of large quantities of heat into vital structures such as the spinal chord and heart, or other specific heat sinks. Since the pioneering work of Smith (1964) these aspects of brown fat regulation have received little attention. However, in the human foetus, Merklin (1974) also noted the close association of the posterior cervical and interscapular bodies with the major cervical blood vessels and posterior vertebral plexus, but he also described direct vascular connections of the perirenal bodies with the kidneys and of the anterior abdominal body wall with the liver. He noted that this suggests a direct association of brown fat with the metabolism of these organs.

To give a quantitative idea of the total amount of brown fat which can be dissected, the following figures are given as a percentage of body weight: 0.8% and 1.4% in adult warm and cold-adapted rats respectively (Foster and Frydman, 1979) and 1.4% in human foetus (Merklin, 1974). Other mammals values can be found in Afzelius (1970). In adult, non-rodent, non-hibernating mammals, the distinction between brown and white fat cannot be made macroscopically and it is impossible to give an estimate of how much brown fat they retain from neonatal life.

3.3.2 Light microscopy

The functional distinction between white and brown fat tissues is that only brown fat is equipped for the rapid oxidation of the products of lipolysis of its fat stores. Indirect, morphological clues of this difference have to be found. The first one is a granular cytoplasm, which indicates a large number of mitochondria in the cell and correlates with the high oxidative power of the cell. The second, much more indirect, is its plurivacuolar aspect which results from the multiple cytosolic fat depots. This is in contrast with the white fat cell, which is normally occupied by a single fat droplet, pushing the nucleus to the periphery. The multilocularity of brown adipocytes is thought to be a consequence of the rapid oxidation of fat, but the evidence is circumstantial. In conditions where it can be thought that the tissue is active (neonates, cold-exposed subjects) the cell is found to be multilocular, whereas after denervation of a brown fat-pad (to inactivate it), many unilocular cells can be found. Often, the multilocularity is said to derive from the breakdown of lipid droplets but since small droplets have a higher surface tension than large ones, such a dispersion of large globules into smaller ones requires energy and cannot proceed spontaneously. Using time-lapse films of cultivated brown fat-cells Dyer (1968) observed only a coalescence of fat droplets and never a dispersion, which conforms to what is expected on physicochemical considerations. In Dyer's (1968) morphological data, from both *in vivo* and *in vitro* systems, lipid synthesis is initially associated with the formation of small droplets, which subsequently increase in size and finally fuse with one another by the process of coalescence. The point is that multilocularity is probably a morphological correlate for rapid multifocal lipid synthesis. When thermogenesis is stimulated, fatty acid synthesis in brown fat is dramatically increased in the rat (McCormack and Denton, 1977; Trayhurn, 1979) and in the mouse (Rath *et al.*, 1979) and thus the multilocular aspect is probably an indirect but valid clue of the thermogenic activity. Finally, two other features have been claimed to characterize brown fat tissue; the size of the cells and the location of the tissue sample. Brown fat is composed of adipocytes, 25–40 μm in diameter, which are, as a rule, smaller than those of white adipose tissue (Feyrter, 1973; Hassi, 1977) and is found in characteristic locations.

Applying all these various criteria it is found that all kinds of evolving forms of brown fat cells occur simultaneously in human fat tissue and Feyrter (1973) proposes the scheme of brown fat evolution illustrated in Fig. 3.1. Fat tissue with the above described cytological appearance can be found in

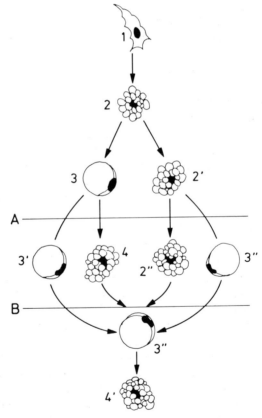

Fig. 3.1 Development pathways of the multilocular brown adipocyte in man. Diagram based on the evaluation of autopsy records. 1: mesenchymal mother cell; 2 ,2′, 2″: primary multilocular brown fat cells; 3, 3′, 3″, 3‴: secondary unilocular brown fat cells; 4, 4′: secondary multilocular brown fat cells. A: time of birth. B: time of maturity (18–20 years old). Redrawn from Feyrter (1973).

human subjects up to the eighth decade of age (Heaton, 1972) close to the neck vessels and muscles, under the clavicles and axillae, around intercostal vessels, between trachea and oesophagus, in the para-aortic region as well as in the perirenal and suprarenal regions. Curiously, brown fat in the interscapular area disappears, gradually, up to 30 years of age, and sharply thereafter. These results have been confirmed by Hassi (1977), who noted that even though, with advancing age from childhood to adulthood, the presence of human brown fat declined, its enzyme intensity on histochemical preparations fell only slightly. On samples taken at autopsy, he studied, among others, succinic dehydrogenase, cytochrome oxidase and a series of enzymes implicated in triglyceride synthesis and degradation. In another large series of autopsies, performed on subjects from 1 month to 86 years, Tanuma *et al.* (1976) found that brown fat in the perirenal region persisted in variable amounts. These authors give some functional clue. They observed in cases of death from burning, drowning, bleeding, etc., cytoplasm-rich multilocular cells, thought to be fat-depleting or consuming cells, together with fat-depleted cells with granular cytoplasm, suggesting that active oxidation of fat had taken place as a result of sympathetic stimulation. Teplitz *et al.* (1974) have described, in cases of severe chronic hypoxaemia, the direct transformation of periadrenal white fat into brown adipose tissue in adult man.

Physiopathological conditions show that brown fat can be reactivated in human adults by circulating catecholamines. Phaeochromocytomas are tumours developing from chromaffin cells that secrete catecholamines so actively that the mean plasma catecholamine levels are about 20 times above normal. These tumours are often associated with brown adipose tissue hyperplasia (Melicow, 1957) and sometimes with brown fat pseudotumours (Rona, 1964; Leiphart and Nudelman, 1970; English *et al.*, 1973; Feyrter, 1973). Utilizing this pathological situation in order to get fresh adult human multilocular fat in sufficient amounts for biochemical studies, Ricquier *et al.* (1982) demonstrated that this tissue fulfills the biochemical criteria for brown fat. These criteria are loose respiratory coupling, specific mitochondrial guanosine diphosphate (GDP) binding and high mitochondrial anion conductance sensitivity to GDP (see Chapter 2). This fills an important missing link between cytological characteristics and function. Incidentally, in case reports on phaeochromocytomas, there are observations which dramatically illustrate two characteristics of the tissue: (*a*) the surgical manipulation of the brown fat mass can trigger a surge of hypertension followed, after clamping of the vascular pedicle, by profound hypotension (Barneon *et al.*, 1975). This is a consequence of the high catecholamine content of brown fat; (*b*) the brown adipose tissue mass is so extensively vascularized that on angiographic records it can be confused with the tumour itself (Angerwall *et al.*, 1964).

3.3.3 Ultrastructure

The ultrastructure of brown adipose tissue has been reviewed by Suter (1969), Afzelius (1970) and Barnard and Skala (1970). In thin sections viewed in the electron microscope the most distinct morphological criterion is the presence of numerous large mitochondria with straight cristae, which are usually tightly packed and traverse the whole width of the mitochondria (Fig. 3.3).

Since the appearance of these comprehensive reviews, new morphological data has been obtained, which documents the syncitial nature of cell organization in brown adipocytes. Three functional categories of intercellular junctions have been described: (a) impermeable junctions (tight junctions) that enable an organ to maintain an internal environment distinct from its surroundings, (b) adhering junctions (desmosomes) that promote adhesion between the cells of an organ, reinforcing its physical integrity and (c) communicating junctions (gap junctions) that enable cells to exchange nutrients and/or signal molecules. Only the latter are found in brown fat. One might recall that gap junctions are patches of intramembranous proteins containing intercellular connecting channels, approximately 2 nm in diameter; thus molecules with a molecular weight of up to 1000 can readily pass from one cell to the other. Most sugars, amino acids, nucleotides and 'messenger' molecules such as cyclic AMP and steroid hormones fall in this range. Obviously, small inorganic ions can exchange, and thus gap junctions form low-resistance electrical pathways between adipocytes. The existence of these junctions and electrical coupling in brown fat was shown by Revel and Sheridan (1968) and Girardier and Seydoux (1971b). In electron micrograph sections, the ultrastructural correlate is regions of apparent cell fusion (Figs 3.2 and 3.3). In freeze-fracture replicas, gap junctions appear as close aggregates of intramembranous particles on the cytoplasmic half-membrane face (P-face) and a corresponding array of pits on the external half-membrane face (E-face) of the underlying cell (Figs 3.4 and 3.5).

The role of gap junctions in brown fat activity is still conjectural. A correlation between parameters of tissue activity and the capacity for intercellular communication has been found. An index for cell communication was determined (Schneider-Picard *et al.*, 1980a) by measuring gap junction frequency and area, and the cell diameter at different ages in the postnatal period. This allows the calculation of the ratio of total gap junction area over the cell volume and represents what Sheridan (1971) calls the 'capability for communication'. On the other hand, an index of the metabolic capacity has been measured on tissue from rats of the same postnatal period. The steady-state shift in redox potential of the flavoprotein complex implicated in the first step of the β-oxidation was chosen (Schneider-Picard and Girardier, 1982). In Fig. 3.6, the amplitude (A) of the flavoprotein reduction in response to a fixed (1 Hz) electrical stimulation of the afferent nerve supply and the index of the capacity for intercellular communication are plotted as a function of age. It can be seen that there is an obvious correlation between both indices. Both are low at birth, increase sharply during the two first postnatal days, stay elevated until to about the 17th day, and then slowly decline thereafter. What can be the function of this cell communication? The answer can only be putative. It could be the spreading of

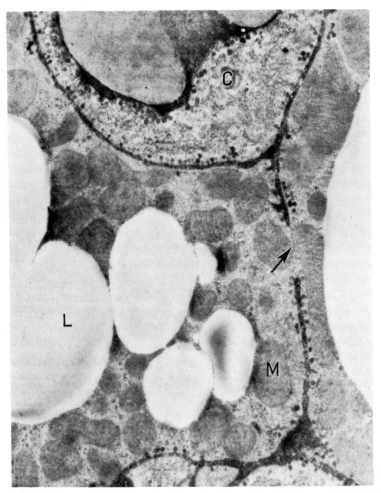

Fig. 3.2 Electron micrograph of interscapular brown fat of rat reared at 23°C. The black precipitate reveals the presence of peroxidase, a tracer of the extracellular space which was injected a few minutes before the sacrifice of the animal. Note the numerous newly-formed micropinocytotic vesicles, and the apparent disappearance of the extracellular space in the region indicated by the arrow showing the probable presence of a gap junction. C: capillary; L: lipid droplet; M: mitochondrion; magnification: ×14 400. From Girardier and Seydoux (1971b).

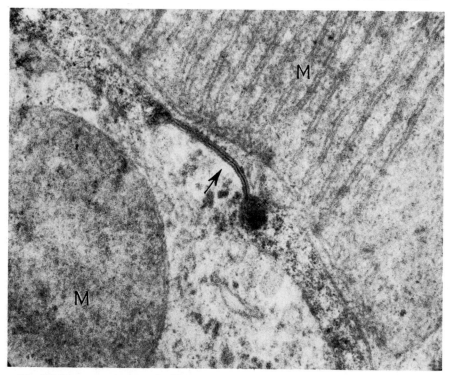

Fig. 3.3 Electron micrograph of interscapular brown fat of a rat reared at 23°C. In this preparation, stained with uranyl acetate, the gap junction appears as a trilamellar structure (arrow). Note the parallel course of the mitochondrial inner membrane where the section is perpendicular to the infoldings. Magnification: ×58 800. From Girardier and Seydoux (1971*b*).

the activating signal and/or the pooling of some important metabolic intermediates. This is still an open field for investigation.

Another potentially interesting ultrastructural feature is clearly visible in Fig. 3.2. This is the pinocytotic activity. The large number of newly formed pinocytotic vesicles both in adipocytes and endothelial cells is striking. In freeze-fracture replicas, which expose large membrane regions, the density of the pinocytotic activity can be appreciated (Figs 3.4 and 3.7). In white adipose tissue, Barrnett and Ball (1960) have shown that the pinocytotic activity requires the presence of insulin and that the metabolic changes induced by the hormone and the vesicular formation seem to be interdependent events. They calculated that, in terms of transmembrane mass transport, pinocytosis is probably negligible but they reason that this vesicular formation results in a continuous loss of surface membrane. This loss would require the cell to synthesize a new membrane, the permeability of which, to glucose and other small molecules, could be modified. Another

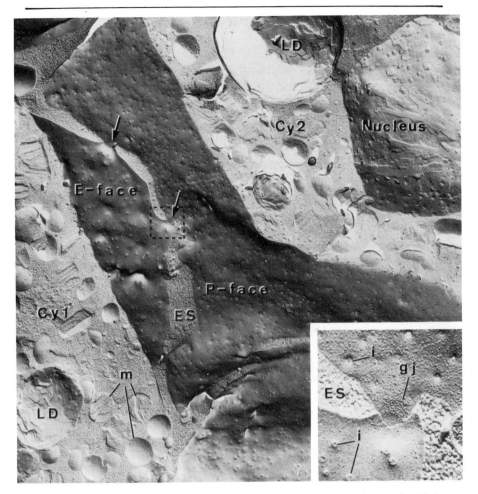

Fig. 3.4 Freeze-fracture replica of brown adipose tissue of rat two hours after birth. The exposed plasma membrane (E-face of cell Cy 1; P-face of cell Cy 2), displays numerous punctiform invaginations (i). The extracellular space (ES) separating the two cells is reduced at two focal regions (arrows) one of which is shown at higher magnification in the inset, which demonstrates a typical gap junctional aggregate (g–j). Magnification: ×11 000. Inset ×58 000. From Schneider-Picard *et al.* (1980*b*).

not exclusive possibility is that, through pinocytosis, large hydrophilic molecules gain access to the cytoplasm of the adipocyte. In a tissue implicated in diet-induced thermogenesis, it might be of interest to know if physiological concentrations of insulin significantly alter this kind of membrane activity.

Blanchette-Mackie and Scow (1981) have described, in white fat, brown fat and liver, continuities between the extracellular space and intracellular

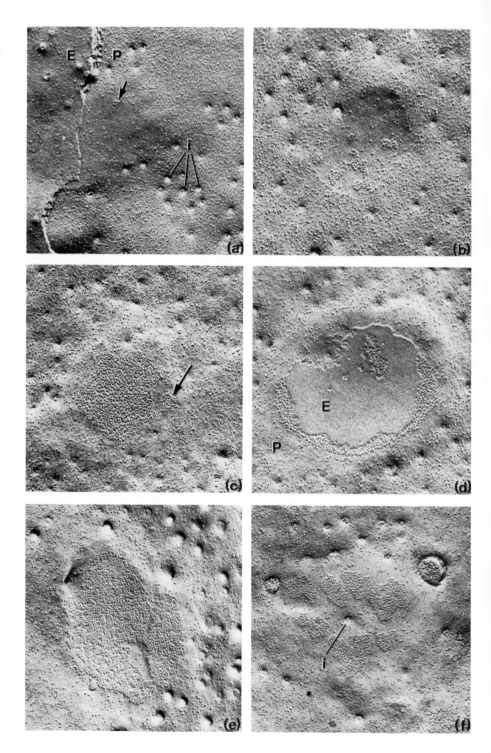

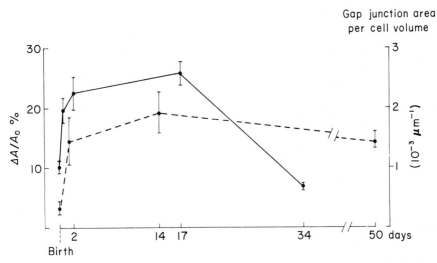

Fig. 3.6 Correlation between indices of metabolic capacity and capability for intercellular communication. Dotted line: steady-state shift in reduction (as a percentage of basal state $\Delta A/A_0$) of the flavoproteins linked to β-oxidation in response to electrical stimulation of the nerve supply (stimulation frequency: 1 Hz). Full line: ratio of total gap junction area per cell volume. From Schneider-Picard *et al.* (1980*b*).

channels. Structures marked C in Fig. 3.7 are compatible with this description. It has been proposed (Blanchette-Mackie and Scow, 1982) that the channels, which are probably elements of endoplasmic reticulum, are implicated in the transport of free fatty acids and other strongly amphiphasic substances between the adipocyte and the capillary lumen.

3.4 THE CAPACITY FOR HEAT PRODUCTION

3.4.1 In vivo studies

In their well-known experiments on the newborn rabbit, Dawkins, Hull and Segall (Dawkins and Hull, 1965) demonstrated the remarkable heat-dissipating power of brown fat. They showed that removal of the few grams

Fig. 3.5 Variations in gap junctions shape and size throughout the stages of development of brown fat. Besides typical gap junctions (d) in a P–E fracture face transition, plaques containing small particle aggregates are also observed 2 hours after birth (a, b) and at 90 days after birth (f). Large gap junction accompanied by a small satellite, 2 hours after birth (c). Gap junction containing particle-free islands, 17 days after birth (e). The above proposed sequence is compatible with a pattern of formation, growth and decay of gap junctions. Magnification about ×50 000. From Schneider-Picard *et al.* (1980*a*).

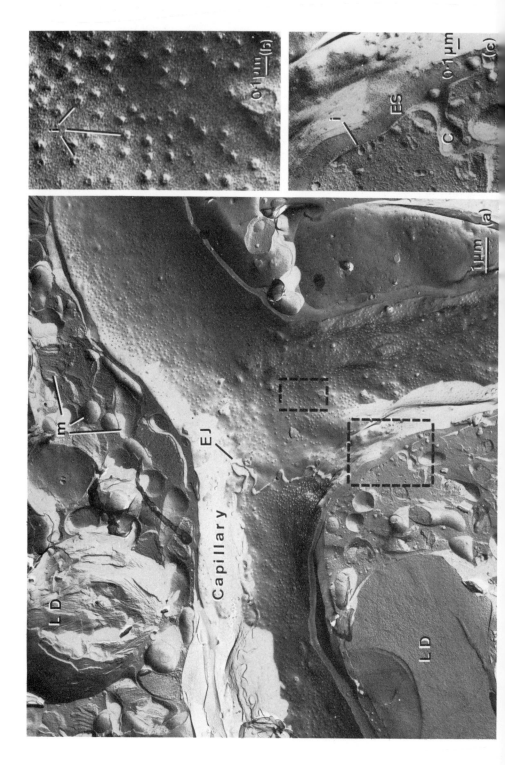

of this specific tissue practically abolished the animal's ability to multiply its oxygen consumption, and hence its heat production, by more than threefold in response to cold. Thus, the brown adipose tissue is entirely, or almost entirely responsible for this metabolic response to cold in the newborn of this species. These data allow an estimate of the maximal specific heat production of the tissue. At an environmental temperature of 35°C, newborn rabbits have a minimum oxygen consumption of 23 ml min^{-1} kg^{-1}. When their environment was cooled to 20°C, their oxygen consumption reached a peak of 85 ml min^{-1} kg^{-1}. If, as a first approximation, it is assumed that the extra oxygen consumption represents the extra heat production of brown fat (5–6% of body weight) one can calculate a dissipative power of about 350 W (kg of tissue)$^{-1}$.

In newborn lambs, Alexander and Bell (1975) found, by careful macroscopic dissection, that brown adipose tissue amounted to approximately 1.5% of the body weight (mean 3.5 kg). Given that 40% of the thermogenic response to cold is due to non-shivering thermogenesis in this tissue, and that summit and whole body oxygen consumption are respectively 3600 and 1000 ml min^{-1} kg^{-1}, the oxygen consumption of brown fat was calculated to reach 70 ml g^{-1} h^{-1}. This corresponds to a heat dissipating power of 381 W (kg of brown fat)$^{-1}$.

In adult rats, Foster and Frydman (1979) have found that decreasing the environmental temperature of a 370 g cold-adapted rat from 21 to -19°C increased its oxygen consumption by 13.8 ml O_2 min^{-1}. They estimated from fractional blood flow measurements that at least 60% of the extra oxygen is utilized by the 5.25 g of brown fat of the animal. This represents a dissipative power of more than 500 W (kg of brown fat)$^{-1}$, that is enough power to heat the tissue, if it is isolated, at a rate of 5–10°C per minute, depending on the fat composition. Clearly, the heat must be rapidly exported and indeed the perfusion rate of the tissue is extremely high. The value of blood flow to brown fat obtained with the use of radioactive microspheres is 8.0 litres kg^{-1} min^{-1} in the newborn lamb (Alexander *et al.*, 1973) and 10.8 litres kg^{-1} min^{-1} in the cold-adapted rat (Foster and Frydman, 1979). This flow is large enough to be able to export the heat evolved with an a–v temperature gradient as low as 0.7°C, in both examples, if it is assumed that thermal equilibration time is small compared with blood transit time.

Thus, three independent groups working on three animal species, one on adult and two newborn, have found a heat dissipating power of brown fat close to about 400 W kg^{-1}.

Fig. 3.7 Freeze-fracture replica of rat interscapular brown adipose tissue. Note the numerous micro-invaginations (i) on both endothelial cells (compare with Figs 3.2 and 3.4). ES: extracellular space; EJ: endothelial junction; LD: lipid droplet; m: mitochondria; C: Intracellular channel. Horizontal bars = 1 μm in (a); 0.1 μm in (b) and (c).

3.4.2 *In vitro* studies

(a) Isolated cell preparations

The oxygen consumption of incubated tissue slices activated by addition of catecholamines in the perifusion medium, yield respiratory rates well below what would be expected from the dissipating power measured *in vivo*. This point will be dealt with below, but isolated cell preparations yield a much higher oxygen consumption. For instance, Bukowiecki *et al.* (1980) found that brown adipocytes from rats gave a maximal utilization of 410 nmol O_2 min^{-1} 10^6 $cells^{-1}$, which would correspond to a heat dissipating capacity of about 300 W (kg of tissue)$^{-1}$. Nedergaard and Lindberg (1982) working on hamsters have found 380 nmol O_2 min^{-1} 10^6 $cell^{-1}$.

These *in vitro* measurements can be considered as being in fairly good agreement with measurements *in vivo* and the value of 400 W kg^{-1} can be reasonably used for rough estimations. For instance, it is interesting to evaluate how long the cell fat store could sustain its summit heat dissipation, or how much brown fat could be required to contribute significantly, when maximally activated, to the whole body heat dissipation of a 75 kg human subject.

If we assume a lipid content of the cell of 50% and suppose that the tissue is stimulated at full capacity without external supply of substrate, then the stores will last for about 13 h. Thus a rapid depletion of lipid stores is expected upon cold exposure of the animal and this has been reported by Barnard and Skala (1970). Suter (1969) observed in rats that within a few hours after birth, substrate stores were largely depleted, but by 1 day post-partum the lipid stores have been reconstituted. Part of this recovery is certainly due to lipogenesis in the brown adipocyte. The brown adipocyte has extremely high activities of pyruvate dehydrogenase and acetyl-CoA carboxylase (McCormack and Denton 1977) and has a high rate or lipogenesis *in vivo*, as measured by the incorporation of ^3H from injected ^3H_2O into fatty acids in the tissue. This lipogenesis is increased during cold exposure (McCormack and Denton, 1977; Trayhurn, 1979; Agius and Williamson, 1980; Rath *et al.*, 1979).

Incidentally, Saggerson and Carpenter (1982) have shown that brown adipose tissue carnitine palmitoyltransferase is extremely sensitive to inhibition by malonyl-CoA and they note: 'clearly there is a conundrum here. Malonyl-CoA must be present in brown fat cells in order to sustain the observed high rates of lipogenesis; how can this be compatible with a simultaneous high rate of β-oxidation'? This is another open question.

How much brown fat is needed to increase by 50% the resting metabolism of a 75 kg human being? This would require about 50 W. If the tissue has the same specific power as in rat and newborn lamb or rabbit, 125 g of brown fat would be needed. Taking a reasonable 14% of body weight for total fat, which amounts to little above 10 kg of fat, it follows that less than 2% of fat would have to be of the brown type to account for such an increase. It is recognized that the calculation is rough and that it is not proven that in adult man the dissipating power of the tissue is as high as 400 W. The point is that the role of brown fat in diet-induced thermogenesis, for instance, cannot be easily discarded on quantitative calculations.

(b) Tissue slice preparations

Perifused tissue slices of brown adipose tissue yield, as noted above, much lower maximal respiratory rates, about an order of magnitude smaller, than isolated cell preparations. These preparations are thus not suitable for measuring the maximal oxidative capacity of brown adipocyte. However, they are useful for several interesting problems, such as the functional role of intercellular gap junctions, membrane potential measurements or intracellular ionic activity measurements, since impalement of isolated cells with microelectrodes is extremely difficult. The study of the neuro-metabolic synapse also necessitates the use of whole tissue preparations. Thus, the cause of the limitation of respiration in slices is of interest to those working on such problems. An explanation for these limitations could be an inhibitory effect of free fatty acids accumulating during stimulation, and/or O_2-diffusion-limited respiration and we have tested these two hypotheses. Regarding free fatty acid accumulation, it has been shown, in white adipose tissue, that intracellular free fatty acid accumulation inhibits cAMP production. In tissue slice preparations from brown fat perifused with protein-free solution, such accumulation is expected to occur, thus limiting the maximal response. We have found, however, that increasing the albumin content of the perifusing solution up to 7 g litre^{-1} had no effect on resting or stimulated heat production (Chinet, unpublished result). This albumin concentration should have drastically decreased the concentration of cytosolic free fatty acid and renders unconvincing the explanation of the respiratory limitation by cytosolic fatty acid accumulation. The oxygen diffusion limitation at maximal neural stimulation is beyond doubt. Seydoux *et al.* (1977) have shown that the partial pressure of oxygen on the tissue surface sinks to near zero at high stimulation rate. However, at submaximal stimulation the respiration is not diffusion limited. This point was investigated by Friedli *et al.* (1977). The respiration rate of two threadlike pieces from the same sample of rat brown adipose tissue were recorded simultaneously, with two different methods, and compared: first, in a respirometer with very efficient mixing in order to decrease to a minimum the unstirred layer around the small tissue sample; secondly, in a laminar flow respirometer surrounded by thermal gradient layers in order to measure heat production and O_2 uptake simultaneously. A caloric equivalent of oxygen can thus be calculated. Both preparations were submitted to noradrenaline stimulation and the same values were obtained in stirred and laminar flow chambers for oxygen uptake at rest and under moderate stimulation. Oxygen diffusion became rate limiting in the laminar flow chamber for stimulations leading to higher metabolic rates. The respiration rate was more than three times higher in the stirred as compared with the unstirred chamber (Fig. 3.8). If the linear velocity of the laminar flow is increased, the difference between O_2 uptake in stirred and unstirred media decreases markedly.

This figure also shows that the caloric equivalent as measured in the laminar flow chamber, remains constant at a value of 105 kcal per mol O_2, compatible with lipid oxidation, even at a high level of stimulation, when respiration rate was severely diffusion limited.

This leads to three conclusions:

(a) Tissue slices which are not suitable for measurement of maximum respiratory capacity of the tissue *in vitro* can validly be used for submaximal stimulation, when this model is irreplaceable for the problem of interest. For surface cells, those explored by microelectrode impalement or surface reflection spectroscopy, the problem of oxygen limitation arises only for high levels of stimulation.

(b) Anaerobic glycolysis cannot sustain noradrenaline-induced calorigenesis to any measurable extent, which agrees well with the findings of Heim and Hull (1966) *in vivo*. Brown adipose tissue is thus a strictly aerobic heat dissipator.

(c) Brown fat does not show signs of autolysis when hypoxic. The catabolic heat production should have increased the apparent caloric equivalent

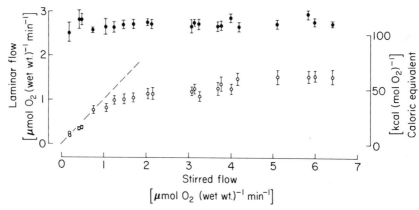

Fig. 3.8 Oxygen uptake of tissue slices measured in stirred or unstirred (laminar flow) perifusing solution. Open circle O, each point corresponds to a paired experiment in which two slices of tissue taken from the same brown fat samples are tested in either laminar (ordinate) or stirred (abscissa) medium in the presence of the same noradrenaline concentration. Dotted line ------, represents the equality line. Filled circles ●, caloric equivalent measured in the unstirred chamber (for details see text). From Friedli *et al.* (1977).

of oxygen in conditions where oxygen supply is diffusion limited. Thus, this tissue seems to be able to adapt to large reductions in blood flow.

3.5 EVIDENCE FOR A CONCERTED REGULATION BETWEEN SUBSTRATE SUPPLY AND RESPIRATION RATE

The level of tissue stimulation determines the rate of lipolysis and, through β-oxidation, the rate of reduced equivalent supply to the respiratory chain. Failure of perfusion to increase correspondingly would bring about a mismatch between supply and utilization and this can be expected to result in a rapid accumulation of reduced equivalents as soon as the oxygen supply becomes limiting. However, this accumulation is hardly ever achieved and it is as if the supply pathway was inhibited by some feedback mechanism when the perfusion rate, hence the oxygen supply, becomes inadequate. A perifused tissue preparation responding to a high frequency burst of stimulation is a workable model. In this preparation the balance between substrate supply and oxidation by the respiratory chain can be monitored by the oxido-reduction state of the flavoproteins of the electron-transport chain. An excess substrate supply relative to oxidation will be recorded as a reduction of the flavoproteins. In the experiments described in Fig. 3.9, the redox states of flavoproteins were monitored spectrometrically: the acyl-CoA dehydrogenase system implicated in the first step of β-oxidation (A), and the NADH dehydrogenase system (F) (for method see Schneider-Picard and Girardier, 1982). The evolution of the redox states of the flavoproteins in response to a train of 100 pulses applied on the afferent nerves of an *in vitro* neuro-adipose preparation is illustrated. Two stimulation frequencies were used, 1 and 4 Hz. For the lower stimulus, the acyl-CoA dehydrogenase (A1) is reduced, as a result of the increased substrate supply, whereas the NADH dehydrogenase (A2) is oxidized, indicating that the increase in respiration rate pulls out more hydrogen than is supplied to this flavoprotein. It is remarkable that both effects are attenuated, instead of increased, at higher stimulation frequencies, during which the oxygen supply would most probably be limiting. The precipitous reduction expected during a phase of relative hypoxia is not observed. A new stimulation at the low rate of the first stimulation reproduced the same higher response. Thus, the cell seems to be protected against respiratory substrate overload. This effect is rapid in its on-off action and can be seen in Fig. 3.9, (B1 and B2). Here the stimulation was changed without any discontinuity between *high* and *low* frequency. This resulted in a large *increase* in the reduction of both flavoprotein systems with a latency that indicates an inhibitory half-time of less than 2–3 min, too short to dispose, for instance, of an excess of free fatty acid in the albumin-free medium utilized. We have no mechanism to propose, but whatever it is, it is interesting to see that the tissue seems to be endowed with a mechanism which permits an adaptation to large variations

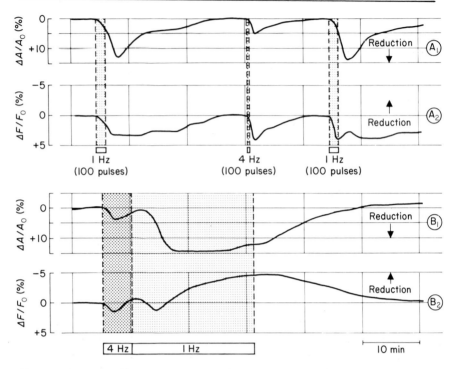

Fig. 3.9 Changes of redox potential of the flavoproteins linked to acyl-CoA dehydrogenase ($\Delta A/A_o$) and NADH dehydrogenase ($\Delta F/F_o$) systems in response to electrical nerve stimulation. For details see text. Schneider-Picard (unpublished).

in perfusion rate by providing an effective impedance matching between substrate release and utilization.

3.6 TISSUE BLOOD FLOW

Unstimulated, isolated, perifused brown fat possesses a stable steady-state metabolism after an initial equilibration period. Immediately after removal and installation in the organ chamber, values of 20–30 W kg^{-1} are found, which slowly decrease, to settle down to a mean dissipation rate of about 5–10 W kg^{-1}. Tissue from reserpinized rats do not show the initial high metabolic phase, which probably is a consequence of the sympathetic discharge occurring during sacrifice and dissection of the tissue sample (Barde et al., 1975). Since, in the unstimulated state, oxygen supply of *in vitro* preparations is not limiting, it can be reasonably inferred that this energy dissipation represents the resting metabolism of the unstimulated tissue *in situ*. The resting metabolic rate in a 300 g rat kept at 25°C is about 5 W kg^{-1}, and therefore it can be estimated that the contribution of brown adipose tissue (less than 1% of body weight) would represent 1–2% of the basal metabolism. This is consistent with the finding of a blood flow of a little less than 2% of the cardiac output found by Foster and Frydman (1979) in the resting rat at 25°C. In tissue dissected from cold-acclimated rats, the resting heat output per unit weight of tissue is *smaller* than that measured in tissue from rats reared at 25°C, but, when calculated as heat output per whole organ (which is 1.7 times larger in cold- than in warm-acclimated rat) it is found that total brown fat of both cold- and warm-acclimated rats have the same resting heat output. Here again, this is in agreement with blood flow measurements. From the values published by Foster and Frydman (1979), one calculates that the blood flow in brown adipose tissue, from warm-adapted rats, is 0.8 ml min^{-1} g^{-1} whereas it is only 0.5 ml min^{-1} g^{-1} in cold-acclimated rats measured at 21°C, but here again, the flow per whole organ is the same in both cold- and warm-acclimated rats at this ambient temperature range. Thus, in rats close to or above thermoneutrality, brown fat represents a low thermic load of less than 1% of basal metabolism. Now, when the perifused tissue is maximally stimulated, its heat output increase is a function of the perifusion flow. For instance, in the presence of a laminar flow at linear velocity of 0.5 mm s^{-1}, the heat output increases to 240% basal and it reaches 640% at a perifusion velocity of 1–1.5 mm s^{-1} (Chinet, unpublished data). What is remarkable is that, in both cases, the response is perfectly reversible and reproducible, and yields the same caloric equivalent for oxygen. Thus, this tissue is extremely resistant to ischaemic conditions, a property resulting most probably from the concerted regulation of substrate supply and respiration.

These data indicate that for a given stimulating drive of the cell, its heat output *will be proportional to the perfusion flow* and consequently the latter must be, in some way, adapted to each stimulation level. Curiously, little is known about the mechanism of regulation of the haemodynamic resistance of the vascular bed of brown fat. There is ample morphological evidence of an adrenergic innervation of the blood vessels supplying brown fat. Nor-adrenaline-containing nerves, identified according to Falk *et al.* (1962) by their greenish fluorescence in preparations exposed to formaldehyde vapour, were described innervating arteries and arterioles (Cottle and Cottle, 1970). Observations with the electron microscope have confirmed the presence of sheathed axons along the arterioles and capillaries (Bargmann *et al.*, 1968). However, the effect of noradrenaline on vascular smooth muscle is not known. There are indications that these vascular smooth muscles possess vasoconstrictive α receptors. Alexander and Stevens (1980), studying lambs chemically sympathectomized *in vitro* with 6-hydroxydopamine have shown that, under thermoneutral conditions, a dose of noradrenaline that stimulates maximum non-shivering thermo-genesis in control lambs failed to stimulate thermogenesis in sympathectomized animals unless accompanied by a dose of α-blocker. This failure was apparently due to vasoconstriction induced by α-receptors, unmasked by the hypersensitivity of denervation.

It is well documented that, in animals with intact sympathetic innervation, noradrenaline infusion increases blood flow in brown adipose tissue from newborn rabbit (Heim and Hull, 1966; Hardman and Hull, 1970) and in rat (Evonuk and Hannon, 1963; Kuroshima *et al.*, 1967; Laury *et al.*, 1971; Hirata, 1982; Foster *et al.*, 1980). No β-vasodilating receptors have been described as yet in this vascular bed, and the effect of the mediator could be indirect, being secondary to the metabolic activation of the tissue. Foster and Depocas (1980) showed evidence in the rat of such an indirect action. In cold-adapted rats, infused with noradrenaline, experimental reductions in the O_2 concentration of arterial blood resulted in marked increases in blood flow without corresponding changes in the calorigenic response of the animal, the oxygen consumption in interscapular brown fat deposit or in the concentration in plasma noradrenaline. They suggest that vasodilation during calorigenesis is regulated by some unidentified substance, the production of which would be a function of the O_2 requirements of the adipocyte. They noted that 'if so, there arises the question of the function of the extensive, vascular, adrenergic innervation of BAT'. Indeed, at this stage of our knowledge, this role is still obscure. Incidentally, it should be borne in mind that there might be differences between the effects of neurally-released and blood-borne noradrenaline on the adrenergic receptors in a given arterial bed. For instance, Glick *et al.* (1967), studying the effect of noradrenaline released reflexly as a result of carotid sinus hypotension, as compared with the effect of intra-arterially administered noradrenaline, have obtained results compatible with the idea that only α-receptors are 'innervated' in vessels of striated muscles. The vasomotor response evoked reflexly is monotonously a constriction, attenuated by α-blockers and not modified by β-blockers, whereas the vasoconstrictive effect of intra-arterially injected noradrenaline was reversed to a dilation by an α-blocker. Suffice to say for our present purpose that the accessibility of vascular receptors could be different for circulating or for neurally released noradrenaline.

In white adipose tissue also, there are data consistent with the hypothesis that vasodilation is secondary to metabolic events. Bulow (1981) has found that the threefold increase in blood flow of human subcutaneous fat observed during prolonged exercise was suppressed when lipolysis was blocked by nicotinic acid, which is reported to be without effect on vasoconstriction elicited by sympathetic nerve stimulation in adipose tissue. Thus, the absence of vasodilation during exercise under the influence of nicotinic acid is compatible with the idea that some product of lipolysis is acting as one of the mediators of the accompanying hyperaemia. However, other metabolic changes which occur in active tissues can be proposed. In striated muscles, hydrogen, potassium and hyperosmolality have been implicated in vasodilation occurring in working muscle. The mechanism is complex because these metabolic correlates are involved both as modulators of the adrenergic neuroeffector transmission and as inhibitors of the contractile mechanism of the vasomotor muscles. In contracting skeletal muscles, the response of the resistance vessels to sympathetic nerve stimulation is reduced (Burcher and Garlick, 1975; Kjellmer, 1965; Rowlands and Donald, 1968). This inhibition of adrenergic neurotransmission can be mimicked by concentrations of the putative mediators lower than those required for their direct effect on the vascular smooth muscle itself (Verhaeghe *et al.*, 1978). The generation of metabolites restraining the sympathetic activity results in a vasodilation, the importance of which, during mild exercise, depends on the balance between metabolic and sympathetic activity. At high rates of work, the excess metabolite completely inhibits the sympathetic vasomotor effect by decreasing further the amount of transmitter release and affecting the smooth muscle cells directly (Strandell and Shepherd, 1967). Now, considering brown fat, at least two of these alleged metabolic modulators are present in increased concentration when the tissue is activated: hydrogen and potassium. During a period of stimulation a large wave of acidosis can be easily recorded with a pH microelectrode applied at the surface of brown fat samples bathed in a Krebs solution containing 25 mM bicarbonate buffer (unpublished from this laboratory). This is not surprising since the CO_2 production is abruptly increased by 5 to 10 times during stimulation. The acidosis of the extracellular fluid, could, as in muscle, contribute to the vasodilation. The potassium concentration also increases in the extracellular space of brown fat during activation. Indeed, a correlate of the tissue's thermogenic response is a graded membrane depolarization (see below) of unkown origin. Whatever the latter may be, this depolarization brings the membrane potential below the equilibrium potential for K^+ ions. This in turn will cause an increase in potassium efflux from the cell to the extracellular space. This effect can be demonstrated with radio-isotopes (Girardier *et al.*, 1968). If the resistance vessels of brown fat have properties comparable to those of striated muscles, the K^+ efflux could also contribute to vasodilation.

The effect of temperature on tissue perfusion of this thermogenic organ deserves some attention. Cooling increases the affinity of the α-receptor for noradrenaline in cutaneous veins (Janssens and Vanhoutte, 1978) and inhibits the neuronal uptake of the transmitter (Vanhoutte and Shepherd, 1969). This favours vasoconstriction despite a progressive depression by cold of the contractile process within the smooth muscle cell. If the same properties prevail in brown fat, heat, the major metabolic product, could decrease the affinity of α-receptors and thereby contribute to the massive increase in blood supply required to sustain the 10- to 40-fold increase in metabolism. To my knowledge, nothing has been done on this aspect of brown fat physiology.

When considering the role of vascular innervation, one should remember the most interesting anatomical studies of Smith and Roberts (1964) on the vascularization of the cervical and interscapular brown fat. They showed a dual venous return system. On one hand, the bilateral arterial supply lies in close apposition to corresponding veins while passing largely through an overlay of brown fat. This constitutes what Smith and Roberts (1964) called a 'reverse type of countercurrent heat exchange system'. Indeed, the temperature in these paired vessels will tend to be higher in venous than in arterial blood, and thus heat will flow from the efferent venous blood to the afferent arterial stream; the more the latter is warmed, the higher will be the rise in metabolism of the fat-pad by the added direct effect of temperature on reaction rate. This positive feedback would obviously help to produce the rapid 'on' effect of the thermogenic reaction, but could also lead to a critical rise in temperature. However, a second venous system, unattended by arteries, the central drainage by the Sulzer vein, can damp the temperature increase, thus providing control over the feedback system without necessarily shutting down the arterial input. This fine sequence of thermogenic control requires, in turn, a fine degree of vascular regulation in which the blood vessel innervation might play an important integrative role.

3.7 THE ADAPTIVE RESPONSE TO COLD EXPOSURE

The brown adipose tissue in the rat, after a period of development, undergoes a slow involution, starting between 17 and 30 days post-partum, and continuing for the remainder of the animal's life. However, this trend can be reversed upon exposure of the animal to a low environmental temperature. The changes in morphology and chemical composition have been reviewed by Barnard and Skala (1970). The most striking adaptive response to chronic cold exposure is an increase in wet weight, which can be several fold after 6 weeks of acclimation. The total amounts of nitrogen, DNA, RNA and phospholipids are increased, but Thomson *et al.* (1969) note that 'the relative contribution to tissue growth from hyperplasia, i.e. undifferentiating cells developing into mature brown fat cells, and from hypertrophy of mature brown fat cells is not easy to assess'. Proliferative activity is maximum after 2–4 days of cold exposure and then declines (Hunt and Hunt, 1967; Cameron and Smith, 1964). The brown adipose tissue *concentration* of DNA and RNA per unit nitrogen reaches a maximum after three days, and then declines, although the total *content* of each continues to rise (Thomson *et al.*, 1969). To our knowledge, no evidence of cell division in mature brown adipocytes has ever been found so the large, initial increase in DNA presumably takes place in precursor cells (probably derived from endothelial cells, according to Barnard and Skala, 1970).

The lipid droplets decrease in size during the first hours of cold exposure, but their diameter is almost restored after 24 h, though they remain smaller and more numerous than in animals maintained at a normal temperature (Suter, 1969). The total lipid concentration is somewhat decreased.

At the level of mitochondria, cold acclimation results in the production and maintenance of a large mitochondrial volume and area of inner membrane. The cytochrome c oxidase and the proportion of a polypeptide of molecular weight 32 000, known to be associated with the thermogenic proton conductance pathway, are markedly increased in rats, mice and guinea pigs (but not the polypeptide in hamsters) and the phospholipid fatty acid composition is altered (Ricquier *et al.*, 1979).

The growth of the sympathetic innervation seems to keep pace with the growth of the tissue since the *relative* noradrenaline content of the tissue, after a sharp decrease at the onset of cold exposure, returns to values close to normal. The tyrosine hydroxylase (a limiting enzyme of catecholamine biosynthesis) is increased upon cold acclimation, and the turnover rate of noradrenaline is more rapid, indicating a raised sympathetic tone (see Barnard *et al.*, 1980). The relative contribution of nerve growth and of increased storing and synthesizing capacity of pre-existing nerve fibres remains to be evaluated.

The exposure of adult rats to cold leads to a general stimulation of sympathetic activity (Leduc, 1961; Tedesco *et al.*, 1977) and there is a large increase in the urinary excretion of catecholamines (Leduc, 1961). But the adaptive changes in brown fat just described cannot be attributed exclusively to the sympathetic drive. Rats injected for several weeks with noradrenaline show tissue hyperplasia, but do not exhibit the changes in mitochondrial protein content and protein conductance pathway seen after cold exposure (Mory *et al.*, 1980). However in chemically sympathectomized rats, the trophic response to cold exposure in brown fat is inhibited (Mory *et al.*, 1982). It thus appears that the sympathetic drive is a necessary, but not sufficient, condition for the adaptive response of the tissue.

More interestingly, the morphological and biochemical changes observed in brown fat upon cold acclimation reverse completely in three weeks after returning to normal temperature (Desautels and Himm-Hagen, 1980). Brown fat slices and isolated cells taken from cold-adapted rats show a decreased sensitivity to noradrenaline (Friedli *et al.*, 1977; Nedergaard, 1982). This is probably a facet of a more general phenomenon since a reduced cardiovascular responsiveness to α-adrenergic stimulation and possibly a reduced cardiac responsiveness to β-adrenergic stimulation in chronically cold-exposed rats has been found (Kikta *et al.*, 1982).

3.8 THE ACUTE THERMOGENIC RESPONSE

A problem still much debated is the elucidation of the mechanism of the concerted regulation of substrate supply and its oxidation. The catabolic cascade starting with the binding of the hormone to the β-adrenergic receptor is well documented. This binding results in the activation of adenylate cyclase, which, by increasing the intracellular cAMP concentration, stimulates a hormone-sensitive lipase and increases the subsequent lipolysis of intracellular fat stores. On the other hand, the increase in conductance of the mitochondrial protein short-circuit pathway leading to graduated release of respiration is now established (see Chapter 2). Now, the cell can increase its respiration rate 40 times with a time constant within one minute, and decrease it at the termination of the stimulation with a level of cell-free fatty acids either unchanged or even decreased (Bieber *et al.*, 1975; Williamson, 1970). As Nicholls remarks, the simplest mechanism for this concerted regulation would be that the 'uncoupling mechanism' is controlled by a normal component of the reaction sequence from the β-receptor to the mitochondria. The level of the messenger would be automatically synchronized with the induction and termination of lipolysis. A candidate could be the free fatty acids themselves, which seem to possess the capacity for increasing proton conductance and hence uncoupling respiration (see Chapter 2).

This would be in line with the fact that added fatty acids can stimulate brown adipocyte respiration to almost its full capacity (Bukowiecki *et al.*, 1981). It is still not easy to visualize the intimate mechanisms of this regulation, based on a single substance whose concentration is reported not to change appreciably. The activating signal (the change in free fatty acid concentration) would be very small and this would require a correspondingly large open loop gain of the regulating mechanism (with its intrinsic risk of oscillations). It is not unreasonable to look for a complementary activator which would give to the system the differential characteristics needed for a stable regulation. In searching for this hypothetical activator, one has to look for a short-lived messenger, produced close to the β-receptor, for a rapid on and off effect. It has long been known that brown adipocytes depolarize in response to both exogenous noradrenaline and electrical stimulation of the nerves (Girardier *et al.*, 1968; Horwitz *et al.*, 1969; Krishna *et al.*, 1970; Williams and Matthews, 1974a; Fink and Williams, 1976). This decrease in membrane potential is accompanied by an increase in membrane conductance (Horowitz *et al.*, 1971). This has led to the suggestion that membrane depolarization, or the related ionic events, may play a role in the regulation of thermogenesis. A great deal of work has been performed in this field, leaving it encumbered with the ghosts of many dead hypotheses and this is certainly due to the fact that the problem is much more complex than often realized. Firstly, the control of the resting potential of brown adipocytes is not yet understood. It is far more complex than in nerve or striated muscle. In the latter, the resting potential is adequately described as an exchange diffusion of Na^+ inward for K^+ outward current with no net current across the membrane capacity. Since the resting conductance of striated muscle membrane is much higher for the smaller hydrated K^+ ion than for Na^+ ion, the resting potential is close to the equilibrium potential for potassium. Thus, when the potassium concentration is increased in the extracellular medium, the cell depolarizes with a slope close to 60 mV per decade of potassium concentration increase. Furthermore the resting membrane potential shows little sensitivity to changes in Na^+ and the contribution to the steady-state potential by other ions is insignificant. In brown adipocytes, on the other hand, increasing K^+ in the extracellular medium depolarizes the membrane with a slope of about 20 mV (for potassium concentrations close to the normal value), indicating a relatively low conductance for this ion. More puzzling is that the replacement of the bicarbonate buffer by phosphate or Tris buffers also results in depolarization of the cell (Girardier and Seydoux, 1971*b*; Williams and Matthews, 1974*b*). This latter effect is accompanied by a sharp decrease in K^+ conductance as measured by $^{42}K^+$ efflux changes (Girardier and Seydoux, unpublished observation). The temperature sensitivity of the brown fat resting potential is much higher than expected for a simple diffusion potential (Girardier *et al.*, 1968; Williams and Matthews, 1974*b*) and, taken together, these observations suggest that an important electrogenic component of an ion-pumping mechanism is contributing to the resting membrane potential of brown adipocytes.

This complicates considerably the interpretation of experimental data. Indeed, whereas in nerve and striated muscle for instance, a rapid depolarization induced by some agents can be readily interpreted as a direct effect of the agent on some membrane ionic conductance, in brown fat the conductance change may be secondary to a metabolic effect of the agents. In such a case, it is evident that depolarization could not be the cause, but rather the consequence of the metabolic effect. We shall come back to this point below.

Another complexity worth mentioning stems from the fact that, for practical reasons, the electrophysiological measurement must be performed on tissue fragments, since microelectrode impalement of an isolated cell is technically too difficult. Apart from adipocytes, tissue fragments contain, among other things, surviving nerve terminals, and this can cause problems. For instance, in order to test Rasmussen's (1970) hypothesis, the effect of depolarization of brown adipocytes on respiration was studied. Rasmussen (1970) proposed a general scheme of activation of target cells by non-steroidal hormones. According to this theory, the hormone, as primary signal, would induce both membrane depolarization and activation of adenylate cyclase. The depolarization would cause a dissociation of the membrane bound calcium which would in turn affect the permeability to Na^+, K^+ and Ca^{2+}. The resulting change in ionic content of the cell is seen as a prerequisite for the hormonal response. Depolarization of brown fat can be achieved by submitting the tissue to one of the following procedures: K^+-free medium, ouabain-containing medium, low-Na^+ medium, high-K^+

medium, glucose-free medium or medium containing glycolytic inhibitors (Lasserre *et al.*, 1973). All six methods to induce depolarization resulted in an increase in cell metabolism, giving the impression at this stage, that membrane potential does have some regulatory effect on brown adipocyte heat production. Bearing in mind the fact that the tissue is richly innervated and that some of these effects could be indirect, i.e., via neural depolarization, a verification of this conclusion was attempted by repeating these experiments in the presence of propranolol, a blocker of the catecholamine β-effect, in order to suppress the effect of a possible neural release. The results showed that, for experiments with glucose-free media, glycolytic inhibitors, and high K^+, propranolol completely blocked the increase in respiration. This evidently alters the former interpretation and further evidence suggests that the depolarization *per se* cannot be a signal for the respiratory control system of the cell. This is illustrated in Fig. 3.10 in which the effect of high K^+ on control tissue and on tissue from rats treated with reserpine, a drug which is known to deplete the nerve endings of their catecholamine content, were compared. The reserpine treatment was found to be without effect on the resting membrane potential of adipocytes, on the amplitude of membrane depolarization in response to 40 mEq of K^+ and on the basal respiration of the tissue. It completely prevents, however, the metabolic response induced by high K^+. Figure 3.10 also shows that the

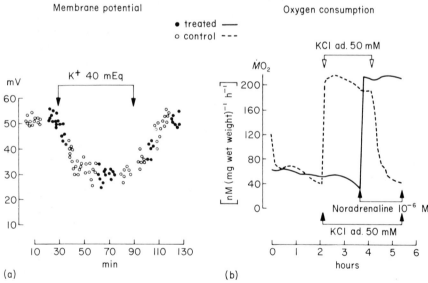

Fig. 3.10 Effect of reserpine on potassium-induced depolarization and oxygen consumption in rat brown adipose tissue. (a) Membrane potential. From Seydoux and Girardier (unpublished). (b) Oxygen consumption; open arrows: K^+ addition on control tissue; filled arrows: K^+ and noradrenaline additions on treated tissue. From Barde *et al.* (1975).

tissue from reserpine-treated rat which is not responding to elevated K^+ exposure by an increase in respiration, is quite normally stimulated by noradrenaline. Thus, in our opinion, the idea that depolarization, by displacing some membrane-bound ion, could trigger reactions leading to the release of respiration is not substantiated. The mechanism of the concerted regulation of substrate supply and its oxidation remains to be found.

3.9 ALPHA- AND BETA-ADRENERGIC MEDIATED MEMBRANE POTENTIAL CHANGES AND METABOLISM

In the initial observation of brown adipocyte depolarization in response to exogenous noradrenaline (Girardier *et al.*, 1968) the potential was measured by successive impalements of different cells. With this technique, which easily misses fast components of membrane potential variations, a slow depolarization was observed which could be blocked by propranolol (a β-antagonist), but not by phentolamine (an α-antagonist). it was concluded that depolarization was a β-adrenergic mediated effect (see also Krishna *et al.*, 1970). Indications of an α-adrenergic mediated depolarization came from the study by Fink and Williams (1976), who showed that an α-antagonist inhibited α-agonist but not β-agonist-mediated depolarization and vice versa. From these agonist and antagonist studies it was not possible to observe the different time courses of the α and β-mediated depolarizations, as the potential was also measured by successive sampling in different cells. The problem was re-investigated by Girardier and Schneider-Picard (1983) with the objective of defining the temporal relationship between membrane potential changes and metabolic activity. As an index of metabolic activity, a spectrometric method was used, which monitors the redox state of the flavoproteins implicated in the first step of β-oxidation of fatty acids (Schneider-Picard and Girardier, 1982). The membrane potential of a single cell was recorded continuously. The membrane potential changes during nerve stimulation were uniform over the region from which the redox state of the flavoprotein was measured, most probably because cells are communicating by gap junctions.

It was found that nerve stimulation evoked two temporally distinct cell depolarizations (Fig. 3.11, upper panel). An initial, rapid depolarization precedes the increase in flavoprotein reduction, and at the time of maximum flavoprotein reduction the cell has repolarized after a transient hyperpolarization. A second, slow depolarization follows flavoprotein reduction. Phentolamine selectively blocks the first depolarization (Fig. 3.11 lower panel). Propranolol delays the first repolarization until the end of nerve stimulation and inhibits the transient hyperpolarization, second depolarization and flavoprotein reduction. Isoproterenol, a β-agonist produces only a transient hyperpolarization and a subsequent slow depolarization. Most interestingly, the addition of the free fatty acid octanoate induced the same complex sequence of potential changes, hyperpolarization followed by slow depolarization, as did the β-agonist and this effect was not blocked by propranolol (Fig. 3.12). It is worth noting that the slow membrane potential depolarization occurs only after the peak in flavoprotein reduction. Thus, it is probable that the electrophysiological effects at the level of the cytomembrane are not due to the fatty acid itself but to a metabolite of its

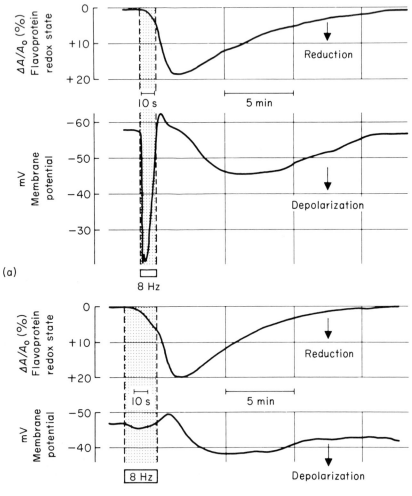

(a)

(b)

Fig. 3.11 Simultaneous recording of redox state of the flavoproteins linked to acyl-CoA dehydrogenase (percent of basal ($\Delta A/A_o$)) and membrane potential. The shaded areas indicate the periods of nerve stimulation at a frequenxy of 8 Hz. During periods of stimulation, the chart recorder speed was increased. (a) Control preparation; (b) Phentolamine, an α-antagonist, was added 40 min prior to the period of nerve stimulation. Schneider-Picard (unpublished).

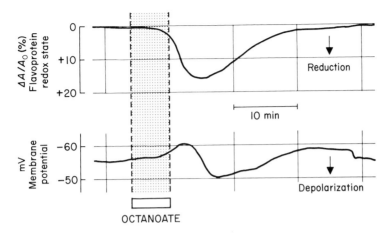

Fig. 3.12 Simultaneous recording of redox state of the flavoprotein linked to acyl-CoA dehydrogenase and membrane potential in rat brown adipose tissue. Shaded area: presence of 5 mM octanoate in the perifusion medium. Schneider-Picard (unpublished).

oxidation. The rapid first depolarization precedes the increase of flavoprotein reduction and its maximum amplitude depends on impulse frequency and not on train duration provided it comprises more than 10–20 impulses. It is not simply related to the amplitude of the subsequent flavoprotein reduction. When it is almost completely blocked by phentolamine, the metabolic response reaches the same amplitude but more slowly. Thus, this membrane effect is probably secondary to changes in membrane permeability due to direct effects of noradrenaline on the α-receptor. The early depolarization codes for the intensity of the neural afferent input and might be involved in the kinetic control of the metabolic response, but is not required for the triggering of the response.

The second depolarization depends on both stimulation frequency and duration and its amplitude can be correlated with the flavoprotein reduction, in contrast to the α-effect. Since it can be reproduced with octanoate, it cannot be related to the binding of noradrenaline to the β-receptor, but rather to some metabolite(s) of fatty acid oxidation leaking back in the cytosol and modifying the cytomembrane. The slow depolarization, via the resulting efflux of potassium could contribute to the regulation of blood flow (see section on blood flow).

3.10 SODIUM PUMP AND THERMOGENESIS

In trying to explain the respiratory release in response to sympathetic stimulation in brown fat, two classes of hypothesis have been proposed:

mitochondrial uncoupling and energy-trap systems. An energy trap can be visualized as a cyclic chain of reactions behaving as an overall ATPase. To be suitable for thermogenesis such a system has to (a) be directly or indirectly stimulated by catecholamines; and (b) have the capacity to turn over ATP at a rate high enough to contribute significantly to the energy dissipation of the stimulated tissue. (For a more detailed discussion on this matter see Chapter 2).

The sodium pump seems to fulfill the first requirement: correlating with noradrenaline action, a depolarization of the cell membrane is recorded, an increase in sodium content of the cell is measured (Girardier and Seydoux, 1971a), and the membrane conductance (Horowitz et al., 1971) and the Na^+K^+-ATPase activity of the tissue are increased (Herd et al., 1970). These results suggest an increase in permeability to Na^+ during catecholamine action and the resulting increase of sodium pumping produces an increase in the overall ATPase activity of the cell.

However, it turned out that the second quantitative requirement cannot be fulfilled. The capacity of the tissue to turn over extra-mitochondrial ATP is limited by the low capacity of the mitochondrial translocase and the rate requirement of the mitochondrial ATP synthetase seems also to be too low (see Chapter 2). But above all, the cost of sodium pumping in brown fat has been directly measured and found to represent only a small fraction of the total metabolism of the cell. The resting heat production and oxygen consumption of brown fat samples from adult rats were measured either in the absence or presence of ouabain, a selective inhibitor of the active NA^+, K^+ transport (Chinet et al., 1977a). Less than 10 minutes after the drug was added, the resting heat output fell by about 5% of the basal rate and oxygen consumption by about 10% of the resting uptake. The difference between the deficit in heat production and oxygen uptake can be accounted for by the heat dissipated by the collapsing free energy stored in the ionic gradients. It can be reasonably assumed that in less than 10 minutes, the cell osmotic balance and the cytoplasmic ionic content are not yet severely perturbed and thus the deficit in energy output is a fair measure of the overall energy cost of sodium pumping. This leads to an estimate of about 500 mW kg^{-1} for the power spent for ion pumping in 1 kg of brown fat (as compared to the 5 W of resting energy dissipation). In order to account for *only 10% of the maximum heat dissipation* of the tissue, the ionic flux would have to increase by 80-fold.

Stimulation by catecholamines of the active sodium transport has been demonstrated by several authors, among which are Horwitz and Eaton (1975) in brown fat of cold-adapted rats, Rogus et al. (1977) and Clausen and Flatman (1977) in striated muscle of rat. These latter authors gave quantitative estimates. They found that adrenaline at a concentration of 6×10^{-6} M increased ^{22}Na efflux by 83% and ^{42}K influx by 34%. These are significant amounts, but out of proportion to what would be required for a significant contribution to the overall thermogenesis.

Ouabain inhibits part of the extra heat production in stimulated brown fat. However, this inhibition develops slowly and takes hours to reach its maximum (Chinet *et al.*, 1977*b*). Moreover the amplitude of the inhibition depends strongly on the pH status of the preparation. The inhibition is barely significant at pH 7.7, amounts to 20–30% at pH 7.4 and to 60–80% at pH 6.8. Furthermore, the effect of the glycoside is larger when administered before the catecholamine than afterwards (Chinet *et al.*, 1977*a*). These effects can hardly be attributed to a *direct* effect on sodium pump.

Even though sodium pumping is quantitatively of minor importance in brown fat thermogenesis, this does not preclude some essential indirect role. Chinet *et al.* (1977*b*) propose that the active Na^+, K^+ transport may, through regulation of intracellular pH exert some control on mitochondrial energy dissipation. Rothwell *et al.* (1981) have shown that there is a linear relation between brown adipose tissue Na^+K^+-ATPase activity and resting oxygen consumption in control and overfed rats, the latter showing a larger oxygen uptake and enzyme activity. This suggests an enzyme induction proportional to tissue stimulation. The Na^+K^+-ATPase activity might be an index of the integrated thermogenic activity of the tissue.

3.11 CONTROL OF BROWN-FAT THERMOGENESIS

3.11.1 Peripheral control

(a) The afferent innervation

Long before brown adipose tissue was recognized as an effector of non-shivering thermogenesis, it was observed that its dense innervation plays a role in the kinetics of utilization of its energy stores. Denervation was found to slow down mobilization of tissue lipids and glycogen deposits (Hausberger, 1934; Clement and Schaeffer, 1947; Sidman and Fawcett, 1954).

The sympathetic origin of a large fraction of the neural afferents to the tissue was established by the finding of a high tissue catecholamine content (Weiner *et al.*, 1962; Stock and Westermann, 1963) and by application of the histochemical fluorescence technique, first by Wirsen (1964), and more recently by Thureson-Klein *et al.* (1976). Using the latter technique, Daniel and Derry (1970) could not demonstrate any catecholamine content in brown fat of newborn rats. The innervation of the parenchyme became discernible at 2–3 days post-partum and that of the arterial blood vessels at only 8–10 days. The full differentiation was reached during the second and third week after birth. However, it was found that electrical stimulation of nerves supplying the tissue produces a maximal response at birth (Schneider-Picard and Girardier, 1982). This indicates that at least the effector limb of the non-shivering thermogenic reflex is functional at birth in the rat and, incidentally, that nerve endings can be functional with a catecholamine content low enough to escape histochemical detection.

The pattern of discharge of the afferent neural supply to the tissue has been little studied as yet. Nouri (1972) has found in newborn rabbits two types of discharge in nerve bundles proximal to the brown fat pad: sudden and very rapid bursts of activity apparently related to the *rate* of cooling of the abdominal skin, and sustained activity, which started when the skin temperature was lowered to about 30°C, with fibres recruited one by one as the temperature was further decreased.

The question as to whether the innervation is exclusively of the adrenergic type remain still open.

(b) The neuro-adipose 'synapse'

In their electron microscopic study of the innervation of brown fat, Bargmann *et al.* (1968) write: 'very thin naked axons can be found closely attached to the fat cells, not infrequently embedded in invaginations of their surface. Those terminal parts of the axons, separated from the plasmalemma of the fat cell by the basement membrane, contain groups of synaptic vesicles. Places where the terminals of the axons come into synaptoid contact with the fat cell are interpreted as sites of liberation of catecholamines.'

There is circumstantial evidence that adrenergic receptors are mainly located in the synaptoid contacts described by Bargmann *et al.* (1968). This evidence is based on the effects of inhibitors of neuronal uptake of noradrenaline on the dose–response curve of the transmitter. If one considers a given steady-state concentration of the amine in the bathing solution and a region close to a nerve ending, as a result of the neuronal uptake, a concentration gradient will develop between the bulk of the solution and this microregion where the nerve terminal acts as a sink for noradrenaline. The adrenoreceptors located in the synaptoid region therefore 'see' a hormone concentration of only a fraction of that found in the bathing medium. Inhibition of neuronal uptake will result in an increased hormone concentration acting on these receptors. Now if a large fraction of the total receptor population is located close to nerve endings, i.e. in the synaptoid region, it is expected that the noradrenaline dose–response curve will be significantly shifted to the left after administration of an inhibitor of neuronal uptake. This was tested by Seydoux and Girardier (1977) and it was found that, in control brown fat preparations perifused *in vitro,* the half-maximum effect of noradrenaline on the steady-state redox potential of NAD(P) was obtained at a concentration of 127 nM. In the presence of desmethylimipramine, an inhibitor of neuronal uptake, the same effect was obtained at 18 nM noradrenaline, that is at a sevenfold lower concentration. This leads to the conclusion that the majority of the adrenoceptors should be located in the synaptoid region of the adipocyte membrane.

The methylation of noradrenaline to normetanephrine by the catechol-*O*-methyltransferase also tends to limit the concentration of the mediator at the adrenoceptor sites (Chinet and Durand, 1978; Depocas *et al.*, 1980), but this effect is much less marked than that of neuronal uptake. A reasonable estimate of the concentration of noradrenaline required at the adrenoceptor level to elicit a half-maximum metabolic response is 10 nM. This value is consistent with that obtained on isolated hamster brown adipocytes by Petterson and Vallin (1976) and by Depocas *et al.* (1980) in rats treated with both neuronal and catechol-O-methyltransferase inhibitors.

It is interesting to compare this concentration of 10 nM noradrenaline, needed to obtain half maximum effect, with the dissociation constant of the β-adrenoceptor for the mediator. Bukowiecki *et al.* (1978) have measured a K_d of 2 μM in crude membrane preparation of rat brown adipocytes. Svoboda *et al.* (1979), using lower ligand concentrations on isolated brown adipocytes of hamster, have found a K_d of 176 nM. Using a wide concentration range of ligand, Seydoux *et al.* (1982a) have found in purified membrane preparations from brown adipocytes of rats two populations of binding sites, one exhibiting a high affinity (low capacity) with a K_d of 180 nM and the second one a low affinity (high capacity) with a K_d of 50 μM. The tissue metabolic response correlates, in normal rats, with binding of the hormone to the high affinity population. Using the K_d value obtained and the dose for half-maximum metabolic effect, Svoboda *et al.* (1979) and Seydoux *et al.* (1982a) found that the high affinity receptor occupancy required for this level of activation was less than 10%. Such a low value might be general for β-receptor-mediated processes (see Levitski, 1978).

Another conclusion can be drawn from these quantitative estimations. In unrestricted, unstressed rats kept at 22°C, the mean plasma concentration of noradrenaline is 1.0 nM and that of adrenaline is 0.2 nM (R. Benzi, personal communication). As seen above, the concentration of catecholamines in the region of the brown adipocyte adrenoceptors is an order of magnitude lower than that of the plasma concentrations due to inactivation mechanisms. Thus the concentration of circulating catecholamines reaching the adrenoceptors is negligible compared with that needed for stimulation of thermogenesis.

To reach the 10 nM required at the receptor level to elicit half maximum response, the plasma catecholamine concentration would have to rise to more than 100 nM. Such concentrations are probably never attained in physiological conditions. In severe insulin-induced hypoglycaemia, plasma adrenaline may increase by 150-fold, reaching a peak concentration of about 30 nM (R. Benzi, personal communication). This increase would result, at the level of brown adipocyte receptors, in a phasic rise in concentration which is probably insufficient to elicit half-maximum calorigenic response. It thus appears that, under physiological conditions, circulating catecholamines have little effect on rat brown fat calorigenesis, and that the tissue's metabolic function is controlled by its sympathetic afferents. It should be noticed, however, that: (a) an indirect or permissive effect of circulating catecholamines, like a trophic action, remains to be investigated; and (b) there are possibly important species differences concerning the relative roles of circulating catecholamines and the direct neural input on sympathetic metabolic effectors. For instance, the *in vivo* lipolysis induced by electrical stimulation of the ventromedial hypothalamus is abolished by bilateral adrenalectomy in the rabbit, but not in the rat (Shimazu, 1981). In the first species, the sympathetic innervation of the white adipose tissue appears to be subdominant and the adrenal medulla is principally involved in ventro-medial hypothalamus-induced lipolysis. The converse is true for the rat.

3.11.2 Central control

In a most interesting series of papers, Brück and Wünnenberg (for review see Brück, 1970) have shown in the guinea pig that the two well-known forms of regulatory heat production, shivering (tremor in skeletal muscle) and non-shivering thermogenesis (a thermoregulatory process that does not involve skeletal muscle contraction) are selectively controlled. An effective influence on non-shivering thermogenesis could be exerted by thermodes placed in the preoptic and supraoptic regions of the hypothalamus. Upon heating of this central region, the oxygen consumption of a cold-acclimated hamster placed at 25°C (without shivering activity) is decreased, suggesting suppression of non-shivering thermogenesis. Upon electrocoagulation of the preoptic area, a dramatic rise in oxygen uptake (without shivering) occurred, followed by hypothermia. These results were obtained from both newborn and cold-acclimated animals, but in animals which had been reared at a thermoneutral temperature, there was little increase in oxygen uptake following electrocoagulation. In the latter animal, the amount of brown fat and the capacity for non-shivering thermogenesis are much diminished. This is in accordance with the fact that brown fat is a major effector of non-shivering thermogenesis. It thus appears that 'the thermosensitive hypo-thalamic structures exert a sustained inhibition upon those portions of the sympathetic system that control the metabolic rate in the brown adipose tissue, and that this inhibition is proportional to the body core temperature' (Brück, 1970).

Brück and Wünnenberg have also shown that the guinea pig possesses, besides the hypothalamic area, a second thermosensitive area, which is located in the cervical spinal cord. The temperature of this region and that of the body surface determine the degree of shivering. The authors draw attention to the fact that the cervical spinal cord receives heat generated from the interscapular and cervical brown adipose tissue (Smith and Roberts, 1964). Thus in a cold-acclimated animal, non-shivering thermo-genesis is activated through temperature sensors located in the skin and in the hypothalamus. The heat generated in the interscapular and cervical brown fat maintains a high temperature in the cervical spinal cord and suppresses shivering as long as non-shivering thermogenesis suffices to compensate for the heat loss in a cold environment. In the rat, the selectivity of the two thermosensitive areas appears not to be as marked as in the guinea pig, since both areas can control both forms of thermogenesis (Banet *et al.*, 1978). But still, the threshold is probably lower in the preoptic area for non-shivering than for shivering thermogenesis, while the spinal cord threshold for shivering is probably lower than that of non-shivering thermogenesis.

In the rat, recent findings are accumulating to show that the ventromedial hypothalamic nucleus (VMH), an area known to be involved in appetite and weight regulation, can modulate the thermogenic activity of brown fat. Shimazu (1981) showed that electrical stimulation of VMH enhanced lipogenesis preferentially in brown adipose tissue, since that of white fat was unmodified and that of liver was decreased. This enhancement of lipogenic activity does not involve insulin since it was also observed in diabetic rats. Perkins *et al.* (1981) have obtained an increase in interscapular brown fat temperature which exceeded core temperature in response to electrical stimulation of VMH. Brown fat isolated from female rats with bilateral VMH lesions and reared at 23°C showed signs of functional disconnection and involution (Seydoux *et al.*, 1981; 1982*b*). In these tissue samples, the number of β-receptors is increased, but the substrate supply to the adipocyte mitochondria and the mitochondrial respiratory capacity are decreased. As an index of the capacity to produce reducing equivalents to the respiratory chain, the steady-state redox level of NAD(P) at different stimulation intensities has been used. It was found that the tissue of VMH-lesioned rat can only mobilize about half (when compared with appropriate controls) of its reducing capacity (Fig. 3.13, left panel). As an index of the thermogenic capacity, the GDP binding to mitochondrial inner membranes has been used and was found to be diminished by half.

All these alterations are reversed by acclimation of the VMH-lesioned animal to cold. Figure 3.13 illustrates the recovery after cold-acclimation of the impaired capacity to produce reducing equivalents in brown adipose tissue of VMH-lesioned rat. The steady-state redox potential of NAD(P) was monitored, by recording the surface-emitted fluorescence, as a function of the frequency of electrical nerve stimulation. In VMH-lesioned rats, reared at 23°C, only about half, as compared to control, of the metabolically mobilizable reducing equivalents were obtained at the higher frequency used. After cold-acclimation, the control shows the usual desensitization to adrenergic stimulation (see above) and in tissues from VMH-lesioned rats the responses to the same stimuli are even slightly greater. When VMH-lesioned rats and control rats are cold-acclimated, both GDP binding and the concentration of the 32 000 molecular weight polypeptide are increased, and the same values are measured in both groups.

These results are compatible with the following working hypothesis. The

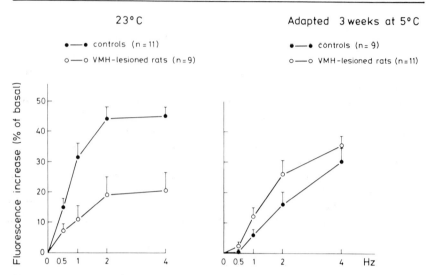

Fig. 3.13 NAD(P) redox state as measured by changes in steady-state surface-emitted fluorescence as a function of electrical stimulation frequency of the afferent nerves to interscapular brown adipose tissue of control and VMH-lesioned rats acclimated at two ambient temperatures. J. Seydoux (unpublished).

sympathetic activation of brown fat represents the sum of two converging inflows: one from the preoptic area, decreasing in intensity with increasing skin and hypothalamic temperatures and the second depending on the ventromedial hypothalamus, increasing its intensity with food intake. This second component would be abolished following bilateral lesion of the ventromedial nucleus. In VMH-lesioned rats, reared at an ambient temperature close to thermoneutrality, the tonic sympathetic input to brown fat would be so low that involution of the tissue would result (the observed *increase* in the number of β-receptors would be a signal of this functional denervation). Keeping the VMH-lesioned animal at a lower ambient temperature, and thus activating skin thermoreceptors, would relieve the inhibition exerted by the preoptic area on the sympathetic inflow to the tissue. Thus 'atrophy of inactivation' could be prevented.

In physiological situations, both central control areas could have a cumulative action on brown fat, since the colder the ambient temperature, the larger the food intake of the animal will be.

ACKNOWLEDGEMENT

Our work is supported by the Swiss National Science Foundation (Berne).

REFERENCES

Afzelius. B.A. (1970) Brown adipose tissue: its gross anatomy, histology and cytology. in *Brown Adipose Tissue* (ed. O. Lindberg). Elsevier. New York. London. Amsterdam. pp. 1–31.

Agius, L. and Williamson, D.H. (1980) Lipogenesis in interscapular brown adipose tissue of virgin, pregnant and lactating rats. *Biochem. J.*, **190**, 477–80.

Alexander, G.. Bell, A.W. and Hales, J.R.S. (1973) Effects of cold exposure on tissue blood flow in the new-born lamb. *J. Physiol. (London)*, **234**, 65–77.

Alexander, G. and Bell. A.W. (1975) Quantity and calculated oxygen consumption during summit metabolism of brown adipose tissue in new-born lambs. *Biol. Neonate*, **26**, 214–20.

Alexander, G. and Stevens. D. (1980) Sympathetic innervation and the development of structure and function of brown adipose tissue. *J. Devel. Physiol.*, **2**, 119–37.

Ambid. L. (1980) Hibernation induite par la simple suppression des protides alimentaires chez le Lérot. *Bull. Soc. Ecophysiol.*, **5**, 171–7.

Angerwall, L., Nilsson, L. and Stener, B. (1964) Microangiographic and histological studies in 2 cases of hibernoma. *Cancer*, **17**, 685–92.

Banet. M., Hensel, H. and Liebermann. H. (1978) Central control of shivering and non-shivering thermogenesis in the rabbit. *J. Physiol.*, **283**, 569–84.

Barde, Y.A., Chinet. A. and Girardier. L. (1975) Potassium-induced increase in oxygen consumption of brown adipose tissue from the rat. *J. Physiol. (London)*, **252**, 523–36.

Bargmann, W., Hehn, G.U. and Lindner, E. (1968) Uber die Zellen des braunen Fettgewebes und ihre Innervation. *Z. Zellforsch. mikroskop. Anat.*, **85**, 601–13.

Barnard, T., Mory, G. and Nechad. M. (1980) Biogenic amines and the trophic response of brown adipose tissue. in *Biogenic Amines in Development* (eds H. Parvez and S. Parvez). pp. 391–439.

Barnard, T. and Skala, J. (1970) The development of brown adipose tissue. in *Brown Adipose Tissue* (ed O. Lindberg). Elsevier. New York, London. Amsterdam, pp. 33–72.

Barneon, G.. Fourcade, J.. Mimran, A., Baldet, P. and Barjon, P. (1975) Association phéochromocytome-pseudo-tumeur de graisse brune. *Nouv. Presse. Méd.*, **4**, 2933–5.

Barrnett. R.J. and Ball, E.G. (1960) Metabolic and ultrastructural changes induced in adipose tissue by insulin. *J. Biophys. Biochem. Cytol.*, **8**, 83–99.

Bernson, V. and Nicholls. D.G. (1974) Acetate, a major end product of fatty-acid oxidation in hamster brown-adipose-tissue mitochondria. *Eur. J. Biochem.*, **47**, 517–25.

Bieber, L.L., Pettersson, B. and Lindberg, O. (1975) Studies on norepinephrine-induced efflux of free fatty acids from hamster brown-adipose-tissue cells. *Eur. J. Biochem.*, **58**, 375–81.

Blanchette-Mackie, E.J. and Scow. R.O. (1981) Lipolysis and lamellar structures in white adipose tissue of young rats: lipid movement in membranes. *J. Ultrastruct. Res.* **77**, 295–318.

Blanchette-Mackie, E.J. and Scow, R.O. (1982) Continuity of intracellular channels with extracellular space in adipose tissue and liver. *Anat. Rec.*, **203**, 205–19.

Bruck, K. (1970) Nonshivering thermogenesis and brown adipose tissue in relation to age, and their integration in the thermoregulatory system. in *Brown Adipose Tissue* (ed O. Lindberg), Elsevier, New York, London, Amsterdam, pp. 117–54.

Bukowiecki, L., Follea, N., Vallieres, J. and Leblanc, J. (1978) Beta-adrenergic receptors in brown-adipose tissue. Characterization and alterations during acclimation of rats to cold. *Eur. J. Biochem.*, **92**, 189–96.

Bukowiecki, L., Follea, N., Paradis, A. and Collet, A. (1980) Stereospecific stimulation of brown adipocyte respiration by catecholamines via β adrenoceptors. *Am. J. Physiol.*, **238**, E552–63.

Bukowiecki, L., Follea, N., Lupien, J. and Paradis, A. (1981) Metabolic relationships between lipolysis and respiration in rat brown adipocytes. *J. Biol. Chem.*, **256**, 12840–8.

Bulow, J. (1981) Human adipose tissue blood flow during prolonged exercise. Effect of beta-adrenergic blockade, nicotinic acid and glucose infusion. *Scand. J. Clin. Lab. Invest.*, **41**, 415–24.

Burcher, E. and Garlick, D. (1975) Effect of exercise metabolites on adrenergic vasoconstriction in gracilis muscle of the dog. *J. Pharmacol. Exp. Ther.*, **192**, 149–56.

Cameron, I. and Smith, R.E. (1964) Cytological responses of brown fat tissue in cold-exposed rats. *J. Cell. Biol.*, **23**, 89–100.

Carey, F.G. (1982) A brain heater in the swordfish. *Science*, **216**, 1327–9.

Chinet, A., Clausen, T. and Girardier, L. (1977a) Microcalorimetric determination of energy expenditure due to active sodium–potassium transport in the soleus muscle and brown adipose tissue of the rat. *J. Physiol. (London)*, **265**, 43–61.

Chinet, A., Friedli, C., Seydoux, J. and Girardier, L. (1977b) Does cytoplasmic alkalinization trigger mitochondrial energy dissipation in the brown adipocyte? in *Effectors of Thermogenesis* (eds L. Girardier and J. Seydoux), Experientia Supplementum, **32**, 25–32.

Chinet, A. and Durand, J. (1978) Control of the brown fat respiratory response to noradrenaline by catechol-*O*-methyltransferase. *Biochem. Pharmacol.*, **28**, 1353–61.

Clausen, T. and Flatman, J.A. (1977) The effect of catecholamines on Na-K transport and membrane potential in rat soleus muscle. *J. Physiol. (London)*, **270**, 383–414.

Clement, G. and Schaeffer, G. (1947) A demonstration of the lipid mobilizing effect of adrenaline. Role of the sympathetic nervous system. *C.R. Soc. Biol.*, **141**, 320.

Conklin, P. and Heggeness, F.W. (1971) Maturation of temperature homeostasis in the rat. *Am. J. Physiol.*, **220**, 333–6.

Cottle, M.K.W. and Cottle, W.H. (1970) Adrenergic fibers in brown fat of cold-acclimated rats. *J. Histochem. Cytochem.*, **18**, 116–9.

Daniel, H. and Derry, D.M. (1970) Sympathetic nerve development in the brown adipose tissue of the rat. *J. Physiol. Pharmacol.*, **48**, 161–8.

Davydov, A.F. and Makarova, A.R. (1964) Changes in heat regulation and circulation in new-born seals on transition to aquatic form of life. *Fed. Proc.*, **24**, T563–6.

Dawkins, M.J.R. and Hull, D. (1965) The production of heat by fat. *Sci. Amer.*, August, 62–7.

Depocas, F., Zarov-Behrens, G. and Lacelle, S. (1980) Noradrenaline-induced calorigenesis in warm- or cold-acclimated rats. *In vivo* estimation of adrenoceptor concentration of noradrenaline effecting half-maximal response. *Can. J. Physiol. Pharmacol.*, **58**, 1072–7.

Desautels, M. and Himms-Hagen, J. (1980) Parallel regression of cold-induced changes in ultrastructure, composition, and properties of brown adipose tissue mitochondria during recovery of rats from acclimation to cold. *Can. J. Biochem.*, **58**, 1057–68.

Dyer, R.F. (1968) Morphological features of brown adipose cell maturation *in vivo* and *in vitro*. *Am. J. Anat.*, **123**, 255–82.

English, J.T., Patel, S.K. and Flanagan, M.J. (1973) Association of pheochromocytomas with brown fat tumors. *Radiology*, **107**, 279–81.

Evonuk, E. and Hannon, J.P. (1963) Cardiovascular and pulmonary effects of noradrenaline in the cold-acclimatized rat. *Fed. Proc.*, **22**, 911–16.

Fairfield, J. (1948) Effects of cold on infant rats: body temperatures, oxygen consumption, electro-cardiograms. *Am. J. Physiol.*, **155**, 355–65.

Falk, B., Hillarp, N.A., Thieme, G. and Torp, A. (1962) Fluorescence of catecholamines and related compounds condensed with formaldehyde. *J. Histochem. Cytochem.*, **10**, 347–54.

Farkas, M., Varnai, I. and Donhoffer, S.Z. (1972) Fallacies in the interpretation of body temperature changes in the newly born. *Acta Physiol. Acad. Sci. Hung.*, **42**, 31–4.

Feyrter, F. (1973) Ein adrenolipoides Syndrom. Normologie und Pathologie des braunen Fettgewebes der Menschen. in *Normale und Pathologische Anatomie* (ed. Bargmann Doerr), Thieme Verlag, Heft 27.

Fink, S.A. and Williams, J.A. (1976) Adrenergic receptors mediating depolarization in brown adipose tissue. *Am. J. Physiol.*, **231**, 700–6.

Foster, D.O. and Frydman, M.L. (1979) Tissue distribution of cold-induced thermogenesis in conscious warm- or cold-adapted rats reevaluated from changes in tissue blood flow. *Can. J. Physiol. Pharmacol.*, **57**, 257–70.

Foster, D.O., Depocas, F. and Frydman, M.L. (1980) Noradrenaline-induced calorigenesis in warm- and in cold-acclimated rats. *Can. J. Physiol. Pharmacol.*, **58**, 915–24.

Foster, D.O. and Depocas, F. (1980) Evidence against noradrenergic regulation of vasodilatation in rat brown adipose tissue. *Can. J. Physiol. Pharmacol.*, **58**, 1418–25.

Friedli, C., Chinet, A. and Girardier, L. (1977) Comparative measurements of *in vitro* thermogenesis of brown adipose tissue from control and cold-adapted rats. in *Effectors of Thermogenesis* (eds L. Girardier and J. Seydoux), Experientia Supplementum **32**, 259–266.

Girardier, L., Seydoux, J. and Clausen, T. (1968) Membrane potential of brown adipose tissue. *J. Gen. Physiol.*, **52**, 925–40.

Girardier, L. and Seydoux, J. (1971a) Cytomembrane phenomena during stimulation of brown fat thermogenesis by norepinephrine. in *Nonshivering Thermogenesis* (ed L. Jansky) Prague Academia, pp. 255–270.

Girardier, L. and Seydoux, J. (1971b) Le contrôle de la thermogénèse du tissu adipeux brun. *J. Physiol. (Paris)*, **63**, 147–86.

Girardier, L. and Schneider-Picard, G. (1983) Alpha and beta adrenergic membrane potential changes and metabolism in rat brown adipose tissue. *J. Physiol. (London)*, **335**, 629–41.

Glick, G., Epstein, S.E., Wechsler, A.S. and Braunwald, E. (1967) Physiological differences between the effects of neurally released and blood born norepinephrine on beta adrenergic receptors in the arterial bed of the dog. *Circ. Res.* **21**, 217–27.

Glick, Z., Teague, R.J. and Bray, G.A. (1981) Brown adipose tissue: thermic response increased by a single low protein, high carbohydrate meal. *Science*, **213**, 1125–7.

Grav, H.J., Blix, A.S. and Pasche, A. (1974) How do seal pups survive birth in arctic winter? *Acta Physiol. Scand.*, **92**, 427–9.

Hardman, M.J. and Hull, D. (1970) Fat metabolism in brown adipose tissue *in vivo*. *J. Physiol. (London)*, **206**, 263–73.

Hassi, J. (1977) The brown adipose tissue in man. *Acta Universatis Ouluensis series D Medica*, **21**, 18–92.

Hausberger, F.X. (1934) Uber die Innervation der Fettorgane. *Z. Mikroscop. Anat. Forsch.*, **36**, 231–66.

Heaton, J.M. (1972) The distribution of brown adipose tissue in the human. *J. Anat.*, **112**, 35–9.

Heim, T. and Hull, D. (1966) The blood flow and oxygen consumption of brown adipose tissue in the new born rabbit. *J. Physiol. (London)*, **186**, 42–55.

Herd, P.A. Horwitz, B.A. and Smith, R. (1970) Norepinephrine-sensitive Na^+/K^+ ATPase activity in brown adipose tissue. *Experientia*, **26**, 825–6.

Hirata, K. (1982) Blood flow to brown adipose tissue and norepinephrine-induced calorigenesis in physically trained rats. *Jap. J. Physiol.*, **32**, 279–91.

Horowitz, J.M., Horwitz, B.A. and Smith, R.E. (1971) Effect *in vivo* of norepinephrine on the membrane resistance of brown fat cells. *Experientia*, **27**, 1419–21.

Horwitz, B.A., Horowitz, J.M. and Smith, R.E. (1969) Norepinephrine-induced depolarization of brown fat cells. *Proc. Natl. Acad. Sci. USA.*, **64**, 113–20.

Horwitz, B.A. and Eaton, M. (1975) The effect of adrenergic agonists and cyclic AMP on the Na^+/K^+ ATPase activity of brown adipose tissue. *Eur. J. Pharmacol.*, **34**, 241–45.

Hunt, T. and Hunt, E. (1967) A radioautographic study of proliferation in brown fat of the rat after exposure to cold. *Anat. Rec.*, **157**, 537–45.

Janssens, W.J. and Vanhoutte, P.M. (1978) Instantaneous changes of alpha-adrenoceptor affinity caused by moderate cooling in canine cutaneous veins. *Am. J. Physiol.*, **234**, H330–7.

Kikta, D.C., Threattle, R.M., Barney, C.C. and Fregly, M.J. (1982) Effect of cold-acclimation on cardiovascular responses of rats. *Fed. Proc.*, **41**, 1238.

Kjellmer, I. (1965) On the competition between metabolic vasodilatation and neurogenic vasoconstriction in skeletal muscle. *Acta Physiol. Scand.*, **63**, 450–9.

Krishna, G., Moskowitz, P., Dempsey, P. and Brodie, B.B. (1970) The effect of norepinephrine and insulin on brown fat cell membrane potentials. *Life Sci.*, **9**, 1353–61.

Kuroshima, A., Konno, N. and Itoh, S. (1967) Increase in the blood flow through brown adipose tissue in response to cold exposure and norepinephrine in the rat. *Jap. J. Physiol.*, **17**, 523–37.

Lasserre, B., Seydoux, J. and Girardier, L. (1973) Energy-supplying metabolism and transmembrane potential of rat brown adipose tissue *in vitro*. *Experientia*, **29**, 745.

Laury, M.C., Bertin, R., Portet, R. and Chevillard, L. (1971) Variation du débit sanguin et de la température de la graisse brune interscapulaire au cours de perfusions i.v. de noradrenaline. *Compt. Rend. Soc. Biol.*, **165**, 535–9.

Leduc, J. (1961) Catecholamine production and release in exposure and acclimation to cold. *Acta Physiol. Scand.*, **53**, suppl. 183, 5–101.

Leiphart, C.J. and Nudelman, E.J. (1970) Hibernoma masquerading as a pheochromocytoma. *Radiology*, **95**, 659–60.

Levitski, A. (1978) Catecholamine receptors. *Rev. Physiol. Biochem. Pharmacol.*, **82**, 1–26.

McCormack, J.G. and Denton, R.M. (1977) Evidence that fatty acid synthesis in the interscapular brown adipose tissue of cold adapted rats is increased *in vivo* by insulin by mechanisms involving parallel activation of pyruvate dehydrogenase and acetyl-CoA carboxylase. *Biochem, J.*, **166**, 627–30.

Melicow, M.M. (1957) Hibernating fat and pheochromocytoma. *Arch. Path.*, **63**, 367–72.

Merklin, R.J. (1974) Growth and distribution of human fetal brown fat. *Anat. Rec.*, **178**, 637–46.

Mory, G., Ricquier, D. and Hemon, P. (1980) Effects of chronic treatment upon the brown adipose tissue of rats. *J. Physiol. (Paris)*, **76**, 859–64.

Mory, G., Ricquier, D., Nechad, M. and Hemon, P. (1982) Impairment of trophic response of brown fat to cold in guanethidine-treated rats. *Am. J. Physiol.*, **242**, C159–65.

Nouri, T.N. (1972) Innervation of brown adipose tissue in new-born rabbits. *J. Physiol. (London)*, **227**, 42P.

Nedergaard, J. (1982) Catecholamine sensitivity in brown fat cells from cold-acclimated hamsters and rats. *Am. J. Physiol*, **242**, C250–7.

Nedergaard, J. and Lindberg, O. (1982) The brown fat cell. Intern review of cytology, **74**, 187–290.

Perkins, M.N., Rothwell, N.J., Stock, M.J. and Stone, T.W. (1981) Activation of brown adipose tissue thermogenesis by the ventro medial hypothalamus. *Nature*, **289**, 401–2.

Pettersson, B. and Valin, I. (1976) Norepinephrine-induced shift in levels of adenosine 3':5' monophosphate and ATP parallel to increased respiratory rate and lipolysis in isolated hamster brown-fat cells. *Eur. J. Biochem.*, **62**, 383–90.

Rasmussen, H. (1970) Cell communication, calcium ion, and cyclic adenosine monophosphate. *Science*, **170**, 404–12.

Rath, E.A., Salmon, D.M.W. and Hems, D.A. (1979) Effect of acute change in ambient emperature on fatty acid synthesis in the mouse. *FEBS Lett.*, **108**, 33–6.

Revel, J.P. and Sheridan, J.D. (1968) Electrophysiological studies of intercellular junctions in brown fat. *J. Physiol. (London)*, **194**, 34P–35P.

Ricquier, D., Mory, G. and Hemon, P. (1979) Changes induced by cold adaptation in the brown adipose tissue from several species of rodents, with special reference to the mitochondrial components. *Can. J. Biochem.*, **57**, 1262–6.

Ricquier, D., Nechad, M. and Mory, G. (1982) Ultrastructural and biochemical characterization of human brown adipose tissue in pheochromocytoma. *J. Clin. Endocrinol. Metab.*, **54**, 803–7.

Rogus, E.M., Cheng, L.A. and Zierler, K. (1977) Beta-adrenergic effect on Na^+-K^+ transport in rat skeletal muscle. *Biochim. Biophys. Acta*, **464**, 347–55.

Rona, G. (1964) Changes in adipose tissue accompanying pheochromocytoma. *Can. Med. Ass. J.*, **91**, 303–5.

Rothwell, N.J., Stock, M.J. and Wyllie, M.G. (1981) Na^+, K^+-ATPase activity and noradrenaline turnover in brown adipose tissue of rats exhibiting diet-induced thermogenesis. *Biochem. Pharmacol.*, **30**, (12), 1709–12.

Rowlands, D.J. and Donald, D.E. (1968) Sympathetic vasoconstrictive responses during exercise- or drug-induced vasodilatation. *Circ.Res.*, **23**, 45–60.

Saggerson, E.D. and Carpenter, C.A. (1982) Sensitivity of brown adipose tissue carnitine palmitoyltransferase to inhibition by malonyl-CoA. *Biochem. J.*, **204**, 373–5.

Schneider-Picard, G., Carpentier, J.L. and Orci, L. (1980*a*) Quantitative evaluation of gap juntions during development of the brown adipose tissue. J. *Lipid Res.*, **21**, 600–7.

Schneider-Picard, G., Seydoux, J. and Girardier, L. (1980*b*) Gap junctional and sympathetic nerve development in the rat brown adipose tissue. in Satellite of 28. *Int. Congr. Physiol. Sci. Pecs.* (eds Z. Szelenyi and M. Szekely).

Schneider-Picard, G. and Girardier, L. (1982) Postnatal development of sympathetic innervation of rat brown adipose tissue reevaluated with a method allowing for monitoring flavoprotein redox-state. *J. Physiol. (Paris)*, **78**, 151–7.

Seydoux, J. and Girarder, L. (1977) Control of brown fat thermogenesis by the sympathetic nervous system. in *Effectors of Thermogenesis Experientia* Suppl. **32**, 153–67.

Seydoux, J., Constandinidis, J., Tsacopoulos, M. and Girardier, L. (1977) *In vitro* study of the control of the metabolic activity of brown adipose tissue by the sympathetic nervous system. *J. Physiol. (Paris)*, **73**, 985–96.

Seydoux, J., Rohner-Jeanrenaud, F., Assimacopoulos-Jeannet, F., Jeanrenaud, B. and Girardier, L. (1981) Functional disconnection of brown adipose tissue in hypothalmic obesity in rats. *Pflugers Arch.*, **390**, 1–4.

Seydoux, J., Giacobino, J.P. and Girardier, L. (1982*a*) Impaired metabolic response to nerve stimulation in brown adipose tissue of hypothyroid rats. *Mol. Cell. Endocrinol.*, **25**, 213–26.

Seydoux, J., Ricquier, D., Rohner-Jeanrenaud, F., Assimacopoulos-Jeannet, F., Giacobino, J.P., Jeanrenaud, B. and Girardier, L. (1982*b*) Decreased guanine nucleotide binding and reduced equivalent production by brown adipose tissue in hypothalamic obesity. *FEBS Lett.*, **146**, 161–4.

Sheridan, J.D. (1971) Electrical coupling between fat cells in newt fat body and mouse brown fat. *J. Cell. Biol.*, **50**, 795–803.

Shimazu, T. (1981) Central nervous system regulation of liver and adipose tissue metabolism. *Diabetologia*, **20**, 343–56.

Sidman, R.L. and Fawcett, D.W. (1954) The effect of peripheral nerve section on some metabolic response of brown adipose tissue in mice. *Anat. Rec.*, **118**, 487–507.

Smith, R.E. (1964) Thermoregulatory and adaptive behavior of brown adipose tissue. *Science*, **146**, 1686–9.

Smith, R.E. and Roberts, J.C. (1964) Thermogenesis of brown adipose tissue in cold-acclimated rats. *Am. J. Physiol.*, **206**, 143–8.

Stock, K. and Westermann, E.O. (1963) Concentration of norepinephrine, serotonine and histamine and of amine-metabolizing enzymes in mammalian adipose tissue. *J. Lipid Res.*, **4**, 297–304.

Strandell, T. and Shepherd, J.T. (1967) The effect in humans of increased sympathetic activity on the blood flow to active muscles. *Acta Med. Scand.* (Suppl.) **472**, 146.

Stucki, W. (1980) The optimal efficiency and the economic degrees of coupling of oxidative phosphorylation. *J. Biochem.*, **109**, 269–83.

Svoboda, P., Svartengren, J., Snochowski, M., Houstek, J. and Cannon, B. (1979) High number of high-affinity sites for (-).[^{3}H]dihydroalprenolol on isolated hamster brown-fat cells. *Eur. J. Biochem.*, **102**, 203–10.

Suter, E.R. (1969) The fine structure of brown adipose tissue. *Lab. Invest.*, **21**, 246–68.

Tanuma, Y., Ohata, M., Ito, T. and Yokochi, C. (1976) Possible function of human brown adipose tissue as suggested by observation on perirenal brown fats from necropsy cases of variable age group. *Arch. Histol. Jap.*, **39**, 117–45.

Teague, R.J., Kanarek, R., Bray, G.A., Orthen-Gambill, N. (1981) Effect of diet on the weight of brown adipose tissue in rodents. *Life Sci.*, **29**, 1531–6.

Tedesco, J.L., Flattery, K.V. and Sellers, E.A. (1977) Effect of thyroid hormones and cold exposure on turnover of norepinephrine in cardiac and skeletal muscle. *Can. J. Physiol. Pharmacol.*, **55**, 515–22.

Teplitz, C., Goss, G., Hammond, R. and Hamolsky, M. (1974) The ultrastructural morphogenesis in the direct transformation of periadrenal white fat into brown adipose tissue in adult man. *Lab. Invest.*, **30**, 405.

Thomson, J.F., Habeck, D.A., Nance, S.L. and Beetham, K.L. (1969) Ultrastructural and biochemical changes in brown fat in cold-exposed rats. *J. Cell. Biol.*, **41**, 312–34.

Thureson-Klein, A., Lagercrantz, H. and Barnard, T. (1976) Chemical sympathectomy of interscapular brown adipose tissue. *Acta Physiol. Scand.*, **98**, 8–18.

Trayhurn, P. (1979) Fatty acid synthesis in brown adipose tissue, liver and white adipose tissue of the cold-acclimated rat. *FEBS Lett.*, **104**, 13–16.

Vanhoutte, P.M. and Shepherd, J.T. (1969) Activity and thermosensitivity of canine cutaneous veins after inhibition of monamine oxidase and catechol-O-methyl transferase. *Circ. Res.*, **25**, 607–16.

Varnai, I., Farkas, M. and Donhoffer, S.Z. (1970) Thermoregulatory effects of hypercapnia in the new born rat. Comparison with the effect of hypoxia. *Acta Physiol. Acad. Sci. Hung.*, **38**, 225–35.

Verhaeghe, R.H., Lorenz, R.R., McGrath, M.A., Shepherd, J.T. and Vanhoutte, P.M. (1978) Metabolic modulation of neurotransmitter release: adenosine, adenine nucleotides, potassium, hyperosmolarity and hydrogen ion. *Fed. Proc.*, **37**, 208–11.

Weiner, N., Perkins, M. and Sidman, R.L. (1962) Effect of reserpine on noradrenaline content of innervated and denervated brown adipose tissue of the rat. *Nature (London)*, **193**, 137–8.

Williams, J.A. and Matthews, E.K. (1974a) Effect of ions and metabolic inhibitors on membrane potential of brown adipose tissue. *Am. J. Physiol.*, **227**, 981–6.

Williams, J.A. and Matthews, E.K. (1974b) Membrane depolarization, cyclic AMP, and glycerol release by brown adipose tissue. *Am. J. Physiol.*, **227**, 987–92.

Williamson, J.R. (1970) Control of energy metabolism in hamster brown adipose tissue. *J. Biol. Chem.* **245**, 2043–50.

Wirsen, C. (1964) Adrenergic innervation of adipose tissue examined by fluorescence microscopy. *Nature, London*, **202**, 913.

Chapter Four

Autonomic Regulation of Thermogenesis

Lewis Landsberg and James B. Young

4.1 INTRODUCTION

Adaptive thermogenesis has been best studied in the context of temperature regulation during cold exposure. It is well established that the maintenance of constant body temperature in the face of environmental cold depends upon adaptive changes in heat production, and that the biological processes involved are under the precise control of the central nervous system. Although the somatic motor system contributes to temperature regulation, the primary importance of the autonomic nervous system in the regulation of thermogenesis is well recognized. The sympathetic nervous system, in particular, and to a lesser extent the adrenal medulla, appear to play the major role in the regulation of mammalian thermogenesis. The parasympathetic system may also participate in the regulation of thermogenic processes, but the role of this portion of the autonomic system is, at present, less clearly defined.

Although the role of the sympathetic nervous system in the regulation of thermogenesis was defined initially in the context of cold exposure, recent evidence has emerged indicating that the sympathetic nervous system may be involved in the regulation of adaptive heat production in situations unrelated to the maintenance of body temperature. A potential role for the sympathetic nervous system, in particular, has been suggested in the regulation of thermogenic processes that accompany changes in dietary intake. This chapter will review the role of the sympathetic nervous system and adrenal medulla, collectively referred to as the sympathoadrenal system, in the regulation of mammalian thermogenesis, with special emphasis on cold exposure and dietary intake.

4.1.1 The functional organization of the sympathoadrenal system

Norepinephrine (NE) is the adrenergic neurotransmitter. Synthesized and stored in peripheral sympathetic nerve endings, NE is released in response to efferent nerve impulses that invade the terminal sympathetic fibres. NE acts principally via stimulation of adrenergic receptors within the immediate vicinity of release and under most circumstances does not function as a circulating hormone. In contrast epinephrine (E) is the circulating hormone of the adrenal medulla. Released in response to preganglionic impulses carried in the splanchnic nerves, E influences physiological events at adrenergic receptors throughout the body. Both the adrenal medulla and the peripheral sympathetic neurons are regulated by preganglionic autonomic neurons originating in the intermediolateral column of the spinal cord. These cholinergic neurons either synapse with postganglionic sympathetic nerve cells in the paravertebral or preaortic sympathetic ganglia from which sympathetic fibres are widely distributed to blood vessels and viscera or pass through the sympathetic ganglia to innervate the adrenal medulla directly.

The activity of the preganglionic neurons is governed by descending bulbo-spinal tracts originating in the reticular formation of the pons and medulla. Although these brainstem sympathetic centres have an intrinsic activity of their own, their functional state is influenced by centres in the hypothalamus, cortex, and limbic lobes. Direct connections between the hypothalamus and the intermediolateral cell column appear to exist as well. These regulatory centres in the hypothalamus and brainstem are responsive to alterations in afferent neural input as well as to change in the physical (temperature, tonicity) and chemical (hormones, substrates) properties of the extracellular fluid. Sympathoadrenal outflow is thus highly integrated and reflects the homeostatic needs of the organism.

The relationship between the sympathetic nervous system and the adrenal medulla is complex. The sympathetic nervous system and the adrenal medulla are often stimulated together, and during periods of intense sympathetic stimulation the adrenal medulla may be progressively recruited so that circulating E reinforces the physiological effects of locally released NE. In other situations, however, the sympathetic nervous system and the adrenal medulla may be activated independently.

4.1.2 Methods of study

A variety of techniques have been utilized to investigate the role of the sympathoadrenal system in the regulation of thermogenesis. Studies based on either chemical or surgical ablation of sympathetic nerves and adrenal medulla, or on the administration of adrenergic blocking agents, have provided much of the basic information about the contribution of the sympathoadrenal system to the production of metabolic heat. These approaches, however, furnish little insight into the mechanisms involved in the thermogenic response and are more suited to situations of increased thermogenesis than to those characterized by hypometabolism. The changes in oxygen consumption and heat production induced by catecholamines or electrical stimulation of autonomic nerves have also been studied in intact animals and man, in tissue preparations *in situ*, and in isolated tissues and cells, particularly brown adipose tissue. While these investigations have generated much useful information about the thermogenic actions of catecholamines, they describe only potential effects; in any given physiological situation the functional state of the sympathoadrenal system must also be considered.

Assessment of sympathoadrenal activity has usually been based on the measurement of catecholamines in urine and plasma. An increase in urinary E excretion or plasma E level is good evidence of adrenal medullary stimulation, although small increments in E secretion may be missed because of the relative insensitivity of the assays at the lower end of the usual physiological range. Plasma and urinary NE levels provide an index of sympathetic activity although the fact that NE is not physiologically important as a circulating hormone, and that the circulating NE pool is only a tangential reflection of NE released at the nerve ending, diminish the sensitivity of these measurements. Under circumstances in which the adrenal medulla is markedly stimulated, moreover, circulating levels of NE may be elevated despite unchanged or even reduced sympathetic nervous system activity. A further drawback to the use of norepinephrine levels in plasma and urine is the lack of information provided about sympathetic activity in different innervated tissues. Since efferent neural impulses are often not distributed uniformly to all sympathetic nerve endings, a single measure of sympathetic activity may not adequately reflect the heterogeneous potential of sympathetic responses. Nonetheless, despite these limitations, plasma NE levels are the best currently available index of sympathetic activity in man.

The use of microelectrodes to record impulse traffic in autonomic nerves in the extremities of human subjects, and in various locations in the experimental animal, has yielded direct information regarding the activity of the instrumented nerves. These techniques, however, are invasive and, at least in animals, frequently involve sectioning of the nerve under study, interventions which may alter the activity of the instrumented nerves. Furthermore, the type of nerve from which the measured impulses derive is not always known with certainty. Under the best of circumstances, moreover, these recordings provide data over relatively short time-intervals. In the experimental animal an additional technique for assessing sympathetic activity involves the measurement of NE turnover rate; this technique has been successfully applied to the study of sympathetic function in a number of sympathetically innervated tissues. When sympathetic activity is increased, the rate of NE turnover is accelerated and when nerve activity is decreased, turnover is diminished.

Although all of the above techniques have limitations, when taken together, the information obtained from studies utilizing a variety of these approaches provides a remarkably comprehensive description of the role of the sympathetic nervous system in the regulation of mammalian thermo-genesis.

Assessment of the functional state of the parasympathetic nervous system is even more difficult. Vagotomy, and the administration of the cholinergic antagonist, atropine, in various physiological situations represent the only techniques that have been employed in the investigation of the role of the parasympathetic nervous system in the regulation of thermogenesis.

4.2 COLD EXPOSURE

4.2.1 Critical role of the sympathetic nervous system in temperature maintenance

Normal temperature maintenance requires an intact autonomic nervous system. Interruption of autonomic function with ganglionic blocking agents renders animals incapable of sustaining body temperature during cold exposure (Fig. 4.1) and leads rapidly to death from hypothermia. Primary involvement of the sympathetic nervous system was suggested over 25 years ago by the classic experiments of Hsieh, Carlson and Gray (1957). As shown in Fig. 4.2, these studies demonstrated that ganglionic blockade abolished the increase in oxygen consumption that normally occurs during cold exposure in curarized, cold-acclimated rats. Since curare eliminated shivering, these experiments indicate a primary effect of the autonomic nervous system on heat production. Involvement of the sympathetic nerves

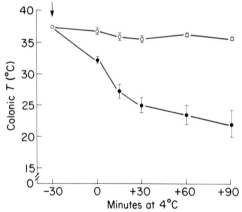

Fig. 4.1 Effect of ganglionic blockade on rectal temperature in mice. Non-cold-acclimated white mice were injected with a ganglionic blocking agent chlorisondamine (dark circles) 30 min before cold exposure at 4°C. In comparison with untreated controls the animals treated with ganglionic blockade fail to maintain body temperature, indicating a critical role for the autonomic nervous system in temperature regulation during cold exposure (Young and Landsberg, unpublished observations).

was suggested by the fact that atropine, in distinction to ganglionic blockade, had no effect on oxygen consumption. The administration of NE, moreover, prevented the fall in oxygen consumption induced by ganglionic blockade (Fig. 4.2). An equivalent dose of E was considerably less effective than that of NE, suggesting a more important role for the sympathetic nervous system, as compared with that of the adrenal medulla. In the years since the experiments of Carlson and co-workers, numerous studies have illuminated the role of the sympathoadrenal system in the response to cold

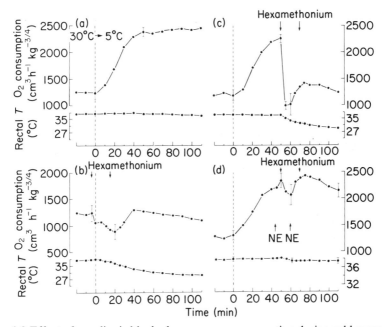

Fig. 4.2 Effect of ganglionic blockade on oxygen consumption during cold exposure in the rat. The effect of cold exposure at 5°C on oxygen consumption and rectal temperature in cold acclimated, curarized rats is shown. Panel A demonstrates the rise in oxygen consumption and maintenance of rectal temperature in intact rats. Panel B demonstrates that ganglionic blockade with hexamethonium prevents the rise in oxygen consumption during cold exposure with subsequent fall in rectal temperature. Panel C demonstrates that ganglionic blockade rapidly diminishes the cold-induced increase in oxygen consumption induced by ganglionic blockade, while panel D demonstrates that the effect of ganglionic blockade is antagonized by the prior administration of NE. Since shivering was inhibited by curarization these studies demonstrate a primary effect of the autonomic nervous system to increase metabolic heat production in the cold (non-shivering thermogenesis). The fact that atropine was without effect on oxygen consumption, and that NE was more potent than E in antagonizing the effect of hexamethonium implies an important role for the sympathetic nervous system. Modified from the classic studies of Hsieh *et al.* (1957), with permission.

exposure and clarified the physiological and biochemical mechanisms involved in the sympathetic regulation of thermogenesis.

4.2.2 Sympathoadrenal activation during cold exposure

(a) Sympathetic nervous system

Stimulation of sympathetic nervous system activity during cold exposure has been convincingly demonstrated in a wide variety of mammalian species including man. Rats exposed to 3°C increase NE excretion four- to five-fold in one day (Leduc, 1961) (Fig. 4.3). Plasma NE levels are increased markedly (as much as five- to ten-fold) during cold-water immersion or acute environmental cold exposure in dogs and man (Bergh *et al.*, 1979; Therminarias, Chirpaz and Tanche, 1979). Lesser degrees of cold exposure are, in general, associated with smaller increases in urinary catecholamine excretion or plasma catecholamine levels (Bergh *et al.*, 1979).

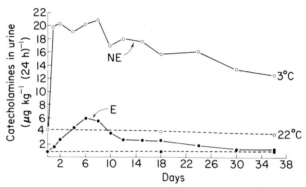

Fig. 4.3 Catecholamine excretion during cold exposure in the rat. Increase in urinary excretion of NE and E in the rat during cold exposure at 3°C (solid line) as compared with 22°C (dashed line). The increase in E excretion is of lesser magnitude and not sustained. From Leduc (1961), with permission.

Measurements of NE turnover (Young and Landsberg, 1979; Jones and Musacchia, 1976; Oliverio and Stjarne, 1965; Bralet, Beley and Lallemant, 1972; Tedesco, Flattery and Sellers, 1977; Johnson, Young and Landsberg, 1981) demonstrate sympathetic activation in several organs of the mouse, rat and hamster during both acute and chronic exposure to cold (Fig. 4.4). Environmental warming on the other hand (30–34°C) is associated with sympathetic suppression (Jones and Musacchia, 1976; Landsberg and Axelrod, 1968). Direct recordings of impulse traffic in cutaneous sympathetic nerves of rabbits and cats also demonstrate enhanced activity during central nervous system cooling (Iriki, Riedel and Simon, 1971; Walther, Iriki and Simon, 1970). NE synthesis in various organs *in vivo* (Beley *et al.*, 1976) and in sympathetic ganglia *in vitro* (Thoenen, 1970) is accelerated during cold exposure as well. In animals exposed to cold for long periods of time sympathetic nervous system activity tends to diminish slowly (Fig. 4.3), although, in the rat, evidence of increased sympathetic activity is present for at least six months (Leduc, 1961).

Although sympathetic stimulation by cold exposure has been demonstrated in multiple organs of mammalian species, current evidence indicates that the sympathetic nervous system response to cold is discriminant rather than generalized. Many studies utilizing different techniques reveal a varied pattern of sympathetic activation in different organ systems or tissues. Measurements of NE turnover in heart, pancreas, lung, spleen, and skeletal muscle are invariably increased in the cold (Young and Landsberg, 1979; Jones and Musacchia, 1976; Tedesco *et al.*, 1977; Beley *et al.*, 1976) while in submaxillary gland, liver, intestine and kidney the effect of cold is much less marked or negligible. The rate of NE biosynthesis, while increased in heart

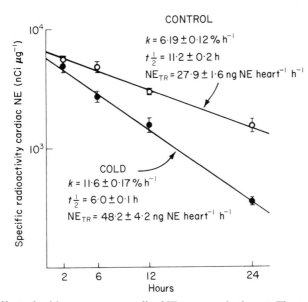

Fig. 4.4 Effect of cold exposure on cardiac NE turnover in the rat. The turnover rate of NE in rat heart is estimated from the rate of disappearance of tracer NE administered at time zero. In this experiment non-cold-acclimated rats were injected with tracer and half the animals placed in the cold at 4°C (dark circles). The rate of decline of specific radio activity is a measure of the fractional turnover rate (k) from which a half time of disappearance ($t_{1/2}$) may be determined. From the product of the slope (k) and the steady-state endogenous NE level a calculated NE turnover rate (NETR) may be computed. The principal determinant of NE turnover is sympathetic activity so that measurement of NE turnover provides an index of sympathetic nervous system activity. Increased sympathetic activity in acutely cold exposed rats is demonstrated in this figure by the line of steeper slope. From Young and Landsberg (1978), with permission.

and spleen during cold exposure, was actually decreased in submaxillary gland (Beley *et al.*, 1976). Nerve impulse traffic in cutaneous sympathetic efferents increased in rabbits exposed to cold, at the same time that it declined in visceral splanchnic nerves (Iriki *et al.*, 1971; Walther *et al.*, 1970). Studies to date, therefore, indicate that cold-induced alterations in sympathetic activity in different tissues are discrete and suggest that the pattern of sympathetic outflow may have important physiological consequences.

(i) Adrenal medulla. Studies from the laboratory of W.B. Cannon over 50 years ago provided evidence that cold exposure stimulated adrenal medullary secretion (Cannon *et al.*, 1979). Subsequent studies have confirmed an increase in E excretion in cold-exposed rats and in plasma E levels in dogs following cold water immersion (Leduc, 1961; Therminarias, Chirpaz and Tanche, 1979). Synthesis of catecholamines within the adrenal medulla is increased by cold exposure in rats and voles (Thoenen, 1970; Guidotti *et al.*, 1973; Feist and Feist, 1978), supporting stimulation of the adrenal medulla by cold in these species. Adrenal medullary stimulation, however, is of lesser degree than sympathetic activation (Leduc, 1961), and not sustained, urinary E excretion returning towards normal as cold exposure progresses despite continued elevation of urinary NE. Not all studies, moreover, have shown an increase in urinary or plasma E during cold exposure (Young and Landsberg, 1981; Jessen, Rabol and Winkler, 1980). When sympathetic nervous system activity is diminished, however, the adrenal medullary response to cold exposure is enhanced (Young and Landsberg, 1981; Himms-Hagen, 1975). In summary, the adrenal medullary response to cold exposure is less pronounced and less reproducible than activation of the sympathetic nervous system. Adrenal medullary secretion is more evident in severe cold exposure, in warm-acclimated as compared with cold-acclimated animals, and in situations in which the response of the sympathetic nervous system is impaired (Young and Landsberg, 1981).

(ii) Parasympathetic nervous system. The function of the parasympathetic nervous system during cold exposure has been less well studied than that of the sympathetic nervous system. As noted above, atropine was without affect on the thermogenic response to cold exposure in cold-acclimated, curarized rats (Hsieh *et al.*, 1957). Vagotomy, however, has more recently been reported to increase oxygen consumption and blood pressure in cold acclimated rats (Leblane and Cote, 1967). Although this latter study suggests a chronic increase in parasympathetic activity during prolonged cold exposure, as well as a suppressive effect of parasympathetic stimulation on the thermogenic response, possible effects of vagotomy on vagal afferents with consequent reflex changes in sympathetic activity were apparently not excluded.

(b) Afferent signals and central neuronal connections

The regulation of body temperature involves peripheral and central sensing mechanisms as well as integrative centres in several regions of the central nervous system. Temperature-sensitive neurons in the skin (peripheral) and hypothalamus (central) increase their firing rate at temperatures below thermoneutrality (Thompson, 1977; Cabanac, 1975; Pierau, and Wurster, 1981; Poulas, 1981; Boulant, 1981; Hellon, 1981). Afferent impulses arising from these areas, as well as from thermosensitive areas in the spinal cord and lower brainstem, are integrated in the preoptic area and the anterior hypothalamus (Thompson, 1977; Gale, 1973). Connections with efferent centres in the posterior hypothalamus establish the appropriate level and distribution of sympathetic outflow. Cold-sensitive spinal neurons appear to contribute importantly to the regulation of shivering thermogenesis (Fuller, 1977), while the hypothalamic neurons are more closely associated with non-shivering thermogenesis, although considerable overlap exists, at least in the rat (Banet, Hensel and Liebermann, 1978). The net result of central integration of the response to cold is a change in sympathetic outflow that acts to maintain a constant body temperature. The interrelationship among various factors that affect central sympathetic outflow is demonstrated by the fact that baroreceptor stimulation diminishes the thermogenic response to cold exposure in monkeys with experimental hypertension (Wasserstrum and Herd, 1977a), presumably by inhibition of sympathetic activity.

The neurotransmitters involved in the central control of thermoregulatory responses have received considerable attention. Evidence in support of a role for acetylcholine, the biogenic amines (including serotonin, NE, dopamine, and histamine), prostaglandins, amino acids, and a variety of neuropeptides has appeared (Hellon, 1975; Blatteis, 1981). The precise contribution of these central neurotransmitters to the regulation of thermogenesis, however, has yet to be determined.

4.2.3 Physiological effects of the sympathoadrenal system during cold exposure

The mammalian response to cold exposure includes mechanisms designed both to conserve and produce heat. Changes in blood flow distribution and in pilo-erection, dependent in part on the sympathetic nervous system (Thompson, 1977), increase the insulating capacity of the skin and diminish heat loss to the environment. Heat production has been traditionally divided into two major types: shivering thermogenesis which occurs as a consequence of muscle contraction, and non-shivering thermogenesis which results from biochemical reactions not coupled to muscular activity. Ultimately this distinction is somewhat artificial since shivering thermogenesis depends upon the synthesis of ATP required to support contraction of muscle elements.

Although catecholamines appear to facilitate shivering (Feist and Feist, 1978; Himms-Hagen, 1975; Banet *et al.*, 1978), shivering thermogenesis depends predominantly on the somatic motor system (Thompson, 1977). Non-shivering thermogenesis (NST), on the other hand, is regulated by the sympathetic nervous system (Fig. 4.2). Two additional aspects of the thermogenic response involve the provision of substrates as a source of fuel for the production of heat and the delivery of oxygen and substrates to metabolizing tissues. Both substrate mobilization and changes in cardiac output and blood flow distribution during cold exposure involve the sympathoadrenal system as well. Thus, with the exception of shivering, the mammalian response to cold exposure is dependent upon the function of the sympathoadrenal system. The various physiological components of the thermogenic response are considered below.

(a) Importance of the sympathetic nervous system as compared with the adrenal medulla

The relative importance of the sympathetic nervous system on the one hand, and of the adrenal medulla, on the other, can be dissected only imperfectly. A variety of pharmacological and surgical techniques for abolishing sympathoadrenal responses have been utilized to address this question. These studies indicate that animals subjected to complete sympathoadrenal ablation fail to increase oxygen consumption in response to cold exposure (Manara *et al.*, 1965; Leduc, 1976; Carlson, 1960) and subsequently die from hypothermia (Johnson, 1963; Pouliot, 1966); unless both sympathetic nerves and adrenal medulla are eliminated, however, the thermogenic response is adequate and body temperature is maintained (Leduc, 1976). Thus, either limb of the sympathoadrenal system appears capable of supporting thermogenesis in the absence of the other. The primacy of the sympathetic nervous system is suggested by the more consistent and greater degree of sympathetic stimulation during acute cold exposure, and more prolonged activation during chronic cold exposure than that of the adrenal medulla. When the function of the sympathetic nervous system is impaired either physiologically (Young and Landsberg, 1981) or by drugs (Himms-Hagen, 1975; Chan and Johnson, 1968), stimulation of the adrenal medulla during cold exposure is increased.

The available evidence thus suggests that the adrenal medulla reinforces the physiological effects of the sympathetic nervous system, and plays a critical role in temperature maintenance when the sympathetic nervous system is suppressed. The thermogenic effects of circulating catecholamines from the adrenal medulla are similar to those of sympathetic nerves, including stimulation of non-shivering thermogenesis (Carlson, 1960), enhancement of substrate supply in support of the increased metabolic rate

(Maickel *et al.*, 1967), and perhaps, facilitation of shivering (Manara *et al.*, 1965).

4.2.4 Non-shivering thermogenesis

(a) Calorigenic effects of catecholamines

At the end of the 19th century, German physiologists, particularly Rubner, suggested that changes in thermogenesis might result from changes in the rate of metabolism ('chemical thermogenesis'). In 1927 W.B. Cannon (Cannon *et al.*, 1927) proposed that cold-induced stimulation of adrenal medullary secretions increased metabolic rate independent of shivering. Subsequent studies clearly demonstrated that catecholamines increased metabolic rate in the absence of shivering (Hsieh and Carlson, 1957; Hemingway, Price and Stuart, 1964), supporting the concept of chemically mediated thermogenesis (Fig. 4.2), and identifying NE as a more potent calorigenic hormone than E. The sympathetic nervous system is now generally recognized to regulate non-shivering thermogenesis with NE as the principal mediator.

(i) Cold acclimation. Animals acclimated to a cold environment over 2–4 weeks display a marked increase in metabolic rate and heat production on

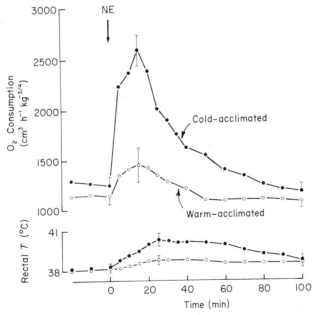

Fig. 4.5 Effect of NE on thermogenesis in the rat. NE injection increases oxygen consumption and rectal temperature in both cold acclimated (dark circles) and warm acclimated (open circles) curarized rats. The effect is markedly enhanced in cold-acclimated animals. From Hsieh and Carlson (1957), with permission.

re-exposure to cold. The capacity for catecholamine-stimulated thermogenesis is markedly exaggerated in these animals (Fig. 4.5) (Hsieh and Carlson, 1957; Hemingway *et al.*, 1964), although the threshold sensitivity for thermogenic effects of catecholamines is apparently not altered (Sellers and Schonbaum, 1963). The mechanisms responsible for this enhancement of the thermogenic response to catecholamines are not known with certainty, although thyroid hormones may contribute (see Chapter 8) and catecholamines themselves may, when present in increased quantities over time, enhance the thermogenic response to subsequent catecholamine administration (Hsieh and Wang, 1971). This latter phenomenon is associated with hypertrophy of brown adipose tissue and indicates that chronic catecholamine administration is able to reproduce several of the elements of the cold-acclimated state. The effects of both cold exposure (Banet *et al.*, 1978) and NE on thermogenesis, at least in small mammals, appear to be mediated by the β-adrenergic receptor (Komaromi, 1977; Schonbaum *et al.*, 1966; LeBlanc, Vallieres and Vachon, 1972) since they are antagonized by β-adrenergic blocking agents, and elicited by β-agonists.

(ii) Studies in man. Human subjects, as well as other large mammals, demonstrate non-shivering thermogenesis. In non-cold-acclimated man a brief period of cold exposure during curarization increases oxygen consumption (Jessen *et al.*, 1980) in association with a rise in the plasma NE level and, despite earlier reports to the contrary (Joy, 1963), NE infusions in normal non-cold-acclimated subjects raise oxygen consumption (Jung *et al.*, 1979*a*). Evidence of cold acclimation due to seasonal variation (Davis and Johnston, 1961; Scholander *et al.*, 1958), geographical climate (Budd and Warhaft, 1966*b*) and experimental manipulation (Davis, 1961) is also available in man. In cold-acclimated human subjects, as in rodents, the thermogenic effect of NE is enhanced (Joy, 1963), although the magnitude of the NE response is considerably less in man.

(b) Sites of non-shivering thermogenesis: brown adipose tissue.

While sympathetic stimulation of non-shivering thermogenesis is well-established, the site or sites of increased heat production in response to catecholamines are less clearly defined. Brown adipose tissue (BAT), an organ highly developed for generating heat, is of recognized importance in thermoregulation in neonates of many mammalian species, during arousal from hibernation, and in cold-acclimated rodents (Smith and Horwitz, 1969). The role of brown adipose tissue in adult, warm-acclimated mammals, particularly of larger species (including man), is still controversial. In all species studied the heat-producing function of BAT is governed by the sympathetic nervous system.

(c) Sympathetic regulation of BAT

BAT is densely innervated with sympathetic nerve endings (Smith and Horwitz, 1969; Cottle and Cottle, 1970; Young, *et al.*, 1982). The tissue content of NE, which reflects the extent of sympathetic innervation, is approximately 1.5 μg per g of tissue in rat interscapular BAT, a concentration greater than that found in heart (Young *et al.*, 1982). During cold acclimation, as the tissue hypertrophies, NE content of interscapular BAT increases (Young *et al.*, 1982), along with histochemical evidence of more extensive innervation (Cottle and Cottle, 1970).

The functional significance of BAT sympathetic innervation has been demonstrated by a variety of experimental techniques. Electrical stimulation of nerves supplying interscapular BAT increases heat production within BAT *in situ* (Smith and Horwitz, 1969; Hull and Segall, 1965). Acute and chronic cold exposure markedly accelerate NE turnover in interscapular BAT (Fig. 4.6) (Young *et al.*, 1982; Cottle *et al.*, 1967), providing evidence of cold-induced sympathetic stimulation in BAT which is of greater magnitude than that noted in heart. Surgical denervation, on the other hand, reduces both lipolysis (Hull and Segall, 1965; Slavin and Bernick, 1974) and glycogen depletion (Slavin and Bernick, 1974) in BAT from cold-exposed rodents and rabbits. In some studies the effects of surgical denervation have only partially impaired metabolic responses in BAT to cold exposure (Steiner *et al.*, 1969; Steiner Loveland and Schonbaum, 1970), although

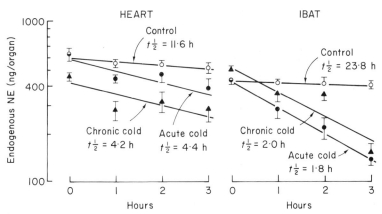

Fig. 4.6 Effect of acute and chronic cold exposure on NE turnover in heart and interscapular brown adipose tissue (IBAT). NE turnover is measured from the rate of fall of endogenous NE after biosynthesis is inhibited with α-methyl tyrosine. The effect of acute (3 h) and chronic (9 days) cold exposure at 4°C is shown in heart and IBAT. Both acute and chronic cold exposure increase NE turnover, the effect being substantially greater in IBAT. This experiment demonstrates increased sympathetic outflow to IBAT during acute and chronic cold exposure. From Young *et al.* (1982), with permission.

more recent evidence indicates that surgical denervation of BAT may often be incomplete (Barde, Chinet and Girardier, 1975). Chemical sympathec- tomy following injection of 6-hydroxydopamine, however, effectively denervated BAT and diminished non-shivering thermogenesis in fetal sheep (Alexander and Stevens, 1980). Current information, therefore, indicates that the sympathetic nervous system plays a prime role in the stimulation of heat production by brown adipose tissue in a variety of experimental animals (see Chapters 2, 3, 8 and 11).

(d) Other potential sites of non-shivering thermogenesis

The participation of organs other than BAT in non-shivering thermogenesis has long been suspected, although, in recent years, recognition of the important role of BAT has diminished the potential importance of other sites (Foster and Frydman, 1978). Nonetheless, in larger, warm-acclimated mammals where the functional role of BAT is uncertain, interest in possible additional sites of adaptive heat production remains active.

Sympathetic stimulation of the vasculature and of substrate mobilization (described in detail below), as well as activation of other metabolic processes not specifically related to thermogenesis, generates a small amount of metabolic heat. These processes, however, would not appear to contribute substantially to the specific, sympathetically mediated increase in non-shivering thermogenesis induced by cold exposure. Based upon measurements of oxygen consumption, the participation of the splanchnic viscera in non-shivering thermogenesis also appears minimal (Jessen *et al.*, 1980; Mejsnar and Jansky, 1976). The thermogenic response to cold exposure and to NE infusions, furthermore, are reasonably well preserved in the rat following functional evisceration of the liver and the gut (Depocas, 1958, 1960 *a,b*). Skeletal muscle, therefore, by virtue of its large mass, has attracted the greatest interest as a potentially important alternative site of non-shivering thermogenesis.

(i) Skeletal muscle. Although a major point in favour of a role for skeletal muscle in non-shivering thermogenesis has been the absence of conclusive data about the participation of other organs, some direct evidence of skeletal muscle involvement in this process is available. Cold exposure produces a modest (approximately twofold) elevation in oxygen con- sumption in the canine hindlimb (Davis, 1967), an increase potentiated by prior cold acclimation, and diminished by section of the nerve supply (Davis, 1967). In the rat cold exposure and NE infusion similarly raise oxygen consumption in isolated leg muscles about twofold (Jansky and Hart, 1963); in the mouse the thermogenic response of skeletal muscle is even less than the small elevation observed in the rat (Dubois-Ferrere and Chinet, 1981).

Skeletal muscle does, in addition, undergo biochemical changes during prolonged cold exposure that may be related to an enhanced capacity for heat production. Skeletal muscle from cold-acclimated rats display increased $Na^+ K^+$-ATPase activity (Guernsey and Stevens, 1977; Horwitz and Eaton, 1977), diminished creatine content (Kurahashi and Kuroshima, 1978) and the presence of loosely coupled mitochondria (Greenway and Himms-Hagen, 1978; Himms-Hagen *et al.*, 1976), findings consistent with enhancement of non-shivering thermogenesis. Adenylate cyclase activity in skeletal muscle of cold-exposed rats is also transiently elevated, but this may be related more to shivering than to non-shivering thermogenesis (Muirhead and Himms-Hagen, 1974).

Despite the existence of these supportive data, serious difficulties remain in attributing a substantial portion of non-shivering thermogenesis to skeletal muscle. As emphasized above, the sympathetic nervous system is the critical regulator of non-shivering thermogenesis. In distinction to BAT, skeletal muscle does not receive a heavy sympathetic innervation (Fuxe and Sedvall, 1965), although this fact does not exclude the possibility that NE released within the interstitial spaces in muscle might exert a local effect on adjacent muscle fibres (Rosell, Kopin and Axelrod, 1963). In fact, low frequency stimulation of the sympathetic chain increases oxygen consumption in innervated skeletal muscles (Duran and Renkin, 1976). The adrenergic receptor mediating the thermogenic response to catecholamines in skeletal muscle adds an additional element of uncertainty in ascribing an important role to muscle. In small mammals non-shivering thermogenesis is blocked by β-receptor antagonists as reviewed above, while the thermogenic response to infused NE in muscle is not clearly β-mediated, and may, in fact, reflect α-adrenergic receptor stimulation (Grubb and Folk, 1977; Nellis *et al.*, 1980).

A contribution of skeletal muscle to sympathetically mediated, non-shivering thermogenesis has, therefore, yet to be established. Although a small change in oxygen consumption in a tissue mass as large as that of skeletal muscle might be difficult to detect, the low levels of oxygen consumption measured in resting muscle, and the modest changes induced by cold exposure or NE administration, make it difficult to attribute to skeletal muscle a significant portion of the large increase in thermogenesis regularly elicited in cold-acclimated animals exposed either to cold or NE.

4.2.5 Regulation of substrate supply and delivery

In addition to directly stimulating thermogenesis, the sympathoadrenal system is importantly involved in supplying fuel to support the increase in metabolic rate. Provision of an adequate energy supply requires substrate mobilization on the one hand, and delivery of mobilized fuels to heat-producing tissues on the other.

(a) Substrate mobilization

The hormonal regulation of intermediary metabolism in the cold is complex and incompletely understood. The effects of catecholamines are exerted directly on liver, skeletal muscle, adipose tissue and other organs, and indirectly by altering the secretion of hormones importantly involved in metabolic regulation, such as insulin and glucagon (see Young and Landsberg, 1983, and below). Several hormones, moreover, including thyroid, glucocorticoids, insulin and glucagon, influence the peripheral responses to catecholamines including, in some cases, the balance between α- and β-adrenergic effects. Glucocorticoids, for example, have a permissive effect on catecholamine-stimulated processes, and are required for normal substrate mobilization in the cold. The hormonal milieu, therefore, exerts an important influence on the physiological expression of catecholamine-induced changes in intermediary metabolism.

During acute and chronic cold exposure the energy requirements of the mammalian organism for shivering and non-shivering thermogenesis are met by breakdown of stored fuels in adipose tissue, liver, and skeletal muscle (Himms-Hagen, 1972). Utilizable substrates in the form of free fatty acids, glucose, ketone bodies, and lactate are released into the circulation and serve as an energy source for metabolizing tissues. The failure of adrenally demedullated, chemically sympathectomized animals to liberate free fatty acids from adipose tissue or to release glucose from liver in response to cold exposure indicates the extent to which this substrate mobilization depends upon catecholamines (Maickel *et al.*, 1967). Although the exact contributions of sympathetic nerves and the adrenal medulla to substrate mobilization during cold exposure are not known with certainty, the role of the sympathetic nervous system in the regulation of lipolysis in white adipose tissue is well defined (Gilgen and Maickel, 1962), while the adrenal medulla has an important effect on hepatic glucose output (Maickel *et al.*, 1967; Gilgen and Maickel, 1962).

(i) Lipid metabolism. Although the adrenergic innervation of white fat is sparse in comparison with that of BAT, the sympathetic nervous system is generally accorded an important role in the regulation of lipolysis within white adipose tissue (Rosell and Belfrage, 1979). NE activates the principal lipolytic enzyme, hormone-sensitive lipase, via interaction with the β-1-adrenergic receptor and stimulation of the adenylate cyclase–cyclic AMP mechanism. Plasma levels of free fatty acids and the activity of adipose tissue lipase increase normally in adrenally demedullated rats acutely exposed to cold, but not in demedullated animals subjected to ganglionic blockade or chemical sympathectomy before cold exposure (Maickel *et al.*, 1963). In the latter group of animals administration of NE coincident with cold exposure restores elevations in FFA and lipase activity toward control values (Maickel *et al.*, 1963).

The changes in lipid metabolism associated with cold exposure and cold acclimation are similar to those seen during NE administration. Exposure to cold leads to a fall in respiratory quotient (RQ; the ration of CO_2 production to O_2 consumption) (Depocas, 1961), an indication of increased fat oxidation. NE infusions lower RQ and accelerate turnover of free fatty acids and ketones (Maekubo, Moriya and Hiroshige, 1977; LaFrance, Lagace and Routhier, 1980) effects that are greater in cold-acclimated as compared with warm-acclimated animals (LaFrance *et al.*, 1980). Plasma triglyceride levels, predominantly from the very low density lipoprotein (VLDL) fraction, also decline with cold exposure in the rat (Himms-Hagen, 1972), consistent with increased lipid utilization and with the known stimulatory effects of NE on lipoprotein lipase in heart and BAT (Radomski and Orme, 1971). Body fat stores, moreover, supply a significant portion of the energy required for cold induced thermogenesis. Even if cold exposure is accompanied by unrestricted access to food, depletion of body fat occurs in both experimental animals and man (Kodama and Pace, 1964; Sugahara *et al.*, 1969; O'Hara *et al.*, 1979). While the energy deficit induced by cold is ultimately balanced by increased caloric intake, mobilization of fat, under the influence of the sympathetic nervous system, plays an important role in support of thermogenesis in the cold until the caloric deficit is restored by increased intake.

(ii) Carbohydrate metabolism. Despite the prime role of fat utilization during cold-induced thermogenesis, carbohydrate metabolism is enhanced as well (Depocas, 1961) and is capable of compensating if mobilization of free fatty acids from adipose tissue is deficient (Himms-Hagen, 1972). Evidence of stimulated carbohydrate metabolism during cold exposure includes diminished hepatic glucogen stores and increased glucose utilization, as determined from kinetic studies of glucose turnover, the latter effect being greater in cold-acclimated, than warm-acclimated, animals (Depocas, 1961). A causal relationship, however, between the cold-induced changes in carbohydrate metabolism, and the increased sympathoadrenal activity has not been established and factors other than catecholamines may also influence carbohydrate metabolism in the cold.

In warm-acclimated dogs exposed to a cold environment sufficient to raise oxygen consumption fivefold, glucose utilization (measured as the rate of disappearance of [U-^{11}C]-glucose) nearly doubles; adrenal demedullation did not diminish either the elevation in oxygen consumption or the increase in glucose utilization (Forichon *et al.*, 1977*a*). When glucose kinetics were examined more carefully in the adrenally demedullated dogs, cold exposure by itself increased endogenous glucose production by 50–60% and the irreversible loss of glucose (glucose utilization) by 70–80%, but did not affect the rate of glucose recycling (a reflection largely of Cori cycle activity) (Forichon *et al.*, 1977*b,c*). Infusion of epinephrine produced its greatest inpact on the rate of glucose recycling which increased 50–100% in these animals depending upon the dose administered; rates of glucose production and of glucose utilization were less affected, but glucose levels, unaffected by cold alone in the demedullated dogs, increased during epinephrine infusion because the change in glucose production always exceeded that of glucose utilization (Forichon *et al.,* 1977*b,c*). These observations are consistent with data indicating that adrenal demedullation blocks the rise in plasma glucose that occurs when warm-acclimated rats are exposed to cold (Maickel *et al.,* 1967). Since warm-acclimated animals are more likely to increase adrenal medullary E secretion during cold exposure, these results support a significant role for E in the regulation of hepatic glucose output in the cold. Although the important alterations in carbohydrate metabolism observed in the adrenally demedullated dogs may be due, in part, to increased hepatic sympathetic activity, the precise role of the sympathetic nerves, in comparison with the effect of other hormones (see below) and substrate levels has not been established. In one study examining sympathetic responses in liver to a milder cold stimulus, only a minimal increase in hepatic sympathetic activity occurred (Young and Landsberg, 1979).

(iii) Role of insulin and glucagon. Effects of catecholamines on the secretion of various peripheral peptide hormones, particularly insulin and glucagon, may also contribute to metabolic regulation during cold exposure. Since the processes of fuel mobilization are opposed by insulin and promoted by glucagon (at least in the liver) suppression of insulin and stimulation of glucagon secretion would facilitate the release of stored fuels for utilization in the cold. In fact, in response to modest degrees of cold exposure glucagon levels rise while insulin remains unchanged (Seitz *et al.,* 1981); with exposure to severe cold suppression of insulin secretion becomes more evident (Kervran *et al.,* 1976; Blackard, Nelson and Labat, 1967; Baum, Dillard and Porte, 1968). Cold-induced activation of the sympathoadrenal system is one potential mediator of these alterations in endocrine pancreatic function. α-Adrenergic blockade partially restores β-cell responsiveness to glucose in animals made hypothermic by cold exposure (Kervran *et al.,* 1976; Baum and Porte, 1971). The catecholamine component of cold-induced inhibition of insulin secretion may represent adrenal medullary activity since cold-stressed, adrenally demedullated dogs displayed no impairment in insulin release (Forichon *et al.,* 1977*b*). On the other hand, the rise in glucagon during cold exposure may reflect increased pancreatic sympathetic activity (Young and Landsberg, 1983) since it occurred in adrenalecto-mized rats under mild cold conditions (Seitz *et al.,* 1981). Thus, sympathetically mediated stimulation of glucagon secretion and adrenal medullary suppression of insulin release may contribute to the direct effects of catecholamines on substrate mobilization during cold exposure.

(iv) Lipoprotein lipase. The activity of lipoprotein lipase, an enzyme importantly involved in the uptake and utilization of circulating triglycerides by extra-hepatic tissues, changes during acute cold exposure. In the rat lipoprotein lipase activity demonstrates tissue-specific alterations, decreasing in epididymal white fat, increasing slightly in heart and markedly in BAT (Radomski and Orme, 1971). Both the sympathetic nervous system and insulin may participate in the regulation of lipoprotein lipase during cold exposure, since NE produces similar, but less-marked effects in warm-acclimated animals and insulin partially reverses the cold-induced changes in enzyme activity (Radomski and Orme, 1971). Thus, both sympathetic stimulation and a fall in circulating insulin during cold exposure may contribute to the observed alterations in lipoprotein lipase activity, changes which would tend to suppress storage of triglyceride in white adipose tissue and favour utilization in BAT and heart.

(b) Cardiovascular system

Sympathetic stimulation of the cardiovascular system contributes to heat conservation and to delivery of oxygen and utilizable substrates to thermogenic tissues during cold exposure. Central integration of peripheral blood flow with temperature regulation resides in the preoptic anterior hypothalamic region (Gilbert and Blatteis, 1977; Lynch, Adair and Adams, 1980). Under the direction of this hypothalamic centre sympathetically mediated subcutaneous vasoconstriction shunts blood away from the skin, thereby increasing the insulating capacity of the subcutaneous tissues and inhibiting heat loss. Despite this peripheral vasoconstriction cardiac output has been shown to double during a brief period of cold exposure in non-cold-acclimated man and oxygen uptake increases almost threefold (Raven *et al.*, 1970). Pulse pressure and cross product (pulse rate × systolic blood pressure) rise in these subjects suggesting that the cardiovascular responses are due to increased sympathetic activity; total peripheral resistance, however, falls 30% which may be a reflection of vasodilation in shivering muscles and increased perfusion of splanchnic viscera. Thus the pattern of alterations in cardiovascular function induced by cold exposure is consistant with the known changes in sympathetic activity in different organs and tissues, as described above (see Section 4.2.2(*a*)). The direct correlation noted between oxygen uptake and cardiac output suggests that the elevation in metabolic rate is the stimulus for the increase in cardiac output.

(i) Vascular responsiveness to NE. The responsiveness of arteries and veins to NE varies with environmental temperature. In intact and isolated vascular preparations from dogs and ducks (Millard and Reite, 1975; Webb-Peploe and Shepard, 1968; Janssens and Vanhoutte, 1978) lower ambient temperatures increase contractile responses to NE. The vasoconstrictor response for a given level of sympathetic activity should, therefore, be

enhanced in superficial vessels, preferentially shunting blood from sub-cutaneous to deeper limb veins (Webb-Peploe and Shepard, 1968). Such a diversion of venous blood flow to deeper *venae comitantes* would contribute to heat conservation through a reduction in skin blood flow and an increase in countercurrent heat exchange; the latter mechanism permits the transfer of heat from arterial blood leaving the body core to cooler venous drainage returning to the central venous pool.

(ii) Cold acclimation. Chronic cold exposure alters the vascular responses to sympathetic nervous system activation. In non-cold-acclimated monkeys (Wasserstrum and Herd, 1977b) and man (Raven *et al.*, 1970; Budd and Warhaft, 1966) acute cold exposure increases blood pressure about 20%, but in cold-acclimated man blood pressure changes much less in response to either a cold stimulus (Budd and Warhaft, 1966a; LeBlanc *et al.*, 1975) or to infused NE (Joy, 1963). Enhancement of vagal parasympathetic tone could account for some of the changes in cardiovascular responses following chronic cold exposure (LeBlanc *et al.*, 1975). In addition, alterations in peripheral sensitivity to catecholamines, including diminished α- and augmented β-adrenergic responses have been demonstrated in cold-acclimated animals (Koo and Liang, 1978; Fregly *et al.*, 1977). Modification of the cardiovascular responses to sympathetic activity following cold acclimation may, therefore, be partially attributable to alterations in the sensitivity of the cardiovascular system to NE.

4.3 DIET

4.3.1 Dietary thermogenesis and the sympathetic nervous system

While adaptive changes in thermogenesis following cold exposures are well established, the possibility that regulatory changes in thermogenesis follow alterations in caloric intake has been less well accepted. The point of controversy is not whether dietary manipulations affect thermogenesis; metabolic rate is clearly decreased during semi-starvation and increased during overfeeding. The debate has centred, rather, on whether these changes in thermogenesis are truly adaptive, or merely reflect differences in the metabolically active tissue mass or in the energy required to assimilate ingested nutrients. Most recent evidence, however, supports the existence of adaptive dietary thermogenesis, or 'luxus consumption', in many laboratory animals and in man (see Chapter 6).

Despite the overriding importance of the sympathetic nervous system in the regulation of non-shivering thermogenesis, only recently has an important role for the sympathetic nervous system been proposed in the regulation of dietary thermogenesis (Young and Landsberg, 1977a,b; Young and Landsberg, 1978; Rothwell and Stock, 1979). Since dietary and non-shivering thermogenesis are now known to share a number of physiological and biochemical features, the relationship between sympathetic activity and heat production in response to cold exposure argues strongly for primary involvement of the sympathetic nervous system in diet-induced alterations in thermogenesis as well.

4.3.2 Effect of dietary intake on sympathoadrenal activity

Recent evidence indicates that caloric restriction suppresses, while increased caloric intake stimulates, the sympathetic nervous system.

(a) Effect of fasting and caloric restriction

In the laboratory rodent measurement of NE turnover initially demonstrated reduction in sympathetic nervous system activity in heart with fasting or caloric restriction (Young and Landsberg, 1977a). An example of such a study is illustrated in Fig. 4.7. Subsequently, similar changes in sympathetic activity have been noted in pancreas, liver (Young and Landsberg, 1979a), kidney (Rappaport, Young and Landsberg, 1981), and interscapular BAT (Young *et al.*, 1982). Sympathetic suppression begins during the first day of

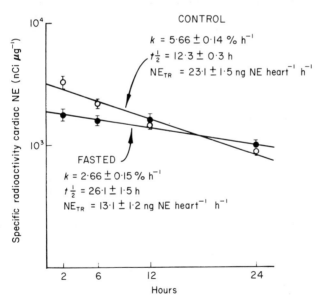

Fig. 4.7 Effect of fasting on cardiac NE turnover in the rat. The effect of a 48 hour fast on NE turnover in heart is shown in comparison with fed controls *ad lib*. Method and notation as described in the legend to Fig. 4.4. The decrease in NE turnover indicates suppression of sympathetic nervous system activity. Reproduced from Young and Landsberg (1977a), with permission of the American Association for the Advancement of Science.

fasting (Rappaport, Young and Landsberg, 1982*a*), partly abates on the first day of refeeding (Young and Landsberg, 1979*a*) and is also demonstrable when caloric intake is reduced to 30% of control (Rappaport, Young and Landsberg, 1982*b*).

(i) Human studies. Data from human studies examining sympathetic nervous system function during caloric restriction are less complete and more contradictory than those obtained in the rat. In obese subjects plasma NE levels fell during a period of controlled semi-starvation in which dietary sodium intake was maintained (Jung *et al.*, 1979*b*; DeHaven *et al.*, 1980). Similarly, patients with anorexia nervosa displayed subnormal levels of plasma NE in the untreated, undernourished state, with restoration to normal following nutritional therapy (Gross *et al.*, 1979). In addition, urinary NE excretion in normal weight men given sodium replacement decreased slightly during a 3-day (R.M. Rosa, J.B. Young and L. Landsberg, unpublished data) or 10-day period of starvation (Palmblad *et al.*, 1977). On the other hand, several studies report elevations in urinary excretion or plasma concentrations of NE with fasting (Januszewicz, Sznajdelman-Ciswicka and Wocial, 1967; Christensen, 1974; Pequignot, Peyrin and Peres, 1980; Galbo *et al.*, 1981), but the interpretation of these data is confounded by concomitant sodium restriction, which increases sympathetic activity. Moreover, the recent observation that the clearance of NE from the circulation is diminished by fasting (O'Dea *et al.*, 1982) alters the relationship between sympathetic activity and plasma NE level, increasing the difficulty of demonstrating a reduction in plasma NE under fasting conditions. Thus, although the current literature is not unequivocal, substantial direct evidence, coupled with physiological data demonstrating blood pressure and pulse rate reductions, is entirely consistent with fasting-induced suppression of sympathetic activity.

(ii) Adrenal medulla. The adrenal medullary response to fasting or caloric restriction is even less well established than that of sympathetic nerves. In the rat evidence of adrenal medullary stimulation is not apparent during a fast unless hypoglycaemia or cold exposure supervenes (Young and Landsberg, 1979*b*; 1981; Landsberg *et al.*, 1980). Data from human studies have either failed to detect a change in plasma E level (Jung *et al.*, 1979*b*; Christensen, 1974; Pequignot *et al.*, 1980; Galbo *et al.*, 1981) or noted an increase in urinary E excretion (Palmblad *et al.*, 1977; Janusezwicz *et al.*, 1967, R.M. Rosa, J.B. Young and L. Landsberg, unpublished data) or in plasma E concentration (Cryer, 1980). Since acute lowering of plasma glucose levels in normal and diabetic human subjects within the normal range (from 95 to 60 mg dl^{-1}) increased plasma E levels (Santiago *et al.*, 1980), the fall in glucose concentration which occurs routinely during fasting may elicit a similar adrenal medullary reaction. Thus, although

fragmentary, the available evidence suggests that the adrenal medullary response to fasting or caloric restriction may be opposite to that of the sympathetic nervous system.

(iii) Physiological role of the sympathoadrenal system during fasting and caloric restriction. Since the sympathetic nervous system is an important regulator of metabolic rate, suppression of sympathetic activity during caloric restriction should diminish thermogenesis. The recent demonstration that sympathetic activity in rat BAT, the major thermogenic tissue in this species, decreases during fasting (Young *et al.*, 1982) adds further support to this contention. An unresolved question, however, is the extent to which withdrawal of sympathetically mediated thermogenesis during fasting is responsible for the overall reduction in metabolic rate. One indication, perhaps, of the potential significance of the sympathetic contribution is provided by the observation that the administration of L-dopa, which induces a mild sympathomimetic effect, prevents the fall in oxygen consumption normally seen in calorically restricted obese subjects (Shetty, Jung and James, 1979). In contrast, neither the absence of thyroid hormone, the other major thermogenic factor, nor its repletion in hypothyroid animals, abolished the fasting-induced decrease in metabolic rate (Wimpfheimer *et al.*, 1979). Thus, the hypothesis that attributes a major part of the decrement in thermogenesis associated with fasting to sympathetic nervous system suppression, has considerable experimental support.

As described in preceding sections, one aspect of sympathetic and adrenal medullary function during cold exposure is the mobilization of substrate in support of increased thermogenesis. Given the reduction in sympathetic activity that occurs with fasting, it is unlikely that substrate mobilization in this setting is mediated by the sympathetic nervous system. The small increase in adrenal E secretion, however, in conjunction with the low circulating level of insulin in the fasting state, may be involved in breakdown of stored fuels. Lipolysis, in particular, is very sensitive to changes in plasma E level within the physiological range (Galster *et al.*, 1981). Earlier observations obtained from *in vitro* studies of isolated human adipocytes had indicated that fasting diminished NE-mediated lipolysis (Arner and Ostman, 1976), apparently through alterations in the β-adrenergic receptor (Burns *et al.*, 1979). Subsequent studies *in vivo*, however, indicate that fasting obese human subjects have increased sensitivity to the lipolytic effect of NE infusions (Arner, Engfeldt and Nowak, 1981). Although an explanation for this discrepancy between *in vivo* and *in vitro* experiments is currently lacking, when taken as a whole, the evidence suggests that small increments in E may facilitate substrate mobilization during fasting. This degree of adrenal medullary activation would be unlikely to affect thermogenesis which, as described in the sections on cold exposure, reflects predominantly sympathetic nervous system activity and responds only to very high circulating levels of catecholamines. Thus, a functional distinction between sympathetic nerves and the adrenal medulla may exist in the fasting state; sympathetic suppression may restrain thermogenesis, while adrenal medullary activation may facilitate substrate mobilization.

(b) Feeding and excess caloric intake

The sympathetic nervous system is stimulated acutely during discrete episodes of feeding, and chronically, in response to sustained increase in caloric intake. The relationship between these sympathetic responses and thermogenesis is complex. The thermogenic response to feeding also has two components: an acute, short-lived meal related effect known as the thermic effect of food, and a sustained increase in resting metabolic rate that reflects antecedent caloric intake, the latter often designated diet-induced thermogenesis.* Involvement of the autonomic nervous system in these two components of the thermogenic response to caloric intake may be different, with perhaps the autonomic nervous system playing a more important role in the thermogenic responses related to antecedent diet.

(i) Acute sympathetic responses to feeding and the thermic effect of food. The potential contribution of sympathetic nervous system stimulation to the physiology of the postprandial state has been recognized for many years. The changes in cardiovascular function noted with feeding, including acute increases in heart rate, blood pressure and cardiac output, along with alterations in blood flow distribution (Grollman, 1929; Aperia and Carlens, 1931; Abramson and Fierst, 1941; Fronek and Stahlgren, 1968; Blair-West and Brook, 1969; Vatner, Franklin and van Citters, 1970a,b; Bloom et al., 1975) are consistent with sympathetic activation. Evidence demonstrating that ganglionic blockade and reserpine attenuate the cardiac response to feeding (Bloom et al., 1975; Ehrlich, Fronkova and Steger, 1958) provides a more direct indication of the sympathetic link between feeding and alterations in cardiovascular function. Recent reports, moreover, describing elevations of plasma NE in man following glucose (Young et al., 1980; Welle, Lilavivathana and Campbell, 1980), but not isocaloric amounts of protein or fat (Welle, Lilavivat and Campbell, 1981), support the notion that sympathetic stimulation occurs with feeding under some experimental circumstances.

In addition to the various parameters of cardiovascular activity which rise with feeding, oxygen consumption increases in the postprandial period. This elevation was observed initially in man following protein ingestion (so-called 'specific dynamic action'), but has since been demonstrated with fat and carbohydrate as well (Welle et al., 1981; Zwillich, Sahn and Weil, 1977). In light of the known thermogenic potential of sympathetic stimulation in

---

*This division into two components may not be so rigid, see Section 4.1.

general, the postprandial increase in sympathetic activity might be expected to account, wholly or in part, for the associated elevation in oxygen consumption. Current evidence, however, does not support this expectation. Although oral glucose administration increases both sympathetic activity and thermogenesis, in one study the rise in oxygen consumption was not antagonized by prior treatment with propranolol (Zwillich, Sahn and Weil, 1977); similarly in sheep β-adrenergic blockade was without effect on oxygen consumption following hay feeding (Webster and Hays, 1968). Furthermore, despite a greater increase in metabolic rate, protein ingestion was not associated with any rise in plasma NE, in contrast to the significant elevation in plasma NE seen following glucose ingestion in association with a lesser increase in metabolic rate (Welle *et al.*, 1981). Thus, in the immediate postprandial state, despite the evidence of sympathetic activation and the indirect suggestions that sympathetic activation mediates some of the cardiovascular responses, the rise in thermogenesis (the 'thermic effect of food') is not clearly related to an increase in sympathetic activity. The possibility remains, however, especially with carbohydrate feeding, that part of the thermic effect of food is mediated by the sympathetic nervous system.

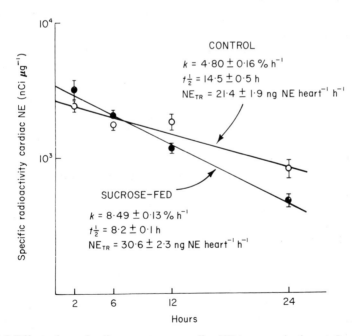

Fig. 4.8 Effect of overfeeding sucrose on cardiac NE turnover in the rat. Increased NE turnover in rats overfed a dilute solution of sucrose for three days. Increased NE turnover indicates stimulation of sympathetic nervous system activity. From Young and Landsberg (1977*b*), with permission.

(ii) Overfeeding. Voluntary overfeeding in rats is associated with increased sympathetic nervous system activity. Rats given *ad lib* access to dilute solutions of sucrose in addition to chow increase their caloric intake approximately 30%. On this regimen NE turnover in heart (Fig. 4.8) (Young and Landsberg, 1977*b*), liver and pancreas (Young and Landsberg, 1979*a*), and kidney (Rappaport *et al.*, 1981), is increased. These effects of sucrose are induced within the first day of overfeeding (Rappaport *et al.*, 1982*a*); the effects of sucrose feeding on NE turnover are sustained through at least eight days of overfeeding and restored to control levels by one day of normal feeding (Rappaport *et al.*, 1982*a*).

Overfeeding a mixed, variable and highly palatable ('cafeteria') diet, a regimen that increases caloric intake from 50 to 100%, for 10 days or longer

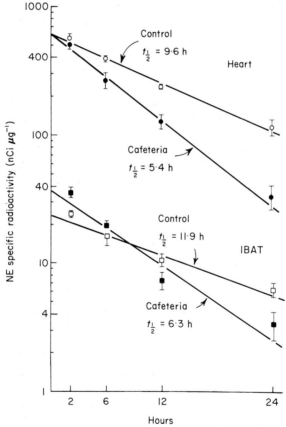

Fig. 4.9 Effect of overfeeding a 'cafeteria' diet on NE turnover in heart and interscapular brown adipose tissue (IBAT). Overfeeding a mixed highly palatable diet for 9 days increases NE turnover in both heart and IBAT indicating increased sympathetic ervous system outflow to these organs. From Young *et al.* (1982), with permission.

also increases NE turnover in heart and interscapular BAT (Young *et al.*, 1982) (Fig. 4.9). Preliminary experiments indicate that fat intake, as well, has an effect on sympathetic nervous system activity. When added to chow, lard increases cardiac NE turnover in rats, an effect that may exceed that of sucrose (Schwartz, Young and Landsberg, 1982).

Thus, in the rat, alterations in dietary intake and the type of nutrient ingested influence sympathetic nervous system activity. The effect of chronic overfeeding on sympathetic nervous system in human subjects has not been clearly established; the effect of overfeeding on the adrenal medulla, in rats and in humans, requires clarification as well.

(c) Role of the sympathetic nervous system in dietary thermogenesis

Chronic overfeeding increases resting metabolic rate in normal weight human subjects (Katzeff and Danforth, 1981) and rats (Rothwell and Stock, 1979; Stirling and Stock, 1968). Since the enhanced thermogenesis (dietary thermogenesis) was noted in the postabsorptive state it is distinct from the acute response to feeding described above (thermic effect of food). In rats chronic overfeeding increases the thermogenic response to NE infusions (Rothwell and Stock, 1979; Stirling and Stock, 1968); the increase in oxygen consumption associated with increased dietary intake is, furthermore, antagonized by β-adrenergic blockade. In human subjects, by comparison, overfeeding does not potentiate the thermogenic effects of infused NE (Katzeff and Danforth, 1981); the effects of β-blockade on oxygen consumption in overfed man have not been clearly defined. The role of the sympathetic nervous system in dietary thermogenesis in man, therefore, needs further investigation. Animal studies, however, that demonstrate physiological and biochemical parallels between non-shivering and dietary thermogenesis provide strong circumstantial evidence for an important sympathetic contribution to dietary thermogenesis (see also Chapter 6).

(d) Physiological role of the sympathetic nervous system during overfeeding

Suppression of sympathetically mediated thermogenesis in response to fasting or caloric restriction is of obvious survival value. The ability to reduce energy expenditure during periods of limited caloric intake would confer obvious selective advantage during famine. Stimulation of sympathetic nervous system activity by overfeeding, on the other hand, is less clearly advantageous since it would lead to the dissipation of nutrient calories as heat unrelated to the demands of environmental temperature. A potential benefit of dietary thermogenesis, however, might accrue to animals ingesting a diet of poor nutritional quality. The capacity to activate sympathetically mediated thermogenesis might allow such animals to satisfy their basic requirements for nitrogen or other essential nutrients by increasing dietary intake of low quality foodstuffs without accumulating excessive amounts of fat. Since individuals are known to vary in the efficiency of utilization of foodstuffs it is conceivable that alterations in the extent of the activation of the sympathetic nervous system (Landsberg and Young, 1981), or alterations in the thermogenic effect of NE, may be involved in the pathogenesis of some cases of obesity. The function of the sympathetic nervous system, including the thermogenic effects of catechol-amines, warrant further study in obese animals and man.

4.3.3 The signal that couples changes in dietary intake with changes in sympathetic nervous system activity

The coupling of sympathetic nervous system activity with dietary intake depends upon continuous assessment of the nutritional status by the central nervous system. Changes in body mass or body composition cannot explain diet-induced changes in sympathetic nervous system activity since the latter are demonstrable long before appreciable changes in these variables could occur (Rappaport *et al.*, 1982*a*). It is not clear, moreover, that body mass or body composition can be readily assessed by the central nervous system. Given the different types of nutrients capable of affecting sympathetic activity, it appears likely that multiple signals are involved. To date, the role of carbohydrate metabolism has been studied most extensively.

(a) Role of glucose: effect of hypoglycaemia

A role for glucose in the relationship between dietary intake and sympathetic nervous system activity is suggested by the potent effect of sucrose on sympathetic nervous system activity, described above, along with the fact that glucose is known to influence the activity of neurons in hypothalamic centres concerned with metabolic regulation (Young and Landsberg, 1980; Oomura and Kita, 1981). Studies examining the effect of hypoglycaemia on sympathetic activity also support a role for glucose. Hypoglycaemia has long been known to stimulate the adrenal medulla; recent evidence, however, from fasting pregnant rats (Young and Landsberg, 1979), phlorizin-treated fasting rats (Landsberg *et al.*, 1980), and from 2-deoxyglucose-treated rats fed *ad lib* (Rappaport *et al.*, 1982*b*), indicates that hypoglycaemia suppresses sympathetic nervous system activity despite concomitant activation of the adrenal medulla. During fasting, therefore, decrements in plasma glucose below postabsorptive levels may contribute to suppression of the sympathetic nervous system.

(i) Effect of 2-deoxyglucose.

Experiments with 2-deoxyglucose (2-DG), which inhibits the metabolism of glucose, reveal that intracellular glucose utilization, within the central nervous system, may be involved in the sympathetic response to glucose deprivation. 2-DG suppresses sympathetic nervous system activity in rats despite an actual increase in dietary intake

(Rappaport *et al.,* 1982*b*), and decreases oxygen consumption as well (Rothwell, Saville and Stock, 1981). These experiments with hypoglycaemia and 2-DG, therefore, suggest that diminished intracellular glucose metabolism within the central nervous system may be involved in suppression of sympathetic nervous system activity with fasting.

(b) Role of insulin

Factors other than plasma glucose concentration, however, may contribute to central nervous system glucose metabolism. Although plasma glucose levels fall during fasting, changes in glucose concentration are relatively small and plasma glucose is maintained within rather narrow limits despite wide variations in dietary intake. During sucrose feeding, moreover, plasma glucose level is elevated only slightly in comparison with rats fed *ad lib* despite an impressive effect on the sympathetic nervous system.

Evidence from a variety of sources suggests the participation of insulin, particularly insulin-mediated glucose metabolism, in the coupling of dietary intake with sympathetic nervous system activity. In distinction of glucose, plasma insulin levels vary widely and reflect, in a general sense, the intake of carbohydrate. Insulin is, furthermore, the important signal to tissues throughout the body that calories have been assimilated; it is logical that insulin serves the same function for the central nervous system as well.

The basis for an insulin effect on specific areas of the brain, particularly the hypothalamus, is well established (Van Houten and Posner, 1981; Oomura and Kita, 1981). Although insulin does not cross the blood–brain barrier, insulin receptors have been identified in the tubero-infundibular region of the hypothalamus and in the area postrema at the floor of the fourth ventricle, both of which lie outside the blood–brain barrier as part of the circumventricular system (Van Houten and Posner, 1981). These areas have direct access to the circulation and appear to be important sites for the initiation of central neural processes involved in regulating homeostasis. In the median eminence insulin appears to interact directly with neuronal axon terminals, thereby providing the basis for an effect of insulin on the hypothalamic centres involved in metabolic regulation (Van Houten and Posner, 1981).

Direct evidence of a stimulatory effect of insulin on sympathetic nervous system activity is provided by studies utilizing glucose and insulin clamp techniques (Rowe *et al.,* 1981). Euglycaemic hyperinsulinism, achieved by simultaneous infusions of glucose and insulin is associated with an increase in plasma NE levels (Rowe *et al.,* 1981). The effect of insulin on sympathetic activity is greater under these circumstances than that of hyperglycaemia. Cardiovascular changes during insulin infusion are also consistent with sympathetic stimulation (Rowe *et al.,* 1981). A separate role for insulin in diet-induced thermogenesis has also been demonstrated since streptozotocin-treated diabetic rats fail to increase oxygen consumption in response to NE after cafeteria feeding, a response which is restored towards normal by insulin (Rothwell and Stock, 1981).

The available data, therefore, suggest that diet-induced changes in sympathetic nervous system activity may be mediated, at least in part, by changes in glucose metabolism within certain critical central neurons; involvement of insulin is likely, presumably by stimulating glucose metabolism within these neurons. Since mixed diets, including fat, also stimulate sympathetic nervous system activity it is likely that other signals are involved as well.

(c) Visceral afferents

Receptors for substrates and hormones in visceral organs, particularly liver, may also initiate afferent neural discharges that convey information to the central nervous system about the nutritional state of the organism (Niijima, 1981). For example, intraduodenal infusion of glucose diminishes the discharge rate of hepatic vagal afferents while 2-DG has the opposite effect (Niijima, 1981). Cholecystokinin (CCK) when given by intravenous infusion, also decreases hepatic vagal afferent discharge (Niijima, 1981). Although much work remains to be done before the physiological significance of visceral afferents in metabolic regulation is established, these results suggest at least a superficial resemblance to the venous and arterial baroreceptor systems that regulate sympathetic nervous system activity. Changes in hormone and substrate levels in association with feeding (increased glucose, increased CCK) may result in increased sympathetic nervous system activity by diminishing inhibitory input into the centres that regulate sympathetic outflow, analogous to the increase in sympathetic outflow that results from diminished baroreceptor impulses from the aorta, carotid sinus and great veins when arterial or venous pressure in these regions is lowered.

4.3.4 Central neuronal integration of diet-induced changes in sympathetic nervous system activity

Diet-induced changes in sympathetic activity appear to be integrated within the hypothalamus. The ventromedial portion of the hypothalamus (VMH), in particular, has been implicated in the regulation of diet-induced changes in sympathetic nervous system activity (Young and Landsberg, 1980; Oomura and Kita, 1981; Shimazu and Takahashi, 1980; Perkins *et al.*, 1981; Shimazu, 1981). Prior treatment of mice with gold thioglucose, a compound that destroys portions of the hypothalamus including the area of the ventromedial nucleus, completely blocks suppression of sympathetic nervous system activity during fasting (Young and Landsberg, 1980). In rats, VMH stimulation has been shown to enhance fatty acid synthesis in BAT (Shimazu and Takahashi, 1980) and to increase thermogenesis in this organ (Perkins *et al.*, 1981). Since the ventromedial portion of the hypothalamus contains neurons sensitive to glucose and insulin, it appears likely that regulation of sympathetic outflow in response to changes in nutritional status is integrated in this area. The fact that VMH destruction with gold thioglucose enhances sympathetic nervous system activity in fasting mice (Young and Landsberg, 1980), while electrical stimulation of the VMH in rats may increase sympathetic activity in brown adipose tissue (Perkins *et al.*, 1981) is not understood, but perhaps reflects the presence of both inhibitory and stimulatory pathways in this region of the brain (Oomura and Kita, 1981).

4.4 SUMMARY AND CONCLUSIONS

The sympathetic nervous system has a critical role in the regulation of adaptive changes in thermogenesis. This role has been best studied in the setting of cold exposure, where the importance of the sympathetic nervous system is clear and unequivocal. Sympathetic nervous system activity is markedly increased during cold exposure; cold sensitive receptors in skin, spinal cord, and hypothalamus initiate the sympathetic nervous system response, which is integrated in the anterior and posterior hypothalamus. The sympathetic nervous system stimulates three major physiological responses to environmental cold: (1) increased metabolic heat production (non-shivering thermogenesis); (2) increased cardiac output and redistribution of blood flow; and (3) substrate mobilization. In small mammals non-shivering thermogenesis is induced by β-adrenergic receptor-mediated stimulation of heat production in BAT. The cardiovascular changes, which depend upon differential patterns of sympathetic outflow, are precisely regulated and result in conservation of body heat while insuring adequate delivery of oxygen and substrate to metabolizing tissues including BAT. Sympathetic stimulation increases substrate supply in the form of free fatty acids mobilized from triglyceride stores in adipose tissue, and, perhaps, in the form of glucose from glycogen stores in liver and from hepatic gluconeogenesis. Catecholamine-mediated changes in insulin and glucagon release contribute to the direct effects of catecholamines on substrate mobilization. The process of cold acclimation is associated with increased sensitivity to the thermogenic effects of catecholamines. As cold acclimation progresses sympathetic nervous system activity diminishes slowly but remains substantially increased in comparison with the level of activity in a thermoneutral environment.

The adrenal medulla is stimulated during cold exposure but to a lesser extent than the sympathetic nervous system. Circulating catecholamines of adrenal medullary origin reinforce the effects of sympathetic stimulation and can compensate, at least in part, for the sympathetic nervous system when the function of the latter is impaired. The adrenal medulla may play a particularly significant role in the regulation of carbohydrate metabolism during cold exposure.

An important role for the sympathetic nervous system in the regulation of dietary-induced thermogenesis is reasonably well established in the laboratory animal. Available evidence suggests an important role in larger mammals, including man. In the rodent, fasting suppresses, and excess caloric intake stimulates, sympathetic nervous system activity in a variety of organs including BAT. In the rat sympathetic stimulation of BAT is importantly involved in the changes in thermogenesis that accompany changes in dietary intake. The increase in metabolic rate following over-feeding in rats is antagonized by β-adrenergic receptor blocking agents; chronic overfeeding, like cold-acclimation, potentiates the thermogenic actions of catecholamines. Glucose utilization within the central nervous system, possibly mediated by insulin, appears to be one signal that couples changes in dietary intake with changes in sympathetic nervous system activity. The ventromedial portion of the hypothalamus may be involved in the relationship between dietary intake and sympathetic activity.

Although catecholamines appear to be of primary importance in the regulation of thermogenesis, synergistic interactions with glucocorticoids and thyroid hormones occur. Adrenal cortical hormones potentiate the effects of catecholamines on substrate mobilization while thyroid hormones potentiate the thermogenic effects of catecholamines.

At present, therefore, the critical role of the sympathetic nervous system in the regulation of non-shivering thermogenesis is well established; evidence for an important role of the sympathetic nervous system in the regulation of dietary thermogenesis is impressive and increasing. It appears likely, moreover, that the sympathetic nervous system, to a lesser or greater degree, may be involved in alterations in thermogenesis that occur in a wide variety of other physiological and pathophysiological states such as hibernation, fever, malignant hyperthermia, tetanus, traumatic injury, shock, and the various withdrawal states involving alcohol, opiates, and clonidine. Evidence for participation of the sympathetic nervous system in these situations is, at present, fragmentary but investigation of the role played by the sympathetic nervous system in these various states may be of major importance in understanding the physiological mechanisms that underlie the characteristic changes in thermogenesis associated with these disorders.

ACKNOWLEDGEMENTS

This work was supported in part by USPHS Grants AM 20378, AM 26455 and HL 24084.

REFERENCES

Abramson, D.I. and Fierst, S.M. (1941) Peripheral vascular responses in man during digestion. *Am. J. Physiol.*, **133**, 686–93.

Alexander, G. and Stevens, D. (1980) Sympathetic innervation and the invalidity of Rb$^+$ based measurements. *Can. J. Physiol. Pharmacol.*, 1978; **56**, 97–109.

Aperia, A. and Carlens, E. (1931) Vergleich zwischen der Wirkung von Fett, Kohlenhydrat und EiweiB auf den Kreislauf des Menschen. *Skand. Arch. Physiol.*, **63**, 151–63.

Arner, P., Engfeldt, P. and Nowak, J. (1981) *In vivo* observations on the lipolytic effect of noradrenaline during therapeutic fasting. *J. Clin. Endocrinol. Metab.*, **53**, 1207–12.

Arner, P. and Ostman, J. (1976) Changes in the adrenergic control and the rate of lipolysis of isolated human adipose tissue during fasting and after re-feeding. *Acta Med. Scand.*, **200**, 273–9.

Banet, M., Hensel, H. and Liebermann, H. (1978) The central control of shivering and non-shivering thermogenesis in the rat. *J. Physiol. (Lond.)*, **283**, 569–84.

Barde, Y.A., Chinet, A. and Girardier, L. (1975) Potassium-induced increase in oxygen consumption of brown adipose tissue from the rat. *J. Physiol.*, **252**, 523–36.

Baum, D., Dillard, D.H. and Porte, D. Jr. (1968) Inhibition of insulin release in infants undergoing deep hypothermic cardiovascular surgery. *N. Engl. J. Med.*, **279**, 1309–14.

Baum, D. and Porte, D. (1971) Alpha-adrenergic inhibition of immunoreactive insulin release during deep hypothermia. *Am. J. Physiol.*, **221**, 303–11.

Beley, A., Beley, P., Rochette, L. and Bralet, J. (1976) Evolution *in vivo* of the synthesis rate of catecholamines in various peripheral organs of the rat during cold exposure. *Pflugers Arch.*, **366**, 259–64.

Bergh, U., Hartley, H., Landsberg, L. and Ekblom, B. (1979) Plasma norepinephrine concentration during submaximal and maximal exercise at lowered skin and core temperatures. *Acta Physiol. Scand.*, **106**, 383–4.

Blackard, W.G., Nelson, N.C. and Labat, J.A. (1967) Insulin secretion in hypothermic dogs. *Am. J. Physiol.*, **212**, 1185–7.

Blair-West, J.R. and Brook, A.H. (1969) Circulatory changes and renin secretion in sheep in response to feeding. *J. Physiol. (Lond.)*, **204**, 15–30.

Blatteis, C.M. (1981) The newer putative central neurotransmitters: roles in thermoregulation. Hypothalamic substances in the control of body temperature: general characteristics. *Fed. Proc.*, **40**, 2735–40.

Bloom, S.R., Edwards, A.V., Hardy, R.N., Malinowska, K. and Silver, M. (1975) Cardiovascular and endocrine responses to feeding in the young calf. *J. Physiol. (Lond.)*, **253**, 135–55.

Boulant, J.A. (1981) Hypothalamic mechanisms in thermoregulation. *Fed. Proc.*, **40**, 2843–50.

Bralet, J., Beley, A. and Lallemant, A.M. (1972) Alterations in norepinephrine turnover in various peripheral organs of the rat during exposure and acclimation to cold. *Pflugers Arch.*, **335**, 186–97.

Budd, G.M. and Warhaft, N. (1966*a*) Body temperature, shivering blood pressure and heart rate during a standard cold stress in Australia and Antarctica. *J. Physiol. (Lond.)*, **186**, 216–32.

Budd, G.M. and Warhaft, N. (1966*b*) Cardiovascular and metabolic responses to noradrenaline in man, before and after acclimatization to cold in Antarctica. *J. Physiol. (Lond.)*, **186**, 233–42.

Burns, T.W., Boyer, P.A., Terry, B.E., Langley, P.E. and Robison, G.A. (1979) The effect of fasting on the adrenergic receptor activity of human adipocytes. *J. Lab. Clin. Med.*, **94**, 387–94.

Cabanac, M. (1975) Temperature regulation. *Ann. Rev. Physiol.*, **37**, 415–39.

Cannon, W.B., Querido, A., Britton, S.W. and Bright, E.M. (1927) Studies on the conditions of activity in endocrine glands. XXI. The role of adrenal secretion in the chemical control of body temperature. *Am. J. Physiol.*, **79**, 466–507.

Carlson, L.D. (1960) Nonshivering thermogenesis and its endocrine control. *Fed. Proc.*, **19**, 25–30.

Chan, W.C. and Johnson, G.E. (1968) Influence of cold exposure on catecholamine depleting actions of hydroxylase inhibitors. *Eur. J. Pharmacol.*, **3**, 40–6.

Christensen, N.J. (1974) Plasma norepinephrine and epinephrine in untreated diabetics, during fasting and after insulin administration. *Diabetes*, **23**, 1–8.

Cottle, M.K.W. and Cottle, W.H. (1970) Adrenergic fibers in brown fat of cold-acclimated rats. *J. Histochem. Cytochem.*, **18**, 116–9.

Cottle, W.H., Nash, C.W., Veress, A.T. and Ferguson, B.A. (1967) Release of noradrenaline from brown fat of cold-acclimated rats. *Life Sci.*, **6**, 2267–71.

Cryer, P.E. (1980) Physiology and pathophysiology of the human sympathoadrenal neuroendocrine system. *N. Engl. J. Med.*, **303**, 436–44.

Davis, T.R.A. (1961) Chamber cold acclimatization in man. *J. Appl. Physiol.*, **16**, 1011–5.

Davis, T.R.A. (1967) Contribution of skeletal muscle to nonshivering thermogenesis in the dog. *Am. J. Physiol.*, **213**, 1423–6.

Davis, T.R.A. and Johnston, D.R. (1961) Seasonal acclimatization to cold in man. *J. Appl. Physiol.*, **16**, 213–4.

DeHaven, J., Sherwin, R., Hendler, R. and Felig, P. (1980) Nitrogen and sodium balance and sympathetic-nervous-system activity in obese subjects treated with a low-calorie protein or mixed diet. *N. Engl. J. Med.*, **302**, 477–82.

Depocas, F. (1958) Chemical thermogenesis in the functionally eviscerated cold-acclimated rat. *Can. J. Biochem. Physiol.*, **36**, 691–9.

Depocas, F. (1960*a*) The calorigenic response of cold-acclimated white rats to infused noradrenaline. *Can. J. Biochem. Physiol.*, **38**, 107–14.

Depocas, F. (1960*b*) Calorigenesis from various organ systems in the whole animal. *Fed. Proc.*, **5**, 19–24.

Depocas, F. (1961) Biochemical changes in exposure and acclimation to cold environments. *Br. Med. Bull.*, **17**, 25–31.

Dubois-Ferrere, R. and Chinet, A.E. (1981) Contribution of skeletal muscle to the regulatory non-shivering thermogenesis in small mammals. *Pflugers Arch.*, **390**, 224–9.

Duran, W.N. and Renkin, E.M. (1976) Influence of sympathetic nerves on oxygen uptake of resting mammalian skeletal muscle. *Am. J. Physiol.*, **231**, 529–37.

Ehrlich, V., Fronkova, K. and Sleger, L. (1958) Die Wirkung des Reserpins auf die Speichelsekretion und den Kreislauf wahrend des unbedingten und bedingten Nahrungsreflexes and wahrend der Differenzierungshemmung beim Hunde. *Arch. Int. Pharmacodyn.*, **115**, 373–96.

Feist, D.D. and Feist, C.F. (1978) Catecholamine-synthesizing enzymes in adrenals of seasonally acclimatized voles. *J. Appl. Physiol.*, **44**, 59–62.

Forichon, J., Jomain, M.J., Dallevet, G. and Minaire, Y. (1977*c*) Effect of cold and epinephrine on glucose kinetics in dogs. *J. Appl. Physiol.*, **43**, (2), 230–7.

Forichon, J., Jomain, M.J., Patricot, M.C. and Minaire, Y. (1977*a*) Tolerance to cold and glucose homeostasis in adrenal demedullated dogs. *Experientia*, **33**, 1070–2.

Forichon, J., Jomain, M.J., Schellhorn, J. and Minaire, Y. (1977*b*) Effect of epine-phrine upon irreversible disposal and recycling of glucose in dogs. *Experientia*, **33**, 1171–3.

Foster, D.O. and Frydman, M.L. (1978) Nonshivering thermogenesis in the rat. II. Measurements of blood flow with microspheres point to brown adipose tissue as the dominant site of the calorigenesis induced by noradrenaline. *Can. J. Physiol. Pharmacol.*, **56**, 110–22.

Fregly, M.J., Field, F.P., Nelson, E.L., Tyler, P.E. and Dasler, R. (1977) Effect of chronic exposure to cold on some responses to catecholamines. *J. Appl. Physiol.*, **42**, 349–54.

Fronek, K. and Stahlgren, L.H. (1968) Systematic and regional hemodynamic changes during food intake and digestion in nonanesthetized dogs. *Circ. Res.*, **23**, 687–92.

Fuller, C.A., Horowitz, J.M. and Horwitz, B.A. (1977) Spinal cord thermo-sensitivity and sorting of neural signals in cold-exposed rats. *J. Appl. Physiol.*, **42**, 154–8.

Fuxe, K. and Sedvall, G. (1965) The distribution of adrenergic nerve fibers to the blood vessels in skeletal muscle. *Acta Physiol. Scand.*, **64**, 75–86.

Galbo, H., Christensen, N.J., Mikines, K.J., Sonne, B., Hilsted, J., Hagen, C. and Fahrenkrug, J. (1981) The effect of fasting on the hormonal response to graded exercise. *J. Clin. Endocrinol. Metab.*, **52**, 1106–12.

Gale, C.C. (1973) Neuroendocrine aspects of thermoregulation. *Ann. Rev. Physiol.*, **35**, 391–430.

Galster, A.D., Clutter, W.E., Cryer, P.E., Collins, J.A. and Bier, D.M. (1981) Epinephrine plasma thresholds for lipolytic effects in man. *J. Clin. Invest.*, **67**, 1729–38.

Gilbert, T.M. and Blatteis, C.M. (1977) Hypothalamic thermoregulatory pathways in the rat. *J. Appl. Physiol.*, **43**, 770–7.

Gilgen, A., Maickel, R.P., Nikodijevic, O. and Brodie, B.B. (1962) Essential role of catecholamines in the mobilization of free fatty acids and glucose after exposure to cold. *Life Sci.*, **12**, 709–15.

Greenway, D.C. and Himms-Hagen, J. (1978) Increased calcium uptake by muscle mitochondria of cold-acclimated rats. *Am. J. Physiol.*, **234**, C7–C13.

Grollman, A. (1929) Physiological variations in the cardiac output of man. III. The effect of ingestion of food on the cardiac output, pulse rate, blood pressure, and oxygen consumption of man. *Am. J. Physiol.*, **89**, 366–70.

Gross, H.A., Lake, C.R., Ebert, M.H., Ziegler, M.G. and Kopin, I.J. (1979) Catecholamine metabolism in primary anorexia nervosa. *J. Clin. Endocrinal. Metab.*, **49**, 805–9.

Grubb, B. and Folk, G.E. (1977) The role of adrenoceptors in norepinephrine-stimulated VO_2 in muscle. *Eur. J. Pharmacol.*, **43**, 217–23.

Guernsey, D.L. and Stevens, E.D. (1977) The cell membrane sodium pump as a mechanism for increasing thermogenesis during cold acclimation in rats. *Science*, **196**, 908–10.

Guidotti, A., Zivkovic, B., Pfeiffer, R. and Costa, E. (1973) Involvement of 3′, 5′-cyclic adenosine monophosphate in the increase of tyrosine hydroxylase activity elicited by cold exposure. *Naunyn Schmiedebergs Arch. Pharmacol.*, **278**, 195–206.

Hellon, R.F. (1975) Monoamines, pyrogens and cations: their actions on central control of body temperature. *Pharmacol. Rev.*, **26**, 289–321.

Hellon, R.F. (1981) Neurophysiology of temperature regulation: problems and perspectives. *Fed. Proc.*, **40**, 2804–7.

Hemingway, A., Price, W.M. and Stuart, D. (1964) The calorigenic action of catecholamines in warm acclimated and cold acclimated non-shivering cats. *Int. J. Neuropharmacol.*, **3**, 495–503.

Himms-Hagen, J. (1972) Lipid metabolism during cold-exposure and during cold-acclimation. *Lipids*, **7**, 310–23.

Himms-Hagen, J. (1975) Role of the adrenal medulla in adaptation to cold. in *Handbook of Physiology, Section VII: Endocrinology* (eds R.O. Greep and E.B. Astwood) American Physiological Society, Washington, D.C., pp. 637–65.

Himms-Hagen, J., Behren, W., Hbous, A. and Greenway, D. (1976) Altered mitochondria in skeletal muscle of cold acclimated rats and the adaptation for nonshivering thermogenesis. in *Regulation of Depressed Metabolism and Thermogenesis* (eds L. Jansky and X.J. Musacchia) Charles C. Thomas, Springfield, pp. 243–260.

Horwitz, B.A. and Eaton, M. (1977) Ouabain-sensitive liver and diaphragm respiration in cold-acclimated hamster. *J. Appl. Physiol.*, **42**, 150–3.

Hsieh, A.C.L. and Carlson, L.D. (1957) Role of adrenaline and noradrenaline in chemical regulation of heat production. *Am. J. Physiol.*, **190**, 243–6.

Hsieh, A.C.L., Carlson, L.D. and Gray, G. (1957) Role of the sympathetic nervous system in the control of chemical regulation of heat production. *Am. J. Physiol.*, **190**, 247–51.

Hsieh, A.C.L. and Wang, J.C.C. (1971) Calorigenic responses to cold of rats after prolonged infusion of norepinephrine. *Am. J. Physiol.*, **221**, 335–7.

Hull, D. and Segall, M.M. (1965) Sympathetic nervous control of brown adipose tissue and heat production in the new-born rabbit. *J. Physiol. (Lond.)*, **181**, 458–67.

Iriki, M., Riedel, W. and Simon, E. (1971) Regional differentiation of sympathetic activity during hypothalamic heating and cooling in anesthetized rabbits. *Pflugers Arch.*, **328**, 320–31.

Jansky, L. and Hart, J.S. (1963) Participation of skeletal muscle and kidney during nonshivering thermogenesis in cold-acclimated rats. *Can. J. Biochem. Physiol.*, **41**, 953–64.

Janssens, W.J. and Vanhoutte, P.M. (1978) Instantaneous changes of alpha-adreno-ceptor affinity caused by moderate cooling in canine cutaneous veins. *Am. J. Physiol.*, **234**, H330–7.

Januszewicz, W., Sznajdelman-Ciswicka, M. and Wocial, B. (1967) Urinary excretion of catecholamines in fasting obese subjects. *J. Clin. Endocrinol. Metab.*, **27**, 130–3.

Jessen, K., Rabol, A. and Winkler, K. (1980) Total body and splanchnic thermo-genesis in curarized man during a short exposure to cold. *Acta Anaesth. Scand.*, **24**, 339–44.

Johnson, G.E. (1963) The effect of cold exposure on the catecholamine excretion of adrenalectomized rats treated with reserpine. *Acta Physiol. Scand.*, **59**, 438–44.

Johnson, T.S., Young, J.B. and Landsberg, L. (1981) Norepinephrine turnover in lung: effect of cold exposure and chronic hypoxia. *J. Appl. Physiol.*, **51**, 614–20.

Jones, S.B. and Musacchia, X.J. (1976) Norepinephrine turnover in heart and spleen of 7-, 22-, and 34 C-acclimated hamsters. *Am. J. Physiol.*, **230**, 564–8.

Joy, R.J.T. (1963) Responses of cold-acclimatized men to infused norepinephrine. *J. Appl. Physiol.*, **18**, 1209–12.

Jung, R.T., Shetty, P.S., James, W.P.T., Barrand, M.A. and Callingham, B.A. (1979a) Reduced thermogenesis in obesity. *Nature*, **279**, 322–3.

Jung, R.T., Shetty, P.S., Berrand, M., Callingham, B.A. and James, W.P.T. (1979b) Role of catecholamines in hypotensive response to dieting. *Br. Med. J.*, **1**, 12–13.

Katzeff, H.L. and Danforth, E. Jr. (1981) The thermogenic response to norepine-phrine, food and exercise in lean man during overfeeding. *Clin. Res.*, **29**, 663A.

Kervran, A.A., Gilbert, M., Girard, J.R., Assan, R. and Jost, A. (1976) Effect of environmental temperature on glucose-induced insulin response in the newborn rat. *Diabetes*, **25**, 1026–30.

Kodama, A.M. and Pace, N. (1964) Effect of environmental temperature on hamster body fat composition. *J. Appl. Physiol.*, **19**, 863–7.

Komaromi, I. (1977) Effects of alpha- and beta-adrenergic blockers on the actions of noradrenaline on body temperature in the newborn guinea-pig. *Experientia*, **33**, 1083–4.

Koo, A. and Liang, I.Y.S. (1978) Microvascular responses to norepinephrine in skeletal muscle of cold-acclimated rats. *J. Appl. Physiol.*, **44**, 190–4.

Kurahashi, M. and Kuroshima, A. (1978) Creatine metabolism in skeletal muscle of cold-acclimated rats. *J. Appl. Physiol.*, **44**, 12–16.

LaFrance, L., Lagace, G. and Routhier, D. (1980) Free fatty acid turnover and oxygen consumption. Effects of noradrenaline in nonfasted and nonanes-thetized cold-adapted rats. *Can. J. Physiol. Pharmacol.*, **58**, 797–804.

Landsberg, L. and Axelrod, J. (1968) The effect of elevated temperature on the retention of 3-H-norepinephrine in the hearts of normal and thyroidectomized rats. *Life Sci.*, **7**, 1171–5.

Landsberg, L., Greff, L., Gunn, S. and Young, J.B. (1980) Adrenergic mechanisms in the metabolic adaptation to fasting and feeding: Effects of phlorizin on diet-induced changes in sympathoadrenal activity in the rat. *Metabolism*, **29**, 1128–37.

Landsberg, L. and Young, J.B. (1981) Diet-induced changes in sympathoadrenal activity: Implications for thermogenesis. *Life Sci.*, **28**, 1801–17.

LeBlanc, J. and Cote, J. (1967) Increased vagal activity in cold-adapted animals. *Can. J. Physiol. Pharmacol.*, **45**, 745–8.

LeBlanc, J., Dulac, S., Cote, J. and Girard, B. (1975) Autonomic nervous system and adaptation to cold in man. *J. Appl. Physiol.*, **39**, 181–6.

LeBlanc, J., Vallieres, J. and Vachon, C. (1972) Beta-receptor sensitization by repeated injections of isoproterenol and by cold adaptation. *Am. J. Physiol.*, **222**, 1043–6.

Leduc, J. (1961) Catecholamine production and release in exposure and acclimation to cold. *Acta Physiol. Scand.*, **53**, 1–101.

Leduc, J. (1976) Effect of sympathectomy on heat production in rats exposed to cold. *Israel J. Med. Sci.*, **12**, 1099–102.

Lynch, W.C., Adair, E.R. and Adams, B.W. (1980) Vasomotor thresholds in the squirrel monkey: effects of central and peripheral temperature. *J. Appl. Physiol.: Resp. Environ. Exercise Physiol.*, **48**, 89–96.

Maekubo, H., Moriya, K. and Hiroshige, T. (1977) Role of ketone bodies in nonshivering thermogenesis in cold-acclimated rats. *J. Appl. Physiol.*, **42**, 159–65.

Maickel, R.P., Matussek, N., Stern, D.N. and Brodie, B.B. (1967) The sympathetic nervous system as a homeostatic mechanism. I. Absolute need for sympathetic nervous function in body temperature maintenance of cold-exposed rats. *J. Pharmacol. Exp. Ther.*, **157**, 103–10.

Maickel, R., Sussman, H., Yamada, K. and Brodie, B. (1963) Control of adipose tissue lipase activity by the sympathetic nervous system. *Life Sci.*, **3**, 210–4.

Manara, L., Costa, E., Stern, D.N. and Maickel, R.P. (1965) Effect of chemical sympathectomy on oxygen consumption by the cold-exposed rat. *Int. J. Neuropharmacol.*, **4**, 301–7.

Mejsnar, J. and Jansky, L. (1976) Mode of catecholamine action during organ regulation of nonshivering thermogenesis. in *Regulation of Depressed Metabolism and Thermogenesis* (eds L. Jansky and X.J. Musacchia), Charles C. Thomas, Springfield, pp. 225–42.

Millard, R.W. and Reite, O.B. (1975) Peripheral vascular response to norepinephrine at temperatures from 2 to 40°C. *J. Appl. Physiol.*, **38**, 26–30.

Muirhead, M. and Himms-Hagen, J. (1974) Changes in the adenyl cyclase system of skeletal muscle of cold-acclimated rats. *Can. J. Biochem.*, **52**, 176–80.

Nellis, S.H., Flaim, S.F., McCauley, K.M. and Zelis, R. (1980) Alpha-stimulation protects exercise increment in skeletal muscle oxygen consumption. *Am. J. Physiol.*, **238**, H331–9.

Niijima, A. (1981) Visceral afferents and metabolic function. *Diabetologia*, **20**, 325–30.

O'Dea, K., Esler, M., Leonard, P., Stockigt, J.R. and Nestel, P. (1982) Noradrenaline turnover during under- and over-eating in normal weight subjects. *Metabolism*, **31**, 896–9.

O'Hara, W.J., Allen, C., Shephard, R.J. and Allen, G. (1979) Fat loss in the cold—a controlled study. *J. Appl. Physiol.: Resp. Environ. Exercise Physiol.*, **46**, 872–7.

Oliverio, A. and Stjarne, L. (1965) Acceleration of noradrenaline turnover in the mouse heart by cold exposure. *Life Sci.*, **4**, 2339–43.

Oomura, Y. and Kita, H. (1981) Insulin acting as a modulator of feeding through the hypothalamus. *Diabetologia*, **20**, 290–8.

Palmblad, J., Levi, L., Burger, A., Melander, A., Westgren, U., von Schenck, H. and Skude, G. (1977) *Acta Med. Scand.*, **201**, 15–22.

Pequignot, J.M., Peyrin, L. and Peres, G. (1980) Catecholamine-fuel inter-relationships during exercise in fasting men. *J. Appl. Physiol.*, **48**, 109–13.

Perkins, M.N., Rothwell, N.J., Stock, M.J. and Stone, T.W. (1981) Activation of brown adipose tissue thermogenesis by the ventromedial hypothalamus. *Nature*, **289**, 401–2.

Pierau, F.K. and Wurster, R.D. (1981) Primary afferent input from cutaneous thermoreceptors. *Fed. Proc.*, **40**, 2819–24.

Pouliot, M. (1966) Catecholamine excretion in adreno-demedullated rats exposed to cold after chronic guanthidine treatment. *Acta Physiol. Scand.*, **68**, 164–8.

Poulos, D.A. (1981) Central processing of cutaneous temperature information. *Fed. Proc.*, **40**, 2825–9.

Radomski, M.W. and Orme, T. (1971) Response of lipoprotein lipase in various tissues to cold exposure. *Am. J. Physiol.*, **220**, 1852–6.

Rappaport, E.B., Young, J.B. and Landsberg, L. (1981) Impact of age on basal and diet-induced changes in sympathetic nervous system activity of Fischer rats. *J. Gerontol.*, **36**, 152–7.

Rappaport, E.B., Young, J.B. and Lansberg, L. (1982*a*) Initiation, duration and dissipation of diet-induced changes in sympathetic nervous systema activity in the rat. *Metabolism*, **31**, 143–6.

Rappaport, E.B., Young, J.B. and Landsberg, L. (1982*b*) Effects of 2-deoxy-D-glucose on the cardiac sympathetic nerves and the adrenal medulla in the rat: Further evidence for a dissociation of sympathetic nervous system and adrenal medullary responses. *Endocrinology*, **110**, 650–6.

Raven, P.B., Nikki, I., Dahms, T.E. and Horvath, S.M. (1970) Compensatory cardiosvascular responses during an environmental cold stress, 5°C. *J. Appl. Physiol.*, **29**, 417–21.

Rosell, S. and Belfrage, E. (1979) Blood circulation in adipose tissue. *Physiol. Rev.*, **59**, 1078–104.

Rosell, S., Kopin, I.J. and Axelrod, J. (1963) Fate of H³-noradrenaline in skeletal muscle before and following sympathetic stimulation. *Am. J. Physiol.*, **205**, 317–21.

Rothwell, N.J., Saville, M.E. and Stock, M. (1981) Acute effects of food, 2-deoxy-D-glucose and noradrenaline on metabolic rate and brown adipose tissue in normal and atropinised lean and obese (fa/fa) Zucker rats. *Pflugers Arch.*, **392**, 172–7.

Rothwell, N.J. and Stock, M.J. (1979) A role for brown adipose tissue in diet-induced thermogenesis. *Nature*, **281**, 31–5.

Rothwell, N.J. and Stock, M.J. (1981) A role for insulin in the diet-induced thermogenesis of cafeteria-fed rats. *Metabolism*, **30**, 673–8.

Rowe, J.W., Young, J.B., Minaker, K.L., Stevens, A.L., Pallotta, J. and Landsberg, L. (1981) Effect of insulin and glucose infusions on sympathetic nervous system activity in normal man. *Diabetes*, **30**, 219–25.

Santiago, J.A., Clarke, W.L., Shah, S.D. and Cryer, P.E. (1980) Epinephrine, norepinephrine, glucagon, and growth hormone release in association with physiological decrements in the plasma glucose concentration in normal and diabetic man. *J. Clin. Endocrinol. Metab.*, **51**, 877–83.

Scholander, P.F., Hammel, H.T., Lange Anderson, K. and Loyning, Y. (1958) Metabolic acclimatization to cold in man. *J. Appl. Physiol.*, **12**, 1–8.

Schonbaum, E., Johnson, G.E., Sellers, E.A. and Gill, M.J. (1966) Adrenergic beta-receptors and non-shivering thermogenesis. *Nature*. **210**, 426.

Schwartz, J., Young, J.B. and Landsberg, L. (1983) Effect of dietary fat on sympathetic nervous system activity in the rat. *J. Clin. Invest.* (in press).

Seitz, H.J., Krone, W., Wilke, W. and Tarnowski, W. (1981) Rapid rise in plasma glucagon induced by acute cold exposure in man and rat. *Pflugers Arch.*, **389**, 115–20.

Sellers, E.A. and Schonbaum, E. (1963) Catecholamines in acclimation to cold: historical survey. *Fed. Proc.*, **22**, 909–10.

Shetty, P.S., Jung, R.T. and James, W.P. (1979) Effect of catecholamine replacement with levodopa on the metabolic response to semistarvation. *Lancet.*, **1**, 77–9.

Shimazu, T. (1981) Central nervous system regulation of liver and adipose tissue metabolism. *Diabetologia*, **20**, 343–56.

Shimazu, T. and Takahashi, A. (1980) Stimulation of hypothalamic nuclei has differential effects on lipid synthesis in brown and white adipose tissue. *Nature*, **284**, 62–3.

Slavin, B.G. and Bernick, S. (1974) Morphological studies on denervated brown adipose tissue. *Anat. Rec.*, **179**, 497–506.

Smith, R.E. and Horwitz, B.A. (1969) Brown fat and thermogenesis. *Physiol. Rev.*, **49**, 330–425.

Steiner, G., Johnson, G.E., Sellers, E.A. and Schonbaum, E. (1969) Nervous control of brown adipose tissue metabolism in normal and cold-acclimated rats. *Fed. Proc.*, **28**, 1017–22.

Steiner, G., Loveland, M. and Schonbaum, E. (1970) Effect of denervation on brown adipose tissue metabolism. *Am. J. Physiol.*, **218**, 566–70.

Stirling, J.L. and Stock, M.J. (1968) Metabolic origins of thermogenesis induced by diet. *Nature*, **220**, 801–2.

Sugahara, M., Baker, D.H., Harmon, B.G. and Jensen, A.H. (1969) Effect of ambient temperature and dietary amino acids on carcass fat deposition in rats. *J. Nutr.*, **98**, 344–50.

Tedesco, J.L., Flattery, K.V. and Sellers, E.A. (1977) Effects of thyroid hormones and cold exposure on turnover of norepinephrine in cardiac and skeletal muscle. *Can. J. Physiol. Pharmacol.*, **55**, 515–22.

Therminarias, A., Chirpaz, M.F. and Tanche, M. (1979) Catecholamines in dogs during cold adaptation by repeated immersions. *J. Appl. Physiol.*, **46**, 662–8.

Thoenen, H. (1970) Induction of tyrosine hydroxylase in peripheral and central adrenergic neurons by cold-exposure of rats. *Nature*, **228**, 861–2.

Thompson, G.E. (1977) Physiological effects of cold exposure. in *International Review of Physiology. Environmental Physiology* II. (ed. D. Robertshaw) Baltimore: University Park Press, pp. 29–69.

Van Houten, M. and Posner, B.I. (1981) Cellular basis of direct insulin action in the central nervous system. *Diabetologia*, **20**, 255–67.

Vatner, S.F., Franklin, D. and van Citters, R.L. (1970a) Mesenteric vascoactivity associated with eating and digestion in the conscious dog. *Am. J. Physiol.*, **219**, 170–4.

Vatner, S.F., Franklin, D. and van Citters, R.L. (1970b) Coronary and visceral vasoactivity associated with eating and digestion in the conscious dog. *Am. J. Physiol.* **219**, 1380–5.

Walther, O.E., Iriki, M. and Simon, E. (1970) Antagonistic changes of blood flow and sympathetic activity in different vascular beds following central thermal stimulation. II. Cutaneous and visceral sympathetic activity during spinal cord heating and cooling in anesthetized rabbits and cats. *Pflugers Arch.*, **319**, 162–84.

Wasserstrum, N. and Herd, J.A. (1977a) Baroreflexive depression of oxygen consumption in the squirrel monkey at 10°C. *Am. J. Physiol.*, **232**, H451–8.

Wasserstrum, N. and Herd, J.A. (1977b) Elevation of arterial blood pressure in the squirrel monkey at 10°C. *Am. J. Physiol.*, **232**, H459–63.

Webb-Peploe, M.M. and Shepard, J.T. (1968) Responses of the superficial limb veins of the dog to changes in temperature. *Circ. Res.*, **22**, 737–46.

Webster, A.J.F. and Hays, F.L. (1968) Effects of beta-adrenergic blockade on the heart rate and energy expenditure of sheep during feeding and during acute cold exposure. *Can. J. Physiol. Pharmacol.*, **46**, 577–83.

Welle, S., Lilavivathana, U. and Campbell, R.G. (1980) Increased plasma norepinephrine concentrations and metabolic rates following glucose ingestion in man. *Metabolism*, **29**, 806–9.

Welle, S., Lilavivat, U. and Campbell, R.G. (1981) Thermic effect of feeding in man: Increased plasma norepinephrine levels following glucose but not protein or fat consumption. *Metabolism*, **30**, 953–8.

Wimpfheimer, C., Saville, E., Voirol, M.J., Danforth, E. Jr and Burger, A.G. (1979) Starvation-induced decreased sensitivity of resting metabolic rate to triiodothyronine. *Science*, **205**, 1272–3.

Young, J.B. and Landsberg, L. (1977a) Suppression of sympathetic nervous system during fasting. *Science*, **196**, 1473–5.

Young, J.B. and Landsberg, L. (1977b) Stimulation of the sympathetic nervous system during sucrose feeding. *Nature*, **269**, 615–7.

Young, J.B. and Landsberg, L. (1978) Fasting, feeding, and the regulation of sympathetic activity. *New Engl. J. Med.*, **298**, 1295–301.

Young, J.B. and Landsberg, L. 1979a) Effect of diet and cold exposure on norepinephrine turnover in pancreas and liver. *Am. J. Physiol.*, **236**, E524–33.

Young, J.B. and Landsberg, L. (1979b) Sympathoadrenal activity in fasting pregnant rats: Dissociation of adrenal medullary and sympathetic nervous system responses. *J. Clin. Invest.*, **64**, 109–16.

Young, J.B. and Landsberg, L. (1980) Impaired suppression of sympathetic activity during fasting in the gold thioglucose-treated mouse. *J. Clin. Invest.* **65**, 1086–94.

Young, J.B. and Landsberg, L. (1981) Effect of concomitant fasting and cold exposure on sympathoadrenal activity in rats. *Am. J. Physiol.*, **240**, E314–9.

Young, J.B. and Landsberg, L. (1983) Adrenergic influence on peripheral hormone secretion. in *Adrenoceptors and Catecholamine Action,* part B. (ed. G. Kunos) John Wiley & Sons, New York, pp. 157–217.

Young, J.B., Rowe, J.W., Pallotta, J.A., Sparrow, D. and Landsberg, L. (1980) Enhanced plasma norepinephrine response to upright posture and glucose administration in elderly human subjects. *Metabolism,* **29,** 532–9.

Young, J.B., Saville, E., Rothwell, N.J., Stock, M.J. and Landsberg, L. (1982) Effect of diet and cold exposure on norepinephrine turnover in brown adipose tissue in the rat. *J. Clin. Invest.,* **69,** 1061–71.

Zwillich, C.W., Sahn, S.A. and Weil, J.V. Effects of hypermetabolism on ventilation and chemosensitivity. *J. Clin. Invest.,* **60,** 900–6.

Chapter Five

Thyroid Hormones and Thermogenesis

Jean Himms-Hagen

5.1 INTRODUCTION

5.1.1 Categories of thermogenesis that may be influenced by hormones

In any discussion of thermogenesis in mammals it is useful to distinguish between obligatory thermogenesis and facultative thermogenesis (Table 5.1). Obligatory thermogenesis is associated with those metabolic reactions essential simply to cell existence (essential thermogenesis), with those metabolic reactions necessary for the warm-blooded state (endothermic thermogenesis) (Girardier, 1977) and, in a steady state, with those reactions associated with the processing of food (post-prandial thermogenesis). The primary endocrine regulator of obligatory thermogenesis is thyroid hormone, which controls endothermic thermogenesis. In contrast to obligatory thermogenesis, at any given time facultative thermogenesis may or may not be occurring, according to the circumstances obtaining at that time. It may be brought on by cold, by exercise or by diet. Cold-induced thermogenesis occurs only at temperatures below thermoneutrality. Diet-induced thermogenesis (DIT)* depends on the amount of food ingested and is associated primarily with overeating. Exercise-associated thermogenesis depends on voluntary activity of the animal. The primary control of facultative thermogenesis is exerted by the nervous system, a control mediated by acetylcholine for processes involving muscle activity and by noradrenaline for processes involving thermogenesis in brown adipose tissue. Numerous other hormones play a permissive or modulating role in the regulation of facultative thermogenesis by the nervous system and in this chapter attention will be largely focused on the role of thyroid hormones.

5.1.2 Target organs for endocrine influences on thermogenesis

Most organs of the body contribute to endothermic thermogenesis, a process on which thyroid hormone has a major influence (Smith and Edelman, 1979). The brain, a significant contributor to obligatory thermogenesis, seems to be the major tissue not subject to this influence of thyroid hormone

*See Chapter 1 and Chapter 7 for discussion of terminology and definition of diet-induced thermogenesis.

Table 5.1 Categories of thermogenesis that may be influenced by hormones

Category of thermogenesis	Major site(s)	Major hormones	
		Direct	Permissive
Obligatory (or essential)			
(1) Essential thermogenesis	All organs	None	
(2) Endothermic thermogenesis	Most organs	Thyroid	
(3) Post-prandial thermogenesis	Intestine, liver, white adipose tissue		Insulin
Facultative (or optional)			
(4) Diet-induced thermogenesis	Brown adipose tissue	Noradrenaline	Thyroid Glucocorticoids (?) Insulin Sex hormones(?)
(5) Cold-induced non-shivering thermogenesis	Brown adipose tissue	Noradrenaline Glucagon (?)	Thyroid Glucocorticoids Insulin Melatonin (?)
(6) Cold-induced shivering thermogenesis	Skeletal muscle	Acetylcholine	Glucocorticoids Catecholamines
(7) Exercise-induced thermogenesis	Skeletal muscle	Acetylcholine	Glucocorticoids Catecholamines

(Smith and Edelman, 1979). Endocrine-mediated changes in endothermic thermogenesis are usually quite slow, requiring hours or days for the full change to be seen.

In contrast, facultative thermogenesis resides principally in two organs, skeletal muscle and brown adipose tissue; it may be switched on and off fairly rapidly, in minutes rather than hours. Facultative thermogenesis in skeletal muscle, in the form of exercise-induced thermogenesis or cold-induced shivering thermogenesis, is switched on by acetylcholine, itself released from motor nerves under central nervous control. Other endocrine influences can, however, modify the capacity of skeletal muscle for contraction and/or for metabolic processes that support contraction, and can thus influence the capacity of skeletal muscle for thermogenesis. These include the glucocorticoids and the hormones of the sympathoadrenal system. In skeletal muscle, thermogenesis is usually a byproduct of another functional activity, namely mechanical work (even though under certain circumstances it may be the primary aim of the activity, as in shivering thermogenesis). That facultative thermogenesis in muscle plays an important role in overall energy balance is obvious (James and Trayhurn, 1981; Jéquier and Schutz, 1981).

Facultative thermogenesis in brown adipose tissue, in the form of diet-induced thermogenesis or cold-induced non-shivering thermogenesis, is switched on by noradrenaline released from sympathetic nerve endings within the tissue. Other hormones can modify the thermogenic capacity of brown adipose tissue, by modifying its sensitivity to noradrenaline, by promoting its growth or regression and/or by promoting changes in its capacity for certain metabolic processes. These hormones include the glucocorticoids, the pancreatic hormones insulin and glucagon, thyroid hormones, possibly sex hormones, and, in some species, the pineal hormone, melatonin. Facultative thermogenesis in brown adipose tissue has only recently been recognized as playing an important role in overall thermogenesis. Although this tissue has been recognized as having the specific function of heat production for over twenty years (Smith and Horwitz, 1969), the extent of its contribution to the large amount of cold-induced non-shivering thermogenesis of which some animals are capable remained controversial until the conclusive demonstration by Foster and Frydman (1978a,b, 1979) that brown adipose tissue is the major site of heat production in cold acclimated rats, in which the metabolic rate may be increased three to four-fold either by exposure to cold or by infusion of noradrenaline in a thermoneutral environment. The cold acclimated rat, living at 4°C and maintaining its body temperature primarily by increasing heat production in brown adipose tissue, is hyperphagic, eating twice as much as a rat at 28°C (Hardeveld, Zuidwijk and Kassenaar, 1979a; Leung and Horwitz, 1976) and is producing the extra heat by metabolizing in its brown adipose tissue products derived from the extra food. The cold acclimated rat remains very lean, despite its large food intake; thus the sympathetic-mediated thermo-genesis in brown adipose tissue allows the diversion of energy from fat deposition to regulatory heat production.

Evidence for a direct role of brown adipose tissue as an energy buffer has been provided by research on the cafeteria-fed rat, which, despite overeat-ing, remains relatively lean, and the genetically obese mouse, which can gain excess weight even when its energy intake is restricted to that eaten by its lean counterpart. The cafeteria-fed rat increases not only its food intake but also its metabolic rate (diet-induced thermogenesis). The process is sympathetic-mediated and occurs primarily in brown adipose tissue (Rothwell and Stock, 1979, 1981a,b; Stock and Rothwell, 1981); its periph-eral mechanism, namely, noradrenaline-stimulated thermogenesis in brown adipose tissue, is the same as that involved in cold-induced non-shivering thermogenesis (Rothwell and Stock, 1980) but the afferent pathways and central integrative mechanisms differ. In contrast, genetically obese mice (the ob/ob mouse and the db/db mouse resemble each other in these respects) have a low metabolic rate, and a high metabolic efficiency (Thurlby and Trayhurn, 1979; Trayhurn and Fuller, 1980; Lin *et al.*, 1979c; Romsos, 1981a,b). Both cold-induced non-shivering thermogenesis and diet-induced thermogenesis appear to be defective in these mice; the defect lying either in the brown adipose tissue itself or in the control mechanisms that govern its thermogenic activity (Thurlby and Trayhurn, 1980; Himms-Hagen and Desautels, 1978; Hogan and Himms-Hagen, 1980, 1981; Trayhurn, 1979; Goodbody and Trayhurn, 1981; Trayhurn *et al.*, 1982; Himms-Hagen, 1981).

Brown adipose tissue enlarges when there is an increased need for its thermogenic function, as in animals living at low temperature and in animals that are overeating; or when there is an anticipated need for its thermogenic function, as in certain hibernators during the period of preparation for hibernation. This is a true growth, accompanied by cell proliferation (Bukowiecki *et al.*, 1978, 1982) and its extent can be assessed from the increase in total DNA content, protein content and cytochrome oxidase content (as measures of number of cells, active tissue mass and mito-chondrial mass). Measurement of wet weight alone is not a useful measure of the growth of brown adipose tissue since wet weight may increase solely because of an increase in triglyceride content.

Not only does the mitochondrial mass increase during the growth of brown adipose tissue, but the composition of the mitochondria may also change. Thus, in cold acclimated rats the mitochondria of the enlarged brown adipose tissue have a relative increase in the 32 000 molecular weight (32K) polypeptide associated with the thermogenic proton conductance pathway (see Chapter 2) (Ricquier, Kader, 1976; Desautels *et al.*, 1978; Ricquier *et al.*, 1979a) and a marked change in composition of phospholipid fatty acids, which becomes relatively more saturated (Ricquier, Mory and

Hémon, 1975, 1976, 1979*b*; Ricquier *et al.*, 1978). While some workers suggest that the increase in the 32K polypeptide may underlie an increase in capacity for loose-coupled respiration (Nicholls *et al.*, 1974; Nicholls, 1976), others have been unable to find the latter increase and suggest that the altered composition may rather underlie a change in intracellular regulation of the proton conductance pathway (Desautels and Himms-Hagen, 1981). The change in polypeptide composition occurs in response only to some growth-promoting stimuli (cold acclimation) (Ricquier and Kader, 1976; Desautels, Zaror-Behrens and Himms-Hagen, 1978; Ricquier *et al.*, 1979*a*) and not to others (cafeteria feeding). Since the nature of the presumed hormonal influence which brings about this change in composition is at present unknown, little further discussion of its control will be possible in this chapter. Since it is associated with a selective change in the synthesis of mitochondrial proteins (Himms-Hagen, Dittmar and Zaror-Behrens, 1980), it is presumably regulated by a hormone that acts via a nuclear mechanism.

The thermogenic state of brown adipose tissue can also be assessed by measuring the binding of purine nucleotides (GDP is usually used), known to bind to the 32K polypeptide (Heaton *et al.*, 1978). The extent of binding of purine nucleotides is not directly related to the concentration of the 32K polypeptide but is a function of the concentration of binding sites and of the extent to which the sites are exposed and accessible for binding. A large increase in binding, without any change in composition but associated with an ultrastructural change in the mitochondria and due to an unmasking of sites already present, occurs in rats in response to exposure to cold (Desautels *et al.*, 1978; Desautels and Himms-Hagen, 1979, 1980) and to cafeteria feeding (Brooks *et al.*, 1980; Himms-Hagen, Triandafillou and Gwilliam, 1981). The unmasking response is mediated by noradrenaline and is a sensitive and readily measured indicator of the thermogenic responsiveness of brown adipose tissue to endogenous catecholamines and/or the stimulation of sympathetic activity in brown adipose tissue by external stimuli.

Some workers have proposed an equation of the extent of binding of purine nucleotides with the amount of the 32K polypeptide ('thermogenin') (Cannon, Nedergaard and Sundin, 1981). This however, is not valid because the two can be dissociated, as, for example, in the acutely cold exposed rat, the cafeteria-fed rat and the cold acclimated hamster (increase in binding, unchanged 32K polypeptide) and in the ob/ob mouse (decreased binding, normal amount of 32K polypeptide).

There is no doubt that other factors, particularly thyroid hormones, exert an influence on the growth and thermogenic function of brown adipose tissue and thus on an animal's capacity for thermogenesis. To what extent this influence is exerted directly on brown adipose tissue (or on other tissues) and to what extent it is permissive or exerted indirectly by inducing changes in sympathetic activity or in the responsiveness of brown adipose tissue to sympathetic stimulation will be discussed in the following sections.

5.2 ROLE OF THYROID HORMONES IN THERMOGENESIS

It has been known for many years that administration of thyroid hormones results in an increase in metabolic rate and that thyroidectomy results in a decrease in metabolic rate. Likewise, a role for thyroid hormones in growth and development has long been recognized. There is, however, still considerable argument, not only about the exact mechanism of action of thyroid hormones but also about the extent to which observed changes in thermogenesis, such as occur in response to altered environmental temperature and in response to fasting or feeding, are due to the simultaneous changes in levels of thyroid hormones.

5.2.1 Thyroid–catecholamine interrelationships

A major difficulty in interpreting the results of experiments designed to determine the exact role of thyroid hormones is that these hormones profoundly influence the action of the other major thermogenic hormones, the catecholamines, while, at the same time, attenuating the activity of the sympathetic nervous system (Gibson, 1981; Axelrod, 1975). Thus, the thermogenic effect of catecholamines increases in hyperthyroidism and decreases in hypothyroidism (Swanson, 1956; Hsieh *et al.*, 1966; Fregly *et al.*, 1979). Because the thermogenic effects of both thyroid hormones and catecholamines tend to wax and wane in parallel it is most difficult to tell which hormones are principally involved in thermogenesis. A further complication is that in many reported experiments hypothyroid animals have been unintentionally maintained at temperatures below thermoneutrality and are, in effect, cold-stressed, showing signs of increased sympathetic activity (Sellers *et al.*, 1971; Sellers, Flattery and Steiner, 1974; Tedesco, Flattery and Sellers, 1977).

Another complication is that thyroid hormones influence the concentration of adrenergic receptors in certain tissues and this should also not be overlooked. The influence is, however, tissue specific and variable. In hyperthyroid rats there is usually an increase in the concentration of β-adrenergic receptors in white adipose tissue (Giudicelli, 1978; Malbon *et al.*, 1978; Ciaraldi and Marinetti, 1978), heart (Ciaraldi and Marinetti, 1978; Scarpace and Abrass, 1981*a*) and brown adipose tissue (Sundin, 1981*b*), a decrease in liver (Malbon, 1980) and no change in lymphocytes or lung (Scarpace and Abrass, 1981*b*). On the other hand, the concentration of α-adrenergic receptors has generally been seen to decrease in white adipose tissue (Giudicelli, Lacasa and Agli, 1980) and heart (Williams and Lefkowitz, 1979; Sharma and Bannerjee, 1978). However, since maximum metabolic responses to catecholamines can be obtained with only a very low receptor occupancy, estimated to be, for example, only about 10% of total β-adrenergic receptors in brown adipocytes (Svoboda *et al.*, 1979; Bukowiecki *et al.*, 1980), the relation of thyroid-induced changes to alterations in adrenergic receptor number is far from clear. It is likely that it is the balance between α- and β-adrenergic receptors, the former usually inhibitory, the latter stimulatory, which is of critical importance in determining the responsiveness of tissues to catecholamines (Gibson, 1981), but this has not been determined quantitatively for any given tissue. Both α- (Svartengren, Mohell and Cannon, 1980; Launay and Himms-Hagen, unpublished results) and β-adrenergic receptors (Bukowiecki *et al.*, 1978; Svoboda *et al.*, 1979) are present in brown adipose tissue and are involved in its thermogenic responsiveness to catecholamines (Bukowiecki *et al.*, 1980; Mohell, Nedergaard and Cannon, 1980; Hamilton and Horwitz, 1977; Horwitz, 1977; Flaim, Horwitz and Horowitz, 1977; Cannon *et al.*, 1981), The metabolic consequences of a possible altered balance between α- and β-adrenergic receptors in this tissue are not known.

Not only do thyroid hormones influence the activity and effectiveness of the sympathetic nervous system, but the sympathetic nervous system in turn reciprocally influences the secretion, peripheral formation and action of thyroid hormones. The sympathetic innervation of the thyroid gland appears to exert an inhibitory effect on the stimulatory control by TSH (Pisarev *et al.*, 1981) and can also have a direct stimulatory action on the thyroid follicle cells themselves (Melander *et al.*, 1975*a,b*). Moreover, increased sympathetic activity promotes the peripheral conversion of thyroxine (3,5,3′,5′-tetra-iodothyronine of T_4) into the more thermogenically active 3,5,3′-tri-iodothyronine $(T_3)^*$ (Hardeveld *et al.*, 1979*b*; Scammell, Shiverick and Fregly, 1980; Storm, Hardeveld and Kassenaar, 1981), thus further complicating our understanding of the interrelationship between the two groups of thermogenically active hormones, catecholamines and thyroid hormones.

The best measure of the influence of thyroid hormones on metabolic rate is the thyroid-induced change in minimal oxygen consumption (MOC), a

*The principal active thyroid hormone, 3,5,3′-tri-iodothyronine, is only in part secreted directly by the thyroid gland, most of it being formed in peripheral tissues by deiodination of thyroxine, the principal secretory product of the gland (Schimmel and Utiger, 1977; Eisenstein *et al.*, 1978), a process described as T_3-neogenesis (Balsam, Sexton and Ingbar, 1981*a*). Regulation of the level of T_3 thus involves not only regulation of the secretory activity of the thyroid gland itself, but also of the 5′-deiodinase that converts T_4 into T_3. Since a similar 5′-deiodinase is also responsible for the degradation of 3,3′,5′-tri-iodothyronine (reverse T_3 or rT_3), an inactive form of thyroid hormone also produced peripherally from T_4, any change in the activity of the 5′-deiodinase that produces a changes in T_3 level is also likely to produce a reciprocal change in the level of rT_3, a change of unknown physiological significance.

measurement supposedly uncomplicated by thyroid-associated changes in sympathetic activity and responsiveness (Danforth and Burger, 1981). Of all hormones and hormone ablations studied, only thyroidectomy and hypophysectomy reduce the MOC of rats; these reductions can be reversed by T_4 (Denckla, 1973). Thyroid hormones alone thus appear to be responsible for the control of essential and endothermic thermogenesis (Danforth and Burger, 1981). MOC decreases with age, an effect partly due to altered tissue responsiveness to thyroid hormone (Denckla, 1974). The effect of thyroid hormones *in vivo* on endothermic thermogenesis persists in isolated tissues, unlike the more rapid and transient thermogenic effects of other hormones such as the catecholamines. Most tissues of the body, excluding the brain (Smith and Edelman, 1979), but including brown adipose tissue (Ikemoto, Hiroshige and Itoh, 1967), are responsive to thyroid hormones.

5.2.2 Thermogenic actions of thyroid hormones

It is now well recognized that thyroid hormone does not have a single site and mechanism of action to bring about the variety of developmental and metabolic effects attributable to it, but rather has multiple sites of action within the body, even within cells, and multiple mechanisms of action (Tata, 1975; Sterling, 1979; Oppenheimer, 1979; Oppenheimer *et al.*, 1979). High affinity receptors for T_3 have been identified in nucleus, cytosol, mitochondrion and plasma membrane of a variety of tissues and effects of T_3 have been identified which may be attributable to interactions with these receptors. Since in this discussion the emphasis is on the thermogenic action of T_3, only those effects leading to altered thermogenesis will be considered.

Three basic mechanisms are believed to underlie the thermogenic action of thyroid hormone. One is a change in the properties of the mitochondria such that their rate of respiration increases even though they remain in a coupled state (Shears and Bronk, 1979; Shears, 1980; Hoch, 1977). The second is an increase in the total mitochondrial content of tissues (Courtright and Fitts, 1979; De Leo *et al.*, 1976; Nishiki *et al.*, 1978). The third is an increase in Na^+K^+-ATPase activity in tissues (Smith and Edelman, 1979). These three mechanisms are not mutually exclusive and may all be involved; there is, however, no agreement as to their relative quantitative importance. To these three basic mechanisms should also be added the effect of thyroid hormone to potentiate the thermogenic action of the catecholamines.

Liver mitochondria from thyroid-treated animals have an elevated rate of state 3 respiration (Shears and Bronk, 1979; Sterling, Brenner and Sakurada, 1980; Nishiki *et al.*, 1978) and an increased rate of ADP translocation (Babior *et al.*, 1973; Portnay *et al.*, 1973). They also have a higher rate of state 4 respiration that is not due to uncoupling since it is associated with a higher Δp (proton electrochemical gradient) (Shears and

Bronk, 1979; Shears, 1980). This effect is thought to be associated with an alteration in membrane properties such that a non-ohmic increase in proton conductance occurs at very high levels of Δp. Such an increase has been observed in both liver (Nicholls, 1974) and brown adipose tissue mitochondria (Nicholls, 1977) and appears to be peculiar to flavoprotein-linked substrates (Nicholls, 1977). The thyroid-induced changes in mitochondrial membrane fatty acid composition (an increase in saturation) may be related to this increase (Hulbert, 1978; Hulbert, Augee and Raison, 1976; Hoch, 1977; Chen and Hoch, 1977). The way in which the change in fatty acid composition is brought about is unknown. Since there is also a change in polypeptide composition of liver mitochondria of hypothyroid rats (Bouhnik *et al.*, 1979; Baudry, Clot and Michel, 1975; Sterling, 1979; De Leo *et al.*, 1976; Jakovcic *et al.*, 1978) a change in overall mitochondrial membrane composition may be involved.

The way in which the increase in mitochondrial oxygen consumption is induced by thyroid hormone is not understood. While it may in part be associated with altered mitochondrial composition, as noted above, the rapidity with which effects of very small amounts of T_3 can be demonstrated in perfused liver (30–60 min) (Müller and Seitz, 1980, 1981) and after T_3 injection into intact animals (Sterling *et al.*, 1980) suggests that they may be, at least, in part, a consequence of a direct interaction of T_3 with the high-affinity receptors in mitochondria (Sterling *et al.*, 1977, 1978; Sterling, 1979).

An influence of thyroid hormone on total mitochondrial mass has been observed in muscle (Courtright and Fitts, 1979), liver (De Leo *et al.*, 1976; Wooten and Cascarano, 1980) and heart (Nishiki *et al.*, 1978). A thyroid hormone-induced increase in total mitochondrial mass may also contribute to the changes in overall oxygen uptake of the animal. The way in which this mitochondrial proliferation is brought about is not understood in detail.

That promotion of active sodium transport by the Na^+K^+-ATPase of the plasma membrane might underlie thyroid-induced thermogenesis was first proposed by Ismail-Beigi and Edelman some twelve yeas ago (Ismail-Beigi and Edelman, 1970, 1971). Since then considerable evidence has been acquired showing that thyroid hormone, administered *in vivo,* increases the oxygen uptake of several different target tissues, measured *in vitro,* in a way that is inhibited by ouabain, an inhibitor of the Na^+K^+-ATPase, and to an extent that is related to an increased activity and number of pump sites in the tissue (Edelman, 1974; Smith and Edelman, 1979). The contribution of the sodium pump to the T_3-induced increase in tissue respiration has been calculated to be more than 90% in liver and 84% in muscle (Smith and Edelman, 1979; Asano, Lieberman and Edelman, 1976). The effect of T_3 is also demonstrable in primary monolayer cultures of hypothyroid rat liver, where it is characterized by a similar lag period of 1–2 days for increases in oxygen uptake, Na^+K^+-ATPase and mitochondrial α-glycerophosphate dehydrogenase (Ismail-Beigi *et al.*, 1979); 80–90% of the T_3-induced increase in respiration was abolished by ouabain and attributed to an increase in Na^+K^+-ATPase action.

Several criticisms of the attribution to sodium pumping of such a large proportion of T_3-induced thermogenesis have been published. Thus, a much smaller contribution, 5–13%, has been calculated for perfused rat liver (Folke and Sestoft, 1977), for mouse intact soleus muscle (Biron *et al.*, 1979) and for perfused hind limb of the rat (Hardeveld, and Kassenaar, 1981). The major difference between these studies and those of Edelman and co-workers is that they all involved intact tissues. Inhibition of respiration by ouabain is usually greater in tissue slices than in intact tissues. Ouabain exerts an immediate inhibitory effect on ion movements in liver slices (McLaughlin, 1973) and in perfused liver (Folke and Sestoft, 1977), but only a delayed inhibitory effect on oxygen consumption. Thus, the depression of oxygen consumption is greater the longer ouabain remains in contact with the tissue. Ouabain may either inhibit or stimulate oxygen uptake of muscle, depending on the experimental conditions (Hardeveld and Kassenaar, 1981; Chinet, Clausen and Girardier, 1977; Biron *et al.*, 1979). It has also been observed to inhibit a variety of metabolic processes, such as lipolysis in white adipose tissue (Ho *et al.*, 1967; Fain and Rosenthal, 1971), hormone-stimulated gluconeogenesis in kidney (Saggerson and Carpenter, 1979; Guder, 1979) and noradrenaline-stimulated adenylate cyclase in brown adipose tissue (Fain, Jacobs and Clément-Cormier, 1973). The principal problem with the use of ouabain and of incubation media modified to inhibit the sodium pump in isolated tissues is the change in ionic composition artefactually brought about within the tissue, particular in the concentrations of Na^+, K^+ and Ca^{2+}, and the usually unknown metabolic consequences of these ionic changes (see Himms-Hagen, 1976 for more detailed discussion), as well as the inhibition of other metabolic processes as noted above. Edelman's experiments (Edelman, 1974; Ismail-Beigi and Edelman, 1970, 1971; Asano *et al.*, 1976; Smith and Edelman, 1979) have usually involved the use of sliced or otherwise damaged tissues and a preincubation of the tissues with ouabain. Only in a recent study with hepatocytes in primary culture were undamaged cells used, but, even in this study, a preincubation with oubain was employed (Ismail-Beigi *et al.*, 1979). The reason the rapid and apparently direct stimulatory effect of T_3 on respiration seen by others in perfused liver (Müller and Seitz, 1980, 1981) was not seen in these cultured cells is not clear. The cells did metabolize the added T_3 fairly rapidly (a $t^{1/2}$ of about 3.5 h) and, since measurements of cell respiration were made some hours after replacement of medium containing T_3, it is possible that the direct rapid effect of T_3 had worn off and only the initial stage of the long-term (1–2 days) response was seen. Even the long-term response to T_3, claimed to be almost entirely (90%) due to sodium pump activity (Ismail-Beigi *et al.*, 1979), can be calculated to be larger than expected from the measured increase in Na$^+$K$^+$-ATPase activity. The increase in Na$^+$K$^+$-ATPase activity was less than 10% of the increase expected if it alone were the cause of the increase in ouabain-inhibited respiration. It is therefore probable that some other metabolic process occurring in these cells, such as gluconeogenesis (Bissell, Hammaker and Meyer, 1973), [known to be stimulated by T$_3$, at least in perfused liver (Müller and Seitz, 1980, 1981)] was inhibited by the preincubation with ouabain.

In conclusion, while it is undoubtedly correct to say that some proportion of the thyroid-induced increase in tissue respiration may be due to a stimulation of ion pumping, it is still not possible to assign a numerical value to this proportion. Other thyroid-stimulated processes, such as the direct effect on mitochondrial properties noted previously, also contribute to the increase, but to an unknown extent.

In the intact animal the thermogenic effect of thyroid hormone includes a component attributable to an enhancement of the thermogenic effect of catecholamines (Hsieh *et al.*, 1966). The size of this component will depend on the extent to which sympathetic-mediated facultative thermogenesis is occurring, and thus on nutritional state and on environmental temperature. The mechanism of this effect of thyroid hormone is not understood; it may involve, at least in part, an altered balance between α- and β-adrenergic receptors, as noted above.

5.2.3 Role of thyroid hormones in states of altered thermogenesis

A role for thyroid hormones has frequently been sought in the alteration of thermogenesis that occurs in a number of different states such as in cold exposure and cold acclimation, in fasting, feeding and overfeeding, and in obesity. Each of these states will be reviewed in turn, to try to elucidate what role(s), if any, thyroid hormones play, as distinct from the role(s) of changes in sympathetic activity, also usually implicated in these same states.

(a) Cold exposure and cold acclimation

When first exposed to cold normal rats raise their metabolic rate by a mixture of shivering thermogenesis and non-shivering thermogenesis (Foster and Frydman, 1979). After about two weeks in the cold their capacity for cold-induced non-shivering thermogenesis increases. The continued elevation of their metabolic rate is then almost entirely due to this sympathetic-mediated metabolic process in brown adipose tissue (Foster and Frydman, 1978*a,b*, 1979); such animals are said to be cold acclimated. Hypothyroid rats, on the other hand, are unable to survive in the cold, having an impaired ability to raise their heat production (Sellers *et al.*, 1971, 1974; Johnson, Flattery and Schönbaum, 1967; Hsieh, 1962). This defect in the hypothyroid animal, together with the normally persistently high metabolic rate in the euthyroid animal kept in the cold, has frequently led to the suggestion that thyroid hormones mediate part or all of cold-induced non-shivering thermogenesis (e.g. Edelman, 1976). This suggestion has been reinforced by the finding of elevated levels of T_3, but not usually of T_4, in the blood of cold-acclimated rats (Bernal and Escobar del Rey, 1975a,b; Balsam and Sexton, 1975; Scammell et al., 1980, 1981; Hardeveld et al., 1979a,b; Hefco et al., 1975). The increase in T_3 level is primarily due to accelerated conversion of T_4 into T_3 (Bernal and Escobar del Rey, 1975a,b; Balsam and Sexton, 1975; Scammell et al., 1980; Hardeveld et al., 1979a,b), an effect probably mediated by the cold-activated sympathetic nervous system (Hardeveld et al., 1970a,b; Scammell et al., 1980).

It might also be asked to what extent activation of the sympathetic nervous system by the hyperphagia which accompanies cold acclimation is involved (Jung, Shetty and James, 1980b; Landsberg and Young, 1981a,b). However, since food restriction does not reduce the level of T_3 in cold-acclimated rats to any greater extent than in control rats (Hardeveld et al., 1979a) it would seem that the hyperphagia is involved to only a limited extent, if at all, in elevating the plasma T_3 levels in the cold-acclimated rat.

Not only do cold-acclimated rats have an elevated level of T_3 in their blood, but their MOC has been reported to be increased (Denckla and Bilder, 1975), an increase not seen in thyroidectomized rats (acclimated by a brief daily exposure to cold). However, this increase in MOC is not always seen and appears to depend on the nature of the diet (Héroux, 1968) and the length of time the rat is allowed to switch off the elevated sympathetic activity on removal from the cold (at least 2.5 h at 28°C is required) (Hsieh, 1963) as well as the temperature at which the metabolic rate is measured (Hsieh, 1963). It should also be noted that cold-acclimated rats may have a lower thermoneutral zone than control rats and may, therefore, be partially heat-stressed, even at 25–28°C (Foster and Frydman, 1979); the heat stress may then lead to an increase in metabolic rate.

Thyroidectomized rats survive well in the cold if provided with a low maintenance dose of T_4 (Sellers et al., 1974), lower than the estimated secretion rate of 5–7.5 μg d^{-1} (Ruegamer, Westerfeld and Richert, 1964; Wood and Carlson, 1956; Hsieh, 1962), an amount insufficient to depress the elevated level of TSH in these animals (Sellers et al., 1974). Moreover, the extent to which partially hypothyroid rats can survive cold exposure is not directly related to the level of T_3 in their blood (Nakashima, Taurog and Krulich, 1981). Thyroidectomy of already cold-acclimated rats leads to only a gradual decline in their capacity for cold-induced non-shivering thermogenesis (Hsieh and Carlson, 1957) and the amount of thyroid hormone required by these animals is no greater than that required by animals at room temperature (Hsieh, 1962). Thus, cold-acclimated animals do not appear to be hyperthyroid, despite increased T_3 levels, yet do require at least a relatively small amount of thyroid hormone to be able to survive in the cold.

It is likely that this requirement is for the well-known permissive effect of thyroid hormone on metabolic effects of catecholamines (Fregly *et al.*, 1979), and particularly on their thermogenic effect (Swanson, 1956).

Since brown adipose tissue mitochondria are not activated in the normal way by acute exposure of the thyroidectomized rat to cold, as indicated by the lack of the normal cold-induced increase in purine nucleotide binding (Triandafillou, Gwilliam and Himms-Hagen, 1982) and the lack of the usual lipid mobilization (Mory *et al.*, 1981), it is likely that the failure of the hypothyroid rat to survive in the cold results from a failure of the mechanism for switching on thermogenesis in brown adipose tissue, a mechanism known to be of considerable importance even to the non-cold acclimated rat (Foster and Frydman, 1978*b*, 1979). The noradrenaline concentration in brown adipose tissue of the adult thyroidectomized rat is normal (Kennedy, Hammond and Hamolsky, 1977) and it seems likely that the lack of responsiveness of this tissue to cold in this animal is due to a refractoriness to noradrenaline, perhaps associated with a reduced affinity of its β-adrenergic receptors (Girardier, 1981; Seydoux, Giacobine and Girardier, 1982) That providing the thyroidectomized rat with a low maintenance dose of thyroid hormone permits a normal thermogenic response of its brown adipose tissue mitochondria to cold (Triandafillou *et al.*, 1982) provides clear evidence for the permissive nature of the effect of thyroid hormone in cold-induced non-shivering thermogenesis.

Treatment of newborn mice, which have a very low level of T_3 in their blood, with T_3 has relatively little effect on MOC but brings about a large increase in their capacity for sympathetic-mediated cold-induced non-shivering thermogenesis (Haidmayer and Hagmüller, 1981). Although thyroid hormone is required for the normal increase in noradrenaline content of brown adipose tissue which occurs during the first three weeks of life in the rat, treatment of newborn rats with T_4 does not further increase the content of noradrenaline in brown adipose tissue (Gripois, Klein and Valens, 1980). Thus the T_3-induced increase in capacity for cold-induced non-shivering thermogenesis in newborn mice (Haidmayer and Hagmüller, 1981) is most probably another manifestation of the permissive effect of thyroid hormone.

Although thyroid hormone has no major role, other than a permissive one, in the acute switching on of cold-induced non-shivering thermogenesis, there remains the possibility that T_3 might be involved in bringing about the adaptive growth of brown adipose tissue and the changes in its mitochondria that lead to a greater capacity for non-shivering thermogenesis after acclimation to cold. However, the role of thyroid hormone in bringing about such long-term cold-induced changes in brown adipose tissue appears also to be minor. Thyroidectomized rats treated with a low maintenance dose of T_4 and exposed to cold grow a normal amount of brown adipose tissue; their brown adipose tissue mitochondria undergo the usual increase in their capacity for binding GDP and in the proportion of the 32K polypeptide they contain (Triandafillou *et al.*, 1982). Thus only a small amount of thyroid hormone is required for the normal cold-induced adaptive changes in brown adipose tissue and no increase in the T_3 level in the blood is necessary. Moreover, treatment of rats with an excess of thyroid hormone leads to a hypertrophy of brown adipose tissue that is almost entirely due to accumulation of triacylglycerol (Heick *et al.*, 1977; Ricquier *et al.*, 1975, 1976; Triandafillou *et al.*, 1982). The normal cold-induced changes in mitochondrial polypeptide composition (Triandafillou *et al.*, 1982) and in phospholipid fatty acid composition (Ricquier *et al.*, 1978) do not occur in brown adipose tissue mitochondria of hyperthyroid rats. In keeping with these findings of little or no effect of hyperthyroidism on brown adipose tissue are the different effects of hyperthyroidism and of cold acclimation on the capacity of rats to respond to noradrenaline with an increase in metabolic rate: the increase in maximum capacity to respond is much greater in cold-acclimated rats than in hyperthyroid rats (Hsieh *et al.*, 1966). Not only does chronic treatment with thyroid hormone not mimic to any extent the effect of cold acclimation on growth of brown adipose tissue, but treatment of rats with large amounts of thyroid hormone during acclimation to cold actually prevents the normal cold-induced increase in DNA and protein content and the change in mitochondrial fatty acid composition (Mory, Ricquier and Hémon, 1980; Ricquier *et al.*, 1976). Furthermore, treatment of rats exposed to mild cold (22°C) with thyroid hormone reduces the thermogenic activity of brown adipose tissue, as indicated by a reduction in mitochondrial GDP binding (Sundin, 1981*a*). It seems likely that the thyroid-induced elevation of heat production reduces the amount of non-shivering thermogenesis in brown adipose tissue, presumably via the reduced activity of the sympathetic nervous system known to occur in hyperthyroidism (Gibson, 1981; Axelrod, 1975; Tedesco *et al.*, 1977).

In conclusion, the only role ascribable to thyroid hormone in cold-induced non-shivering thermogenesis is a permissive one. Minimal amounts are required for the catecholamine-induced switching on of this metabolic process in brown adipose tissue. Thyroid hormone also does not mediate the cold-induced growth and mitochondrial adaptations that occur in brown adipose tissue and which lead to an increased capacity for cold-induced non-shivering thermogenesis; it may even impair these responses if it is present in excessive amounts.

Considerable effort has been expended over the last few years to extend the Edelman hypothesis for thyroid thermogenesis to the mechanism of cold-induced non-shivering thermogenesis (Edelman, 1976; Horwitz, 1978, 1979*a,b*; Stevens, 1973). The findings of increased ouabain-sensitive respiration in liver (Videla *et al.*, 1975; Guernsey and Stevens, 1977; Stevens and Kido, 1974; Horwitz and Eaton, 1977) and muscle (Horwitz and Eaton, 1977; Guernsey and Stevens, 1977; Stevens and Kido, 1974; Mokhova and Zorov, 1973) of cold-acclimated rodents, and of increased Na^+K^+-ATPase activity in both liver (Videla *et al.*, 1975) and muscle (Himms-Hagen *et al.*, 1978) have been interpreted as providing support for the concept that cold-induced non-shivering thermogenesis might be mediated by the same mechanism as thyroid-induced thermogenesis. One would not, however, expect a facultative and rapidly switched on and off process such as noradrenaline-mediated, cold-induced non-shivering thermogenesis to persist in isolated tissues and a further argument against a direct action of thyroid hormones in cold-induced non-shivering thermogenesis is based on the lag in onset and the long duration of their thermogenic action. These are in direct contrast to the very rapid switching on and off of heat production which characterizes cold-induced non-shivering thermogenesis.

In view of the conclusion presented above, namely, that the only role of thyroid hormone in cold-induced non-shivering thermogenesis is a permissive one, and in view of the currently accepted conclusion that most of cold-induced non-shivering thermogenesis occurs in brown adipose tissue, it is clear that the changes observed in other isolated tissues are not part of the actual heat-producing mechanism itself but only incidental to it. The relatively small size of such changes and their persistence *in vitro* suggest that they are related to the small and variable increase in resting metabolic rate seen in cold-acclimated animals and possibly directly related to the elevated T_3 levels in their blood.

This does not mean that there is no role for Na^+K^+-ATPase in cold-induced non-shivering thermogenesis in brown adipose tissue. Noradrenaline stimulates the activity of this enzyme at the same time as it stimulates adenylate cyclase, lipolysis and loose-coupled respiration of the mitochondria (Horwitz, 1977, 1978, 1979*a,b*; Horwitz and Eaton, 1975; Rothwell, Stock and Wyllie, 1981). The extent to which the accelerated ATP hydrolysis catalysed by the stimulated enzyme contributes to the stimulation of tissue respiration is uncertain but it is probably minor. Although cold acclimation reduces the concentration of Na^+K^+-ATPase in brown adipose tissue the total amount of the enzyme is actually increased due to the increase in the amount of tissue (Himms-Hagen and Dittmar, unpublished results).

(b) Fasting, feeding and overfeeding

It is now well-established that fasting brings about a reduction in resting oxygen consumption whereas feeding increases it (Danforth and Burger, 1981; Burger, Hughes and Saville, 1981; Danforth *et al.*, 1979, 1981; Jung, Shetty and James, 1980*b*). The parallel decrease in T_3 level during fasting and increase on refeeding has prompted considerable research aimed at demonstrating a causal relationship between the changes in T_3 levels and the changes in resting oxygen consumption induced by diet. The reduction in T_3 level during fasting is mainly due to reduced conversion of T_4 into T_3 by the

5'-deiodinase (Harris *et al.*, 1978; Balsam *et al.*, 1981*a*,*b*; Eisenstein *et al.*, 1978; Suda *et al.*, 1978; Kaplan, 1979).

However, nutritional status influences not only the level of T_3 in the blood but also the activity of the sympathetic nervous system, there being an increase in activity in the fed state and a decrease in the fasting state (see Chapter 4; Jung *et al.*, 1980*a*,*b*; Shetty, Jung and James, 1979). Thus, just as in the case of the cold exposed or cold acclimated animal it is necessary to distinguish between changes in thermogenesis that are secondary to changes in T_3 levels and changes in thermogenesis that are due to altered sympathetic activity. It now appears likely that it is the changes in sympathetic activity that play the dominant role in determining the level of thermogenesis during dietary manipulation rather than the changes in T_3 level.

Thus, fasting does not induce any change in MOC (a measurement of metabolic rate that includes essential and thyroid-sensitive endothermic thermogenesis) whereas it causes a fall in ROC (a measurement that includes MOC plus diet-induced sympathetic-mediated facultative thermogenesis) in intact rats, in thyroidectomized rats and in thyroidectomized rats treated with replacement amounts of either T_1 or T_3 (Danforth and Burger, 1981; Burger *et al.*, 1981; Wimpfheimer *et al.*, 1979). It can be concluded, therefore, that the fasting-induced decrease in T_3 level itself has no direct influence on thermogenesis and is not even required for the fasting-induced decrease in metabolic rate. The reason no decrease in MOC accompanies the decrease in T_3 level in the fasting state appears to be the development during fasting of a partial refractoriness to T_3 (Wimpfheimer *et al.*, 1979), possibly due to altered tissue uptake (Okamura, Taurog and Distefano, 1981) or associated with reduced numbers of T_3 receptors in target tissues. Fasting reduces nuclear T_3 receptors in liver of rats (De Groot *et al.*, 1977; Schussler and Orlando, 1978; Burman *et al.*, 1977); possible changes in nuclear T_3 receptors in other tissues or in mitochondria T_3 receptors appear not to have been studied as yet. The extent to which the stimulatory influence of T_3 on Na^+K^+-ATPase is modified by fasting appears also to be unknown.

Not only is the relationship of plasma levels of T_3 to thermogenesis during fasting not a simple one, but the relationship of T_3 levels to amount and composition of the diet in the fed or overfed state is rather complex. Thus, re-feeding a fasting rat with carbohydrate or protein will elevate T_3 levels fairly rapidly, whereas refeeding with fat has no effect on T_3 levels (Burger *et al.*, 1980, 1981; Harris *et al.*, 1978). Feeding a high fat diet to fed rats actually brings about a reduction in T_3 level (Otten *et al.*, 1980). Moreover, over-feeding a palatable mixed diet raises T_3 levels, whereas overfeeding a high fat diet does not (Danforth *et al.*, 1979). Thus, in considering the metabolic consequences of overfeeding, particularly the alteration in energy balance which leads to obesity, it is necessary to consider the composition of the palatable diet provided and the effect of that composition on both T_3 levels and sympathetic activity. A diet made palatable to rats by mixing chow with fat, and inducing them to overeat and become obese, would be expected to have less effect both on T_3 levels (Danforth *et al.*, 1979; Burger *et al.*, 1980, 1981) and on sympathetic activity (Landsberg and Young, 1981*a*) than a palatable diet of the supermarket or cafeteria kind which also provides excess carbohydrate. The elevated metabolic rate of rats that are overeating a cafeteria diet is associated with elevated T_3 levels (Rothwell, Stock, 1979, 1981*a*, 1981*c*; Tulp, Frink and Danforth, 1982) and refeeding fasting rats a large meal of carbohydrate does result in both a large increase in T_3 and an increase in metabolic rate (Rothwell, Saville and Stock, 1982). However, both the elevated metabolic rate of cafeteria fed rats (Rothwell and Stock, 1979) and the increases in metabolic rate and in T_3 level in refed rats (Rothwell *et al.*, 1982) are abolished by the β-adrenergic blocking agent, propranolol. This effect of refeeding on metabolic rate and T_3 levels can be mimicked with single subcutaneous injection of noradrenaline, suggesting sympathetic mediation of both processes.

The protein content and composition of the diet also has important effects on thermogenesis, which are also associated with altered T_3 levels. A low protein diet results in increased thermogenesis, an increased level of T_3, and a decreased efficiency of energy retention and leanness (Tyzbir *et al.*, 1981; Tulp *et al.*, 1979*a,b*; Glass *et al.*, 1978). Conversely, a high protein diet reduces T_3 levels and promotes metabolic efficiency and obesity (Rabolli and Martin, 1977; Donald, Pitts and Pohl, 1981). The relationships between caloric intake, protein intake, protein composition and T_3 level are complex; even a diet low in a single amino acid (valine) can raise T_3 levels (Glass *et al.*, 1978). By analogy with the effects of cold acclimation and of fasting/feeding/overfeeding on T_3 levels and the relative unimportance of changes in these levels in bringing about the changes in overall thermogenesis, it might be predicted that the protein content and composition of the diet will influence thermogenesis via an alteration in the activity of the sympathetic nervous system, possibly via a modulation of peripheral and brain neurotransmitters (Anderson, 1979; Pardridge, 1977; Agharanya, Alonso and Wurtman, 1981), and that the changes in T_3 levels are only incidental, playing at most a permissive role.

Alterations in thermogenesis in brown adipose tissue play a major role in at least some of the changes in overall thermogenesis brought about by fasting/feeding. This major role has been demonstrated quantitatively for the transition from the fed state to the overfed state, a transition associated with growth of brown adipose tissue (Rothwell and Stock, 1979, 1981*b*). A reduction in brown adipose tissue thermogenesis may be presumed to occur during the transition from the fed state to the fasting state. The increased capacity of rats fed a low protein diet to respond to noradrenaline by an increase in metabolic rate (Stirling and Stock, 1968) suggests a growth of brown adipose tissue under these conditions and this has been confirmed using low protein cafeteria diets which produce much larger increases in brown adipose tissue mass and its mitochondrial proton conductance pathway than normal cafeteria diets (Rothwell, Stock and Tyzbir, 1982). The role of the simultaneous changes in T_3 levels in bringing about diet-induced alterations in thermogenic functioning of brown adipose tissue is uncertain but is probably minor, by analogy with the minor role of an increase in T_3 level in bringing about changes in brown adipose tissue in the cold-acclimated state (Section 5.2.3 (c)). The altered T_3 levels may, however, play some part in determining the responsiveness of brown adipose tissue to catecholamine; this responsiveness would be predicted to be increased in the fed/overfed state and in association with a low protein diet and to be decreased in the fasting state.

There is some evidence for an influence of T_3 on brown adipose tissue in relation to fasting/feeding/overfeeding. Similar increases in the thyroid-sensitive enzyme, Na^+K^+-ATPase, occur in hyperthyroid and in cafeteria fed rats and similar decreases occur in hypothyroid and in fasting rats (Rothwell *et al.*, 1981*a,c*), changes which can be correlated with the altered levels of T_3 in the blood. An increase in hepatic Na^+K^+-ATPase is also reported to occur in overfed mice (Flier, Usher and De Luise, 1981). It is unlikely that these changes in Na^+K^+-ATPase activity contribute appreciably to the altered thermogenesis in the tissue, since even the highest activity measured in brown adipose tissue can be calculated to contribute no more than a few μl of oxygen per minute in an animal consuming several ml per minute (Rothwell *et al.*, 1981*a*). They are most probably indicative of an altered responsiveness of the plasma membrane to noradrenaline, possibly mediated by T_3.

In conclusion, the role of thyroid hormone in the changes in thermogenesis brought out by changes in amount or composition of the diet would appear to be at most a permissive one, modulating the responsiveness, primarily of brown adipose tissue, to catecholamines. Thus, the decreased level of T_3 in the fasting state and the increased level in the fed state do not themselves exert a direct effect on thermogenesis. The observed changes in thermogenesis are, rather, brought about by the simultaneous alteration in activity of the sympathetic nervous system. One might logically then ask what is the function of the changes in T_3 level in response to diet, if it is not to promote alterations in thermogenesis. An interesting possibility, worthy of further investigation, has been proposed by Cahill (1981), who suggests that the reduction in T_3 level during fasting reduces proteolysis in muscle, itself stimulated by the lowered insulin levels, and is thus responsible for the adaptation that conserves muscle protein when the capacity for ketone body production reaches a level sufficient to supply much of the brain's needs and thus partially to replace gluconeogenesis. Indeed, administration of thyroid hormone to fasting obese subjects (Bray, Melvin and Chopra, 1973; Cahill, 1981; Burman *et al.*, 1979) or to obese subjects on a hypocaloric diet (Wilson *et al.*, 1981) in attempts to reverse the decrease in metabolic rate and thus accelerate weight loss, promotes loss of lean body tissue as well as loss of body fat. The role of diet-induced changes in T_3 level can then be viewed as a part of the function of the thyroid gland in the development, growth and maintenance of tissues, rather than as a part of any function it might have in the control of thermogenesis.

(c) Obesity

The frequent association of obesity with a low level of thermogenesis is dealt with elsewhere in this volume (see Chapter 8). Extension of the Edelman hypothesis for thyroid hormone action to the reduced thermogenesis of obesity has led to extensive studies of Na^+K^+-ATPase activity in various tissues of several different animal models of obesity (Romsos, 1981*a,b*; Bray, York, 1979; Bray *et al.*, 1978; Lin *et al.*, 1978, 1979*a*, 1980, 1981; York *et al.*, 1978*a,b*; Knehans, Romsos and Leveille, 1980; Guernsey and Morishige, 1979; Vander Tuig *et al.*, 1979, 1981) as well as in tissues of human subjects (De Luise, Blackburn and Flier, 1980; Bray, Kral and Björntorp, 1981) and to claims that reduced hydrolysis of ATP for the pumping of sodium and potassium is the cause of the reduced thermogenesis in obesity (e.g. De Luise *et al.*, 1980). The discussion in this chapter will be limited mainly to studies of the genetically obese (ob/ob) mouse since this animal has been the most studied from the point of view of thyroid activity and of the role of altered thyroid function in Na^+K^+-ATPase activity of various tissues and in metabolism of brown adipose tissue.

The ob/ob mouse has a lower resting metabolic rate than normal lean mice at all temperatures except thermoneutral (30–33°C for the mouse) (Trayhurn and James, 1978; Boissoneault *et al.*, 1978) and is extremely cold-sensitive, becoming rapidly hypothermic and dying in about three hours when kept at 4°C (Davis and Mayer, 1954; Trayhurn and James, 1978). The lower resting metabolic rate of the ob/ob mouse has suggested that it might be hypothyroid. However, extensive investigations have failed to demonstrate any major hypothalamic-pituitary-thyroid abnormality (see Bray and York, 1979). The level of T_3 in the blood of the ob/ob mouse is normal or above normal throughout most of its life (Mobley and Dubuc, 1979; Ohtake, Bray and Azukizawa, 1977; York *et al.*, 1978*a*; Naltchayan, Bouhnik and Michel, 1981) although a mild hypothyroid state (low T_3 level) may occur at a very early age (Mobley and Dubuc, 1979; Van der Kroon and Boldewijn, 1980; Van der Kroon and Speijers, 1979). The T_3 level of the adult ob/ob mouse decreases in a normal way on fasting (Naltchayan *et al.*, 1981). Whether it changes in response to cold exposure is not known.

It is possible that the ob/ob mouse fails to respond to the normal level of T_3 in its blood by an increase in metabolic rate. These mice do have an exaggerated increase in metabolic rate in response to administration of a dose of thyroid hormone that is without effect in normal lean mice (Lin *et al.*,

1978c; Vander Tuig *et al.*, 1979; Hogan and Himms-Hagen, 1981) suggesting that they may be partly resistant to the thermogenic effect of their own T_3. Treatment of ob/ob mice with thyroid hormone does improve their cold resistance to some extent; however, their survival rate is only slightly prolonged (Ohtake *et al.*, 1977; Bégin-Heick and Heick, 1977; Thenen and Carr, 1980; Mayer and Barrnett, 1953).

An explanation for the thyroid-related thermogenic defect in the ob/ob mouse has been sought via two approaches: one considers the possible role of thyroid hormone in the defective functioning of brown adipose tissue, the other considers the role of defective control of tissue Na^+K^+-ATPases. The major thermogenic defect in the ob/ob mouse can be attributed to a failure of noradrenaline-mediated, cold-induced non-shivering thermogenesis and diet-induced thermogenesis in brown adipose tissue (Himms-Hagen and Desautels, 1978; Hogan and Himms-Hagen, 1980, 1981; Thurlby and Trayhurn, 1980; Trayhurn *et al.*, 1982; see Chapter 8). It might be argued that the ob/ob mouse fails to secrete an appropriate amount of nor-adrenaline in its brown adipose tissue on cold exposure. However, the noradrenaline content of brown adipose tissue is more or less normal in the ob/ob mouse (Knehans and Romsos, 1982; Zaror-Behrens and Himms-Hagen, 1983) and noradrenaline is secreted normally on exposure to 4°C (Zaror-Behrens and Himms-Hagen, 1983). It follows that if thyroid hormone treatment improves cold resistance of the ob/ob mouse it should also improve the defective functioning of its brown adipose tissue. This is indeed the case, since treatment of the ob/ob mouse with thyroid hormone permits a more normal thermogenic response of the mouse to noradrenaline and of its brown adipose tissue mitochondria to cold exposure (as measured by an increase in GDP binding) (Hogan and Himms-Hagen, 1981). In view of the lack of any role for thyroid hormone in the control of brown adipose tissue function and growth, other than one which permits the action of noradrenaline (see Section 5.2.3 (c)), the effect of thyroid hormone on brown adipose tissue of the ob/ob mouse is interpreted as bringing about an improvement of the responsiveness of the tissue to noradrenaline which is secreted in response to cold or to diet.

The reduced activity of Na^+K^+-ATPase in liver (Bray *et al.*, 1978; Lin *et al.*, 1979b; York *et al.*, 1978a,b; Guernsey and Morishige, 1979), muscle (Lin *et al.*, 1978, 1979b, 1980) and brown adipose tissue (Knehans *et al.*, 1980; Rothwell *et al.*, 1981a) suggests a partial resistance of the ob/ob mouse to the normal level of T_3 in its blood. That T_3 fails to increase Na^+K^+-ATPase activity in tissues (Bray *et al.*, 1978; York *et al.*, 1978a,b; Shimomura, Bray and York, 1981) even though it increases the amount of enzyme (as measured by oubain binding sites) (Lin *et al.*, 1979a; Vander Tuig *et al.*, 1979) also suggests an impairment of responsiveness to T_3. The finding of a reduced number of nuclear T_3 receptors in liver (Guernsey and Morishige, 1979) would be in keeping with this suggestion. The significance of the decreased tissue Na^+K^+-ATPase for the reduced thermogenesis of the ob/ob mouse is not clear. As noted above, the defect in the ob/ob mouse is in facultative thermogenesis, specifically the noradrenaline-mediated cold-induced non-shivering thermogenesis and diet-induced thermogenesis which occur primarily in brown adipose tissue and do not involve altered tissue Na^+K^+-ATPase activity except as a small part of the noradrenaline-induced response of the tissue. The exaggerated response of the resting oxygen consumption of the ob/ob mouse to thyroid hormone may, however, be in part due to altered functioning of Na^+K^+-ATPase in other tissues.

An attempt to use ouabain to determine the contribution of the sodium pump to cold-induced thermogenesis in intact mice has shown a similar absolute and proportional reduction in heat production at thermoneutrality but the absolute reduction is greater in lean mice at 14°C than in ob/ob mice (Lin *et al.*, 1981). However, even assuming that in the intact animal ouabain inhibits only the Na pump and causes no other metabolic changes, either directly or as a consequence of ionic changes, it is known that ouabain will inhibit the effect of noradrenaline on brown adipose tissue (Horwitz, 1973). Since brown adipose tissue makes a much greater contribution to cold-induced non-shivering thermogenesis in the lean mouse than in the ob/ob mouse at 14°C (Thurlby and Trayhurn, 1980), the results can be interpreted in terms of an inhibition of noradrenaline-induced thermogenesis in brown adipose tissue, rather than in terms of a thermogenic contribution of the Na^+K^+-ATPases of other tissues.

It can be concluded that the ob/ob mouse is partially refractory to at least some actions of the normal level of T_3 in its blood and that this leads to a refractoriness of its brown adipose tissue to noradrenaline. This would result in a low capacity for sympathetic-mediated facultative thermogenesis (both cold-induced non-shivering thermogenesis and diet-induced thermogenesis) in brown adipose tissue and thus to obesity. The reduction in Na^+K^+-ATPase in many thyroid-sensitive tissues is also indicative of a partial refractoriness to some actions of T_3 but probably has only a small effect on overall thermogenesis. Other actions of T_3, in particular those involved in development and growth, appear to be relatively normal in the ob/ob mouse; this animal is not the equivalent of the thyroidectomized animal, which fails to grow and develop normally. One might ask whether a partial refractoriness T_3 is close to the primary genetic defect in the ob/ob mouse, leading to the other multiple metabolic and endocrinological abnormalities. This seems unlikely in view of the failure of administered thyroid hormone to normalize completely the metabolic abnormalities (Ohtake *et al.,* 1977; Bégin-Heick nd Heick, 1977; Hogan and Himms-Hagen, 1981; Thenen and Carr, 1980). However, it remains possible that the consequences of a partial refractoriness to T_3 might not be reversed to normal by treatment with thyroid hormone because the treated animals would be euthyroid in one sense (the thermogenic effect of T_3) but rendered hyperthyroid in another sense (the growth and developmental effects of T_3) and thus not metabolically normal. The most probable nature of the genetic defect in the ob/ob mouse is one which finds its expression in multiple membrane abnormalities (Chang, Huang and Cuatrecasas, 1975) which manifest themselves in various ways, including refractoriness to certain actions of T_3 and of catecholamines.

In view of the conclusion outlined above about the thyroid-related nature of the metabolic defect in the ob/ob mouse, a brief comparison of the fragmentary evidence available for man is of interest. A recent report that Na^+K^+-ATPase of red blood cells is reduced in obese human subjects (De Luise *et al.*, 1980) and that the extent of its reduction is inversely related to the degree of obesity in Pima Indians (Klimes *et al.*, 1982) suggests a similarity between obesity in man and in the ob/ob mouse. However, reduced Na^+K^+-ATPase in white adipose tissue of obese subjects is inversely related to their insulin levels and an effect of insulin to repress this enzyme appears likely (Belfiore *et al.*, 1981). As noted above, the reduced Na^+K^+-ATPase of tissues of the ob/ob mouse is thought not to be a contributory cause of their obesity and it is probable that the same applies to man. Moreover, since hyperthyroid human subjects also have reduced Na^+K^+-ATPase in their red blood cells (Cole and Waddell, 1976; DeLuise and Flier, 1983) it is clear that in man there is no direct relationship between this enzyme in red cells and thermogenesis. In addition, since obese human subjects may have increased Na^+K^+-ATPase in their liver (Bray *et al.*, 1981), they obviously do not resemble the ob/ob mouse in this respect.

It remains a possibility, however, that a partial refractoriness to T_3 might occur in obese humans, as in the ob/ob mouse, and that a consequent reduced responsiveness their brown adipose tissue to catecholamines might so reduce their capacity for sympathetic-mediated facultative thermogenesis as to lead to a high metabolic efficiency and obesity. (For further discussion of the applicability of the concept of brown adipose tissue thermogenesis as an energy buffer to human obesity see Chapter 8).

5.3 SUMMARY

The major role of thyroid hormones in the regulation of overall heat production lies in the control of endothermic thermogenesis, one of the two components of obligatory thermogenesis. It exerts this control by reacting with both nuclear and mitochondrial T_3 receptors in most body tissues to bring about changes in the properties and overall mass of mitochondria and in the activity of the sodium pump. The physiological role of this action of thyroid hormone is apparently to set obligatory heat production at a suitable level to balance normal heat loss from a warm-blooded animal at thermoneutrality, at rest and in the unfed state.

The maintenance of such an appropriate level of obligatory thermogenesis by thyroid hormone appears to be a necessary background for facultative thermogenesis, influencing in some hitherto unexplained way the responsiveness of tissues, primarily brown adipose tissue, to the thermogenic effect of catecholamines, the secretion of which by the sympathetic nervous system varies directly with and directly influences the thermogenic status during cold exposure, fasting, feeding and overfeeding. The principal role of thyroid hormone in facultative thermogenesis appears then to be a permissive one. The importance of the permissive effect is illustrated by the fact firstly, that in the absence of thyroid hormone animals die when exposed to cold, and secondly, that the genetically obese mouse, in which this permissive effect appears to be specifically defective, becomes obese.

For the present, no explanation can be offered for the significance of the changes in level of circulating T_3 which accompany the thermogenic changes during fasting, feeding, overfeeding and exposure to cold. However, the fact that thyroidectomized rats maintained on a constant minimum daily amount of thyroid hormone increase their heat production in response to cold acclimation indicates that changes in T_3 level are not an essential part of cold-induced facultative thermogenesis. Since thyroidectomized rats maintained on a constant minimum daily amount of T_3 also decrease their heat production in response to fasting, it is also clear that diet-induced changes in T_3 level are not an essential factor in diet-induced facultative thermogenesis.

Many gaps remain in our understanding of endocrine influences on thermogenesis, particularly on thermogenesis in brown adipose tissue and on capacity for thermogenesis in brown adipose tissue. Elucidation of the possible role of defective thermogenesis in brown adipose tissue in the aetiology of obesity will require considerably more information about such endocrine influences and their interactions with the sympathetic nervous control of this tissue.

REFERENCES

Agharanya, J., Alonso, R. and Wurtman, J. (1981) Changes in catecholamine excretion after short-term tyrosine ingestion in normally fed human subjects. *Am. J. Clin. Nutr.,* **34,** 82–7.

Anderson, G.H. (1979) Control of protein and energy intake: role of plasma amino acids and brain neurotransmitters. *Can. J. Physiol. Pharmacol.,* **57,** 1043–57.

Asano, Y., Liberman, U.A. and Edelman, I.S. (1976) Thyroid thermogenesis. Relationships between Na^+-dependent respiration and $Na^+ + K^+$-adenosine triphosphatase activity in rat skeletal muscle. *J. Clin. Invest.,* **57,** 368–79.

Axelrod, J. (1975) Relationship between catecholamines and other hormones. *Recent Prog. Horm. Res.,* **31,** 1–27.

Babior, B.M., Creagan, S., Ingbar, S.H. and Kipnes, R.S. (1973) Stimulation of mitochondrial adenosine diphosphate uptake by thyroid hormones. *Proc. Nat. Acad. Sci. USA,* **70,** 98–102.

Balsam, A. and Sexton, F.C. (1975) Increased metabolism of iodothyronines in the rat after short-term cold adaptation. *Endocrinology*, **97**, 385–91.

Balsam, A., Sexton, F. and Ingbar S.H. (1981a) Effects of dietary manipulation on the *in vitro* generation of 3,5,3'-triiodothyronine from thyroxine in rat liver preparations. *Life Sci.*, **28**, 1727–36.

Balsam, A., Sexton, F. and Ingbar, S.H. (1981b) The influence of fasting and the thyroid state on the activity of thyroxine 5'-monodeiodinase in rat liver: a kinetic analysis of microsomal formation of triiodothyronine from thyroxine. *Endocrinology*, **108**, 472–7.

Baudry, M., Clot, J.-P. and Michel, R. (1975). Influence de l'état thyroidien sur le profil électrophorétique des protéines mitochondriales de foie de rat. *Biochimie*, **57**, 77–83.

Bégin-Heick, N. and Heick, H.M.C. (1977) Increased response of adipose tissue of the ob/ob mouse to the action of adrenaline after treatment with thyroxin. *Can. J. Physiol. Pharmacol.*, **55**, 1320–9.

Belfiore, F., Iannello, S., Rabuazzo, A.M. and Borzi, V. (1981) The activity of sodium and potassium-activated adenosine triphosphatase (NaK-ATPase) in the adipose tissue of obese patients. in *Obesity: Pathogenesis and Treatment* (ed. G. Enzi, G. Crepaldi, G. Pozza and A.E. Renold), Academic Press, pp. 129–34.

Bernal, J. and Escobar Del Rey, F. (1975a) T_3/T_4 Ratios and α-glycerophosphate activity in intact rats exposed to a cold environment. *Horm. Metab. Res.*, **7**, 222–7.

Bernal, J. and Escobar Del Rey, F. (1975b) Effect of the exposure to cold on the extrathyroidal conversion of L-thyroxine to triiodo-L-thyronine, and on intramitochondrial α-glycerophosphate dehydrogenase activity in thyroidectomized rats on L-thyroxine. *Acta Endocrinol.*, **78**, 481–92.

Biron, R., Burger, A., Chinet, A., Clausen, T. and Dubois-Ferrière, R. (1979) Thyroid hormones and the energetics of active sodium-potassium transport in mammalian skeletal muscles. *J. Physiol.*, **297**, 47–60.

Bissell, D.M., Hammaker, L.E. and Meyer, U.A. (1973) Parenchymal cells from adult rat liver in nonproliferating monolayer culture. I. Functional studies. *J. Cell. Biol.*, **59**, 722–34.

Boissonneault, G.A., Hornshuh, M.J., Simons, J.W. Romsos, D.R and Leveille, G.A. (1978) Oxygen consumption and body fat content of young lean and obese (ob/ob) mice. *Proc. Soc. Exp. Biol. Med.*, **157**, 402–6.

Bouhnik, J., Clot, J.-P., Baudry, M. and Michel, R. (1979) Early effects of thyroidectomy and triiodothyronine administration on rat liver mitochondria. *Mol. Cell. Endocrinol.*, **15**, 1–12.

Bray, G.A., Kral, J.G. and Björntorp, P. (1981) Hepatic sodium-potassium-dependent ATPase in obesity. *New Engl. J. Med.*, **304**, 1580–2.

Bray, G.A., Melvin, K.E.W. and Chopra, I.J. (1973) Effect of triiodothyronine on some metabolic responses of obese patients. *Am. J. Clin. Nutr.*, **26**, 715–21.

Bray, G.A. and York, D.A. (1979) Hypothalamic and genetic obesity in experimental animals: an autonomic and endocrine hypothesis. *Physiol. Rev.*, **59**, 719–809.

Bray, G.A., York, D.A. and Yukimura, Y. (1978) Activity of $(Na^+ + K^+)$-ATPase in the liver of animals with experimental obesity. *Life Sci.*, **22**, 1637–42.

Brooks, S.L., Rothwell, N.J, Stock, M.J., Goodbody, A.E. and Trayhurn, P. (1980) Increased proton conductance pathway in brown adipose tissue mitochondria of rats exhibiting diet-induced thermogenesis. *Nature (London)*, **285**, 274–6.

Bukowiecki, L., Collet, A.J., Folléa, N., Guay, G. and Jahjah, L. (1982) Brown adipose tissue hyperplasia: a fundamental mechanism of adaptation to cold and hyperphagia. *Am. J. Physiol.*, **242**, E353–9.

Bukowiecki, L., Folléa, N., Vallières, J. and Leblanc, J. (1978) β-Adrenergic receptors in brown adipose tissue. Characterization and alterations during acclimation of rats to cold. *Eur. J. Biochem.*, **92**, 189–96.

Bukowiecki, L., Folléa, N., Paradis, A. and Collet, A.J. (1980) Stereospecific stimulation of brown adipose adipocyte respiration by catecholamines via β_1-adrenoreceptors. *Am. J. Physiol.*, **238**, E552–63.

Burger, A.G., Berger, M., Wimpfheimer, K. and Danforth, E. (1980) Inter-relationships between energy metabolism and thyroid hormone metabolism during starvation in the rat. *Acta Endocrinologica*, **93**, 322–31.

Burger, A.G., Hughes, J.N. and Saville, E. (1981) Starvation and thyroid function: effects on thermogenesis and serum thyrotropin. *Life Sci.*, **28**, 1737–44.

Burman, K.D., Lukes, Y., Wright, F.D. and Wartofsky, L. (1977) Reduction in hepatic triiodothyronine binding capacity induced by fasting. *Endocrinology*, **101**, 1331–4.

Burman, K.D., Wartofsky, L., Dinterman, R.E., Kesler, P. and Wannemacher, R.W. jr. (1979) The effect of T_3 and reverse T_3 administration on muscle protein catabolism during fasting as measured by 3-methylhistidine excretion. *Metabolism*, **28**, 805–13.

Cahill, G.F., jr. (1981) Role of T_3 in fasted man. *Life Sci.*, **28**, 1721–6.

Cannon, B., Nedergaard, J. and Sundin, U. (1981) Thermogenesis, brown fat and thermogenin. in *Survival in the Cold: Hibernation and Other Adaptations* (eds X.J. Musacchia and L. Jansky), Elsevier North Holland, Inc., Amsterdam, pp. 99–120.

Chang, K-J., Huang, D. and Cuatrecasas, P. (1975) The defect in insulin receptors in obese-hyperglycemic mice: a probable accompaniment of more generalized alterations in membrane glycoproteins. *Biochem. Biophys. Res. Commun.*, **64**, 566–73.

Chen, Y-D.I. and Hoch, F.L. (1977) Thyroid control over biomembranes. Rat liver mitochondrial inner membranes. *Arch. Biochem. Biophys.*, **181**, 470–83.

Chinet, A., Clausen, T. and Girardier, L. (1977) Microcalorimetric determination of energy expenditure due to active sodium-potassium transport in the soleus muscle and brown adipose tissue of the rat. *J. Physiol.*, **265**, 43–61.

Ciaraldi, T.P. and Marinetti, G.V. (1978) Hormone action at the membrane level VIII. Adrenergic receptors in rat heart and adipocytes and their modulation by thyroxine. *Biochim. Biophys. Acta*, **541**, 334–46.

Cole, C.H. and Waddell, R.W. (1976) Alteration in intracellular sodium concentration and ouabain-sensitive ATPase in erythrocytes from hyperthyroid patients. *J. Clin. Endocrinol. Metab.*, **42**, 1056–63.

Courtright, J.B. and Fitts, R.H. (1979) Effects of thyrotoxicosis on mitochondrial enzymes of rat soleus. *Horm. Metab. Res.*, **11**, 304–6.

Danforth, E., jr. and Burger, A.G. (1981) Hormonal control of thermogenesis. in *The Body Weight Regulatory System: Normal and Disturbed Mechanisms* (eds

L.A. Cioffi, W.P.T. James and T.B. van Itallie), Raven Press, New York, pp. 107–14.

Danforth, E., jr., Horton, E.S., O'Connell, M., Sims, E.A.H., Burger, A.G., Ingbar, S.H., Braverman, L. and Vagenakis, A.G. (1979) Dietary-induced alterations in thyroid hormone metabolism during overnutrition. *J. Clin. Invest.,* **64**, 1336–47.

Danforth, E. jr., Horton, E.S. and Sims, E.A.H. (1981) Nutritionally-induced alterations in thyroid hormone metabolism. in *Nutritional Factors: Modulating Effects on Metabolic Processes* (eds R.F. Beers, jr. and E.G. Bassett), Raven Press, New York, pp. 139–53.

Davis, T.R.A. and Mayer, J. (1954) Imperfect homeothermy in the hereditary obese hyperglycemic syndrome of mice. *Am. J. Physiol.,* **177**, 222–6.

Degroot, L.J., Coleoni, A.H., Rue, P.A., Seo, H., Martino, E. and Refetoff, S. (1977) Reduced nuclear triiodothyronine receptors in starvation-induced hypothyroidism. *Biochem. Biophys. Res. Commun.,* **79**, 173–8.

De Leo, T., Di Meo, S., Barletta, A., Martino, G. and Goglia, F. (1976) Modification of nucleic acid levels per mitochondrion induced by thyroidectomy or triiodothyronine administration. *Pflügers Arch.,* **366**, 73–7.

De Luise, M., Blackburn, G.L. and Flier, J.S. (1980) Reduced activity of the red-cell sodium-potassium pump in human obesity. *New Engl. J. Med.,* **303**, 1017–22.

De Luise, M. and Flier, J.S. (1983) Status of the red cell, Na, k-pump in hyper- and hypothyroidism. *Metabolism,* **32**, 25–30.

Denckla, W.D. (1973) Minimal O_2 consumption as an index of thyroid status: standardization of method. *Endocrinology,* **93**, 61–73.

Denckla, W.D. (1974) Role of the pituitary and thyroid glands in the decline of minimal O_2 consumption with age. *J. Clin. Invest.,* **53**, 572–81.

Denckla W.D. and Bilder, G.E. (1975) Investigations into the hypermetabolism of pregnancy, lactation and cold-acclimation. *Life Sci.,* **16**, 403–14.

Desautels, M. and Himms-Hagen, J. (1979) Roles of noradrenaline and protein synthesis in the cold-induced increase in purine nucleotide binding by rat brown adipose tissue mitochondria. *Can. J. Biochem.,* **57**, 968–76.

Desautels, M. and Himms-Hagen, J. (1980) Parallel regression of cold-induced changes in ultrastructure, composition and properties of brown adipose tissue mitochondria during recovery of rats from acclimation to cold. *Can. J. Biochem.,* **58**, 1057–68.

Desautels, M. and Himms-Hagen, J. (1980) Brown adipose tissue mitochondria of cold-acclimated rats: change in characteristics of purine nucleotide control of the proton electrochemical gradient. *Can. J. Biochem.,* **59**, 619–25.

Desautels, M., Zaror-Behrens, H. and Himms-Hagen, J. (1978) Increased purine nucleotide binding, altered polypeptide composition and thermogenesis in brown adipose tissue mitochondria of cold-acclimated rats. *Can. J. Biochem.,* **56**, 378–83.

Donald, P., Pitts, G.C. and Pohl, S.L. (1981) Body weight and composition in laboratory rats: effects of diets with high or low protein concentrations. *Science,* **211**, 185–6.

Edelman, I.S. (1974) Thyroid thermogenesis. *New Engl. J. Med.,* **290**, 1303–8.

Edelman, I.S. (1976) Transition from the poikilotherm to the homeotherm: possible role of sodium transport and thyroid hormone. *Fed. Proc.,* **35**, 2180–4.

Eisenstein, Z., Hagg, S., Vagenakis, A.G., Fang, S.L., Ransil, B., Burger, A., Balsam, A., Braverman, L.E. and Ingbar, S.H. (1978) Effect of starvation on the production and peripheral metabolism of 3,3'5'-triiodothyronine in euthyroid obese subjects. *J. Clin. Endocrin. Metab.*, **47**, 889–93.

Fain, J.N., Jacobs, M.D. and Clément-Cormier, Y.C. (1973) Interrelationship of cyclic AMP, lipolysis, and respiration in brown fat cells. *Am. J. Physiol.*, **224**, 346–51.

Fain, J.N. and Rosenthal, J.W. (1971) Calorigenic action of triiodothyronine on white fat cells: effects of ouabain, oligmycin, and catecholamines. *Endocrinology*, **89**, 1205–11.

Flaim, K.E., Horwitz, B.A. and Horowitz, J.M. (1977) Coupling of signals to brown fat: α- and β-adrenergic responses in intact rats. *Am. J. Physiol.*, **232**, R101–9.

Flier, J.S., Usher, P. and De Luise, M. (1981) Effect of sucrose overfeeding on Na, K-ATPase-mediated ^{86}Rb uptake in normal and obese mice. *Diabetes*, **30**, 975–8.

Folke, M. and Sestoft, L. (1977) Thyroid calorigenesis in isolated, perfused rat liver: minor role of active sodium-potassium transport. *J. Physiol*, **269**, 407–19.

Foster, D.O. and Frydman, M.L. (1978a) Comparison of microspheres and [86]Rb[+] as tracers of the distribution of cardiac output in rats indicates invalidity of [86]Rb[+]-based measurements. *Can. J. Physiol. Pharmacol.*, **56**, 97–109.

Foster, D.O. and Frydman, M.L. (1978b) Nonshivering thermogenesis in the rat. II. Measurements of blood flow with microspheres point to brown adipose tissue as the dominant site of the calorigenesis induced by noradrenaline. *Can. J. Physiol. Pharmacol.*, **56**, 110–22.

Foster, D.O. and Frydman, M.L. (1979) Tissue distribution of cold-induced thermogenesis in conscious warm- or cold-acclimated rats reevaluated from changes in tissue blood flow: The dominant role of brown adipose tissue in the replacement of shivering by nonshivering thermogenesis. *Can. J. Physiol. Pharmacol.*, **57**, 257–70.

Fregly, M.J., Field, F.P., Katovich, M.J. and Barney, C.C (1979) Catecholamine-thyroid hormone interaction in cold-acclimated rats. *Fed. Proc.*, **38**, 2162–9.

Gibson, A. (1981) The influence of endocrine hormones on the autonomic nervous system. *J. Auton. Pharmacol.*, **1**, 331–58.

Girardier, L. (1977) The regulation of the biological furnace of warm blooded animals. *Experientia*, **33**, 1121–2.

Girardier, L. (1981) Brown adipose tissue as energy dissipator: a physiological approach. In *Obesity: Pathogenesis and Treatment* (eds G. Enzi, G. Crepaldi, G. Pozza and A.E. Renold), Academic Press, London, pp. 55–72.

Giudicelli, Y. (1978) Thyroid-hormone modulation of the number of β-adrenergic receptors in rat fat-cell membranes. *Biochem. J.*, **176**, 1007–10.

Giudicelli, Y., Lacasa, D. and Agli, B. (1980) White fat cell α-adrenergic receptors and responsiveness in altered thyroid status. *Biochem. Biophys. Res. Commun.*, **94**, 1113–22.

Glass, A.R., Mellitt, R., Burman, K.D., Wartofsky, L. and Swedloff, R.S. (1978) Serum triiodothyronine in undernourished rats: dependence on dietary composition rather than total calorie or protein intake. *Endocrinology*, **102**, 1925–8.

Goodbody, A.E. and Trayhurn, P. (1981) GDP binding to brown-adipose-tissue mitochondria of diabetic-obese (db/db) mice. Decreased binding in both the obese and pre-obese states. *Biochem. J.*, **194**, 1019–22.

Gripois, D., Klein, C. and Valens, M. (1980) Influence of hypo- and hyperthyroidism on noradrenaline metabolism in brown adipose tissue of the developing rat. *Biol. Neonate*, **37**, 53–9.

Guder, W.G. (1979) Stimulation of renal gluconeogenesis by angiotensin II. *Biochim. Biophys. Acta*, **584**, 507–19.

Guernsey, D.L. and Morishige, W.K. (1979) Na^+ pump activity and nuclear T_3 receptors in tissues of genetically obese (ob/ob) mice. *Metabolism*, **28**, 629–32.

Guernsey, D.L. and Stevens, E.D. (1977) The cell membrane sodium pump as a mechanism for increasing thermogenesis during cold acclimation in rats. *Science*, **196**, 908–10.

Haidmayer, R. and Hagmüller, K. (1981) Influence of ambient temperature on calorigenic action of thyroid hormones in young mice. *Pflügers Arch.*, **391**, 125–8.

Hamilton, J. and Horwitz, B.A. (1977) Adrenergic- and cyclic nucleotide-induced glycerol release from brown adipocytes. *Eur. J. Pharmacol.*, **56**, 1–5.

Hardeveld, C. Van and Kassenaar, A.A.H. (1981) Evidence that the thyroid state influences Ca^{++}-mediated metabolic processes in perfused skeletal muscle. *Horm. Metab. Res.*, **13**, 33–7.

Hardeveld, C. Van, Zuidwijk, M.J. and Kassenaar, A.A.H. (1979*a*) Studies on the origin of altered thyroid hormone levels in the blood of rats during cold exposure. I. Effect of iodine intake and food consumption. Acta Endocrinol., **91**, 473–83.

Hardeveld, C. Van, Zuidwijk, M.J. and Kassenaar, A.A.H. (1979*b*) Studies on the origin of altered thyroid hormone levels in the blood of rats during cold exposure. II. Effect of propranolol and chemical sympathectomy. *Acta Endocrinol.*, **91**, 484–92.

Harris, A., Fang, S.L., Vagenakis, A.G. and Braverman, L.E. (1978) Effect of starvation, nutriment replacement, and hypothyroidism on *in vitro* hepatic T_4 to T_3 conversion in the rat. *Metabolism*, **27**, 1680–90.

Heaton, G.M., Wagenvoord, R.J., Kemp, A., jr, and Nicholls, D.G. (1978) Brown adipose tissue mitochondria: photoaffinity labelling of the regulatory site of energy dissipation. *Eur. J. Biochem.*, **82**, 515–21.

Hefco, E., Krulich, L., Illner, P. and Larsen, P.R. (1975) Effect of acute exposure to cold on the activity of the hypothalamic-pituitary-thyroid system. *Endocrinology*, **97**, 1185–95.

Heick, H.M.C., Vachon, C., Kallai, M.A., Bégin-Heick, N. and Leblanc, J. (1977) The effects of thyroxine and isopropyl-noradrenaline on cytochrome oxidase activity in brown adipose tissue. *Can. J. Physiol. Pharmacol.*, **51**, 751–8.

Héroux, O. (1968) Thyroid parameters and metabolic adaptation to cold in rats fed a low-bulk thyroxine-free diet. *Can. J. Physiol. Pharmacol.*, **46**, 843–6.

Himms-Hagen, J. (1976) Cellular thermogenesis. *Annu. Rev. Physiol.*, **38**, 315–51.

Himms-Hagen, J. (1981) Nonshivering thermogenesis, brown adipose tissue and obesity. in *Nutritional Factors: Modulating Effects on Metabolic Processes* (eds R.F. Beers, jr. and E.G. Bassett), Raven Press, New York, pp. 85–99.

Himms-Hagen, J., Cerf, J., Desautels, M. and Zaror-Behrens, G. (1978) Thermogenic mechanisms and their control. in *Effectors of Thermogenesis* (eds L. Girardier and J. Seydoux), Birkhäuser Verlag, Basel, pp. 119–34.

Himms-Hagen, J. and Desautels, M. (1978) A mitochondrial defect in brown adipose tissue of the obese (ob/ob) mouse: reduced binding of purine nucleotides and a failure to respond to cold by an increase in binding. *Biochem. Biophys. Res. Commun.*, **83**, 628–34.

Himms-Hagen, J., Dittmar, E. and Zaror-Behrens, G. (1980) Polypeptide turnover in brown adipose tissue mitochondria during acclimation of rats to cold. *Can. J. Biochem.*, **58**, 336–44.

Himms-Hagen, J., Triandafillou, J. and Gwilliam, C. (1981) Brown adipose tissue of cafeteria-fed rats. *Am. J. Physiol.*, **241**, E116–20.

Ho, R.J., Jeanrenaud, B., Posternak, T.H. and Renold, A.E. (1967) Insulin-like action of ouabain. II. Primary antilipolytic effect through inhibition of adenyl cyclase. *Biochim. Biophys. Acta*, **144**, 74–82.

Hoch, F. (1977) Adenine nucleotide translocation in liver mitochondria of hypothyroid rats. *Arch. Biochem. Biophys.*, **178**, 535–45.

Hogan, S. and Himms-Hagen, J. (1980) Abnormal brown adipose tissue in obese mice (ob/ob): response to acclimation to cold. *Am. J. Physiol.*, **239**, E301–9.

Hogan, S. and Himms-Hagen, J. (1981) Abnormal brown adipose tissue in genetically obese mice (ob/ob): effect of thyroxine. *Am. J. Physiol.*, **241**, E436–43.

Horwitz, B.A. (1973) Ouabain-sensitive component of brown fat therogenesis. *Am. J. Physiol.*, **224**, 352–5.

Horwitz, B.A. (1977) Adrenergic receptor involvement in brown adipose tissue activation. in *Drugs, Biogenic Amines and Body Temperature* (eds K.E. Cooper, P. Lomax and E. Schönbaum), Karger, Basel, pp. 160–6.

Horwitz, B.A. (1978) Neurohumoral regulation of nonshivering thermogenesis in mammals. in *Strategies in Cold: Natural Torpidity and Thermogenesis* (eds L.C. Wang and J.W. Hudson), Academic Press, New York, pp. 619–53.

Horwitz, B.A. (1979a) Metabolic aspects of thermogenesis: neuronal and hormonal control. *Fed. Proc.*, **38**, 2147–9.

Horwitz, B.A. (1979b) Cellular events underlying catecholamine-induced thermogenesis: cation transport in brown adipocytes. *Fed. Proc.*, **38**, 2170–6.

Horwitz, B.A. and Eaton, M. (1975) The effect of adrenergic agonists and cyclic AMP on the Na^+/K^+-ATPase activity of brown adipose tissue. *Eur. J. Pharmacol.*, **34**, 241–5.

Horwitz, B.A. and Eaton, M. (1977) Ouabain-sensitive liver and diaphragm respiration in cold-acclimated hamster. *J. Appl. Physiol.*, **42**, 150–3.

Hsieh, A.C.L. (1962) The role of the thyroid in rats exposed to cold. *J. Physiol.*, **161**, 175–88.

Hsieh, A.C.L. (1963) The basal metabolic rate of cold-adapted rats. *J. Physiol.*, **169**, 851–61.

Hsieh, A.C.L. and Carlson, L.D. (1957) Role of the thyroid in metabolic response to low temperature. *Am. J. Physiol.*, **188**, 40–4.

Hsieh, A.C.L., Pun, C.W., Li, K.M. and Ti, K.W. (1966) Circulatory and metabolic effects of noradrenaline in cold-adapted rats. *Fed. Proc.*, **25**, 1205–9.

Hulbert, A.J. (1978) The thyroid hormones: a thesis concerning their action. *J. Theor. Biol.*, **73**, 81–100.

Hulbert, A.J., Augee, M.L. and Raison, J.K. (1976) The influence of thyroid hormones on the structure and function of mitochondrial membranes. *Biochim. Biophys. Acta*, **455**, 597–601.

Ikemoto, H., Hiroshige, T. and Itoh, S. (1967) Oxygen consumption of brown adipose tissue in normal and hypothyroid mice. *Jap. J. Physiol.*, **17**, 516–22.

Ismail-Beigi, F., Bissell, D.M., Edelman, I.S. (1979) Thyroid thermogenesis in adult rat hepatocytes in primary monolayer culture. Direct action of thyroid hormone *in vitro. J. Gen. Physiol.*, **73**, 369–83.

Ismail-Beigi, F. and Edelman, I.S. (1970) Mechanism of thyroid calorigenesis role of active sodium transport. *Proc. Nat. Acad. Sci. USA*, **67**, 1071–8.

Ismail-Beigi, F. and Edelman, I.S. (1971) The mechanism of the calorigenic action of thyroid hormone. *J. Gen. Physiol.*, **57**, 710–22.

James, W.P.T. and Trayhurn, P. (1981) Obesity in mice and men. in *Nutritional Factors: Modulating Effects on Metabolic Processes.* (eds R.F. Beers, jr. and E.G. Bassett), Raven Press, New York, pp. 123–38.

Jakovcic, S., Swift, H.H., Gross, N.J. and Rabinowitz, M. (1978) Biochemical and stereological analysis of rat liver mitochondria in different thyroid states. *J. Cell Biol.*, **77**, 887–901.

Jéquier, E. and Schutz, Y. (1981) The contribution of BMR and physical activity to energy expenditure. in *The Body Weight Regulatory System: Normal and Disturbed Mechanisms.* (eds L.A. Cioffi, W.P.T. James and T.B. Van Itallie), Raven Press, New York, pp. 89–96.

Johnson, G.E., Flattery, K.V. and Schönbaum, E. (1967) The influence of methimazole on the catecholamine excretion of cold-stressed rats. *Can. J. Physiol. Pharmacol.*, **45**, 415–21.

Jung, R.T., Shetty, P.S. and James, W.P.T. (1980a) The effect of beta-adrenergic blockade on metabolic rate and peripheral thyroid metabolism in obesity. *Eur. J. Clin. Invest.*, **10**, 179–82.

Jung, R.T., Shetty, P.S. and James, W.P.T. (1980b) Nutritional effects on thyroid and catecholamine metabolism. *Clin. Science*, **58**, 183–91.

Kaplan, M.M. (1979) Subcellular alterations causing reduced hepatic thyroxine-5′-monodeiodinase activity in fasted rats. *Endocrinology*, **104**, 58–64.

Kennedy, D.R., Hammond, R.P. and Hamolsky, M.W. (1977) Thyroid cold acclimation influences on norepinephrine metabolism in brown fat. *Am. J. Physiol.*, **232**, E565–9.

Klimes, I., Nagulesparan, M., Unger, R.H., Aronoff, S.L. and Mott, D.M. (1982) Reduced Na^+, K^+-ATPase activity in intact red cells and isolated membranes from obese man. *J. Clin. Endocrinol. Metab.*, **54**, 721–4.

Knehans, A.W. and Romsos, D.R. (1982) Reduced norepinephrine turnover in brown adipose tissue of ob/ob mice. *Am. J. Physiol.*, **242**, E253–61.

Knehans, A.W., Romsos, D.R. and Leveille, G.A. (1980) Cytochrome oxidase (CO) and Na^+, K^+-ATPase (ATPase) activities in brown adipose tissue (BAT) of obese (ob/ob) mice (OM). *Fed. Proc.*, **39**, 886.

Landsberg, L. and Young, J.B. (1981a) Diet-induced changes in sympathoadrenal activity: implications for thermogenesis. *Life Sci.*, **28**, 1801–19.

Landsberg, L. and Young, J.B. (1981b) Diet-induced changes in sympathetic nervous system activity. in *Nutritional factors: Modulating Effects on Metabolic Processes* (eds R.F. Beers, jr. and E.G. Bassett), (Miles International Symposium Series, No. 13) Raven Press, New York, pp. 155–74.

Leung, P.M.B. and Horwitz, B.A. (1976) Free-feeding patterns of rats in response to changes in environmental temperature. *Am. J. Physiol.*, **231**, 1220–4.

Lin, M.H., Romsos, D.R., Akera, T. and Leveille, G.A. (1978) Na$^+$, K$^+$-ATPase enzyme units in skeletal muscle from lean and obese mice. *Biochem. Biophys. Res. Commun.*, **80**, 398–404.

Lin, M.H., Romsos, D.R., Akera, T. and Leveille, G.A. (1979*b*) Na$^+$, K$^+$-ATPase enzyme units in skeletal muscle and liver of 14-day old lean and obese (ob/ob) mice. *Proc. Soc. Exp. Biol. Med.*, **161**, 235–8.

Lin, M.H., Romsos, D.R., Akera, T. and Leveille, G.A. (1981) Functional correlates of Na$^+$, K$^+$-ATPase in lean and obese (ob/ob) mice. *Metabolism*, **30**, 431–8.

Lin, M.H., Vander Tuig, J.G., Romsos, D.R., Akera, T. and Leveille, G.A. (1979*a*) Na$^+$, K$^+$-ATPase enzyme units in lean and obese (ob/ob) thyroxine-injected mice. *Am. J. Physiol.*, **237**, E265–72.

Lin, M.H., Vander Tuig, J.G., Romsos, D.R., Akera, T. and Leveille, G.A. (1980) Heat production and Na$^+$- K$^+$-ATPase enzyme units in lean and obese (ob/ob) mice. *Am. J. Physiol.*, **238**, E193–9.

Lin, P.-Y., Romsos, D.R., Vander Tuig, J.G. and Leveille, G.A. (1979*c*) Maintenance energy requirements, energy retention and heat production of young obese (ob/ob) and lean mice fed a high-fat or a high-carbohydrate diet. *J. Nutr.*, **109**, 1143–53.

Malbon, C.C. (1980) Liver cell adenylate cyclase and β-adrenergic receptors. Increased β-adrenergic receptor number and responsiveness in the hypothyroid rat. *J. Biol. Chem.*, **255**, 8692–9.

Malbon, C.C., Moreno, F.J., Cabelli, R.J. and Fain, J.N. (1978) Fat cell adenylate cyclase and β-adrenergic receptors in altered thyroid states. *J. Biol. Chem.*, **253**, 671–8.

Mayer, J. and Barrnett, R.J. (1953) Sensitivity to cold in the hereditary obese-hyperglycemic syndrome of mice. *Yale J. Biol. Med.*, **26**, 38–45.

McLaughlin, C.W. (1973) Control of sodium, potassium and water content and utilization of oxygen in rat liver slices, studied by affecting cell membrane permeability with calcium and active transport with ouabain. *Biochim. Biophys. Acta*, **323**, 285–96.

Melander, A., Ericson, L.E., Sundler, F. and Westgren, U. (1975*a*) Intrathyroidal amines in the regulation of thyroid activity. *Rev. Physiol. Biochem. Pharmacol.*, **73**, 39–71.

Melander, A., Ranklev, E.; Sundler, F. and Westgren, V. (1975*b*) Beta$_2$-adrenergic stimulation of thyroid hormone secretion. *Endocrinology*, **97**, 332–6.

Mobley, P.W. and Dubuc, P.U. (1979) Thyroid hormone levels in the developing obese-hyperglycemic syndrome. *Horm. Metab. Res.*, **11**, 37–9.

Mohell, N., Nedergaard, J. and Cannon, B. (1980) An attempt to differentiate between α- and β-adrenergic responses in hamster brown fat cells. in *Contributions to Thermal Physiology* (eds Z. Szelenyi and M. Szekely), Pergamon Press, Hungary, pp. 495–7.

Mokhova, E.N. and Zorov, D.B. (1973) The effects of cold stress on respiration of diaphragm muscle. *J. Bioenergetics*, **5**, 119–28.

Mory, G., Ricquier, D. and Hémon, P. (1980) Effects of chronic treatments upon the brown adipose tissue of rats. II. Comparison between the effects of catecholamine injections and cold adaptation. *J. Physiol. Paris*, **76**, 859–64.

Mory, G., Ricquier, D., Pesquiés, P. and Hémon, P. (1981) Effects of hypothyroidism on the brown adipose tissue of adult rats: comparison with the effects of adaptation to cold. *J. Endocr.*, **91**, 515–24.

Müller, M.J. and Seitz, H.J. (1980) Rapid and direct stimulation of hepatic gluconeogenesis by L-tri-iodothyronine (T₃) in the isolated-perfused rat liver. *Life Sci.*, **27**, 827–35.

Müller, M.J. and Seitz, H.J. (1981) Dose dependent stimulation of hepatic oxygen consumption and alanine conversion to CO_2 and glucose by 3,5,3′-triiodo-L-thyronine (T₃) in the isolated perfused liver of hypothyroid rats. *Life Sci.*, **28**, 2243–9.

Nakashima, T., Taurog, A. and Krulich, L. (1981) Serum thyroxine, triiodo-thyronine, and TSH levels in iodine-deficient and iodine-sufficient rats before and after exposure to cold. *Proc. Soc. Exp. Biol. Med.*, **167**, 45–50.

Naltchayan, S., Bouhnik, J. and Michel, R. (1981) Concentrations des iodothyronine sériques au cours du jeûne chez la souris génétiquement obèse. *C.R. Soc. Biol.*, **174**, 118–20.

Nicholls, D.G. (1974) The influence of respiration and ATP hydrolysis on the proton-electrochemical gradient across the inner membrane of rat-liver mito-chondria as determined by ion distribution. *Eur. J. Biochem.*, **50**, 305–15.

Nicholls, D.G. (1976) The bioenergetics of brown adipose tissue mitochondria. *FEBS Lett.*, **61**, 103–10.

Nicholls, D.G. (1977) The effective proton conductance of the inner membrane of mitochondria from brown adipose tissue. Dependency on proton electro-chemical potential gradient. *Eur. J. Biochem.*, **77**, 349–56.

Nicholls, D.G., Cannon, B., Grav, H.J. and Lindberg, O. (1974) Energy dissipation in non-shivering thermogenesis. in *Dynamics of Energy-Transducing Membranes* (eds L. Ernster, R.W. Estabrook and E.C. Slater), Elsevier, Amsterdam, pp. 529–37.

Nishiki, K., Erecinska, M., Wilson, D.F. and Cooper, S. (1978) Evaluation of oxidative phosphorylation in hearts from euthyroid, hypothyroid, and hyper-thyroid rats. *Am. J. Physiol.*, **235**, C212–9.

Ohtake, M., Bray, G.A. and Azukizawa, M. (1977) Studies on hypothermia and thyroid function in the obese (ob/ob) mouse. *Am. J. Physiol.*, **233**, R110–5.

Okamura, K., Taurog, A. and Distefano, J.J. (1981) Elevated serum levels of T₃ without metabolic effect in nutritionally deficient rats, attributable to reduced cellular uptake of T₃. *Endocrinology*, **109**, 673–5.

Oppenheimer, J.H. (1979) Thyroid hormone action at the cellular level. *Science*, **203**, 971–9.

Oppenheimer, J.H., Dillmann, W.H., Schwartz, H.L. and Towle, H.C. (1979) Nuclear receptors and thyroid hormone action: a progress report. *Fed. Proc.*, **38**, 2154–61.

Otten, M.H., Hennemann, G., Doctor, R. and Visser, T.J. (1980) The role of dietary fat in peripheral thyroid hormone metabolism. *Metabolism*, **29**, 930–5.

Pardridge, W.M. (1977) Regulation of amino acid availability to the brain. in *Nutrition and the Brain* (vol. 1, *Determinants of the Availability of Nutrients to the Brain*) (eds R.J. Wurtman and J.J. Wurtman), Raven Press, New York, pp. 141–204.

Pisarev, M.A., Cardinali, D.P., Juvenal, G.J., Vacas, M.I., Barontini, M. and Boado, R.J. (1981) Role of the sympathetic nervous system in the control of the goitrogenic response in the rat. *Endocrinology*, **109**, 2202–7.

Portnay, G.I., McClendon, F.D., Bush, J.E., Braverman, L.E. and Babior, B.M. (1973) The effect of physiological doses of thyroxine on carrier-mediated ADP uptake by liver mitochondria from thyroidectomized rats. *Biochem. Biophys. Res. Commun.*, **55**, 17–21.

Rabolli, D. and Martin, R.J. (1977) Effects of diet composition on serum levels of insulin, thyroxine, triiodothyronine, growth hormone, and corticosterone in rats. *J. Nutr.*, **107**, 1068–74.

Ricquier, D., Gervais, C., Kader, J.C. and Hémon, P. (1979*a*) Partial purification by guanosine-5'-diphosphate–agarose affinity chromatography of the 32000 molecular weight polypeptide from mitochondria of brown adipose tissue. *FEBS Lett.*, **101**, 35–8.

Ricquier, D. and Kader, J.C. (1976) Mitochondrial protein alteration in active brown fat: a sodium dodecyl sulfate–polyacrylamide gel electrophoretic study. *Biochem. Biophys. Res. Commun.*, **73**, 577–83.

Ricquier, D., Mory, G. and Hémon, P. (1975) Alterations of mitochondrial phospholipids in the rat brown adipose tissue after chronic treatment with cold or thyroxine. *FEBS Lett.*, **53**, 342–6.

Ricquier, D., Mory, G. and Hémon, P. (1976) Effects of chronic treatments upon the brown adipose tissue of young rats. I. Cold exposure and hyperthyroidism. *Pflügers Arch.*, **362**, 241–6.

Ricquier, D., Mory, G. and Hémon, P. (1979*b*) Changes induced by cold adaptation in the brown adipose tissue from several species of rodents, with special reference to the mitochondrial components. *Can. J. Biochem.*, **57**, 1262–6.

Ricquier, D., Mory, G., Néchad, M. and Hémon, P. (1978) Effects of cold adaptation and re-adaptation upon the mitochondrial phospholipids of brown adipose tissue. *J. Physiol. (Paris)*, **74**, 695–702.

Romsos, D.R. (1981*a*) Efficiency of energy retention in genetically obese animals and in dietary-induced thermogenesis. *Fed. Proc.*, **40**, 2524–9.

Romsos, D.R. (1981*b*) Alterations in Na$^+$, K$^+$-ATPase in obese animal models in *Nutritional Factors: Modulating Effects on Metabolic Processes* (eds R.F. Beers, jr. and E.G. Bassett), (Miles International Symposium Series, No. 13) Raven Press, New York, pp. 115–22.

Rothwell, N.J., Saville, M.E. and Stock, M.J. (1982) Sympathetic and thyroid influences on metabolic rate in fed, fasted and refed rats. *Am. J. Physiol.*, **243**, R339–46.

Rothwell, N.J. and Stock, M.J. (1979) A role for brown adipose tissue in diet-induced thermogenesis. *Nature (London)*, **281**, 31–5.

Rothwell, N.J. and Stock, M.J. (1980) Similarities between cold- and diet-induced thermogenesis in the rat. *Can. J. Physiol. Pharmacol.*, **58**, 842–8.

Rothwell, N.J. and Stock, M.J. (1981*a*) Regulation of energy balance. *Annu. Rev. Nutr.*, **1**, 235–56.

Rothwell, N.J. and Stock, M.J. (1981*b*) Influence of noradrenaline on blood flow to brown adipose tissue in rats exhibiting diet-induced thermogenesis. *Pflügers Arch.*, **389**, 237–42.

Rothwell, N.J. and Stock, M.J. (1981*c*) A role for insulin in the diet-induced thermogenesis of cafeteria-fed rats. *Metabolism*, **30**, 673–8.

Rothwell, N.J., Stock, M.J. and Tyzbir, R.S. (1982) Energy balance and mitochondrial function in liver and brown fat of rats fed cafeteria diets of varying protein content. *J. Nutr.*, **112**, 1663–72.

Rothwell, N.J., Stock, M.J. and Wyllie, M.G. (1981) Na^+, K^+-ATPase activity and noradrenaline turnover in brown adipose tissue of rats exhibiting diet-induced thermogenesis. *Biochem. Pharmacol.*, **30**, 1709–12.

Ruegamer, W.R., Westerfeld, W.W. and Richert, D.A. (1964) α-Glycerophosphate dehydrogenase response to thyroxine in thyroidectomized, thiouracil-fed and temperature-adapted rats. *Endocrinology*, **75**, 908–16.

Saggerson, E.D. and Carpenter, C.A. (1979) Ouabain and K^+ removal blocks α-adrenergic stimulation of gluconeogenesis in tubule fragments from fed rats. *FEBS Lett.*, **106**, 189–92.

Scammell, J.G., Barney, C.C. and Fregly, M.J. (1981) Proposed mechanism for increased thyroxine deiodination in cold-acclimated rats. *J. Appl. Physiol.*, **51**, 1157–61.

Scammell, J.G., Shiverick, K.T. and Fregly, M.J. (1980) *In vitro* hepatic deiodination of L-thyroxine to 3,5,3'-triiodothyronine in cold-acclimated rats. *J. Appl. Physiol.*, **49**, 386–9.

Scarpace, P.J. and Abrass, I.B. (1981a) Thyroid hormone regulation of rat heart, lymphocyte and lung β-adrenergic receptors. *Endocrinology*, **108**, 1007–11.

Scarpace, P.J. and Abrass, I.B. (1981b) Thyroid hormone regulation of β-adrenergic receptor number in aging rats. *Endocrinology*, **108**, 1276–8.

Schimmel, M. and Utiger, R.D. (1977) Thyroidal and peripheral production of thyroid hormones. Review of recent findings and their clinical implications. *Ann. Int. Med.*, **87**, 760–8.

Schussler, G.C. and Orlando, J. (1978) Fasting decreases triiodothyronine receptor capacity. *Science*, **199**, 686–8.

Sellers, E.A., Flattery, K.V., Shum, A. and Johnson, G.E. (1971) Thyroid status in relation to catecholamines in cold and warm environment. *Can. J. Physiol. Pharmacol.*, **49**, 268–75.

Sellers, E.A., Flattery, K.V. and Steiner, G. (1974) Cold acclimation of hypothyroid rats. *Am. J. Physiol.*, **226**, 290–4.

Seydoux, J., Giacobino, J.P. and Girardier, L. (1982) Impaired metabolic response to nerve stimulation in brown adipose tissue of hypothyroid rats. *Mol. Cell. Endocrinol.* **25**, 213–26.

Sharma, V.K. and Banerjee, S.P. (1978) α-Adrenergic receptor in rat heart. Effects of thyroidectomy. *J. Biol. Chem.*, **253**, 5277–9.

Shears, S.B. (1980) The thyroid gland and the liver mitochondrial protonic electrochemical potential difference: a novel hormone action? *J. Theor. Biol.*, **82**, 1–13.

Shears, S.B. and Bronk, J.R. (1979) The influence of thyroxine administered *in vivo* on the transmembrane protonic electrochemical potential difference in rat liver mitochondria. *Biochem. J.*, **178**, 505–7.

Shetty, P.S., Jung, R.T. and James, W.P.T. (1979) Effect of catecholamine replacement with levodopa on the metabolic response to semistarvation. *Lancet*, **i**, 77–9.

Shimomura, Y., Bray, G.A. and York, D.A. (1981) Effects of thyroid hormone and adrenalectomy on $(Na^+ + K^+)$ATPase in the ob/ob mouse. *Horm. Metab. Res.*, **13**, 578–82.

Smith, R.E. and Horwitz, B.A. (1969) Brown fat and thermogenesis. *Physiol. Rev.*, **49**, 330–425.

Smith, T.J. and Edelman, I.S. (1979) The role of sodium transport in thyroid thermogenesis. *Fed. Proc.*, **38**, 2150–3.

Sterling, K. (1979) Thyroid hormone action at the cell level. *New Engl. J. Med.*, **300**, 117–23.

Sterling, K., Brenner, M.A. and Sakurada, T. (1980) Rapid effect of triiodothyronine on the mitochondrial pathway in rat liver *in vivo*. *Science*, **210**, 340–2.

Sterling, K., Lazarus, J.H., Milch, P.O., Sakurada, T. and Brenner, M.A. (1978) Mitochondrial thyroid hormone receptor: localization and physiological significance. *Science*, **201**, 1126–9.

Sterling, K., Milch, P.O., Brenner, M.A. and Lazarus, J.H. (1977) Thyroid hormone action: the mitochondrial pathway. *Science*, **197**, 996–9.

Stevens, E.D. (1973) The evolution of endothermy. *J. Theor. Biol.*, **38**, 597–611.

Stevens, E.D. and Kido, M. (1974) Active sodium transport: a source of metabolic heat during cold adaptation in mammals. *Comp. Biochem. Physiol.*, **47A**, 395–7.

Stirling, J.L. and Stock, M.J. (1968) Metabolic origins of thermogenesis induced by diet. *Nature*, **220**, 801–2.

Stock, M.J. and Rothwell, N.J. (1981) Diet-induced thermogenesis: a role for brown adipose tissue. in *Nutritional Factors: Modulating Effects on Metabolic Processes* (eds R.F. Beers, jr. and E.G. Bassett), (Miles International Symposium Series, No. 13), Raven Press, New York, pp. 101–13.

Storm, H., Hardeveld, C. Van and Kassenaar, A.A.H. (1981) Thyroid hormone-catecholamine interrelationships during exposure to cold. *Acta Endocrinol.*, **97**, 91–7.

Suda, A.K., Pittman, C.S., Shimizu, T. and Chambers, J.B. jr. (1978) The production and metabolism of 3,5,3'-triiodothyronine and 3,3',5'-triiodothyronine in normal and fasting subjects. *J. Clin. Endocrinol. Metab.*, **47**, 1311–9.

Sundin, U. (1981*a*) GDP binding to rat brown fat mitochondria: effects of thyroxine at different ambient temperatures. *Am. J. Physiol.*, **241**, C134–9.

Sundin, U. (1981*b*) Brown adipose tissue. Control of heat production. Development during ontogeny and cold adaptation. Ph.D. Thesis, University of Stockholm.

Svartengren, J., Mohell, N. and Cannon, B. (1980) Characterization of [^{3}H]dihydro-ergocryptine binding sites in brown adipose tissue. Evidence for the presence of α-adrenergic receptors. *Acta Chem. Scand.*, **B34**, 231–2.

Svoboda, P., Svartengren, J., Snochowski, M., Houstek, J. and Cannon, B. (1979) High number of high-affinity binding sites for (−)-(^{3}H) dihydroalprenolol on isolated hamster brown fat cells. *Eur. J. Biochem.*, **102**, 203–10.

Swanson, H.E. (1956) Interrelations between thyroxine and adrenalin in the regulation of oxygen consumption in the albino rat. *Endocrinology*, **59**, 217–25.

Tata, J.R. (1975) How specific are nuclear 'receptors' for thyroid hormones? *Nature (London)*, **257**, 18–23.

Tedesco, J.L., Flattery, K.V. and Sellers, E.A. (1977). Effects of thyroid hormones and cold exposure on turnover of norepinephrine in cardiac and skeletal muscle. *Can. J. Physiol. Pharmacol.*, **55**, 515–22.

Thenen, S.W. and Carr, R.H. (1980) Influence of thyroid hormone treatment on growth, body composition and metabolism during cold stress in genetically obese mice. *J. Nutr.*, **110**, 189–99.

Thurlby, P.L. and Trayhurn, P. (1979) The role of thermoregulatory thermogenesis in the development of obesity in genetically-obese (ob/ob) mice pair-fed with lean siblings. *Br. J. Nutr.*, **42**, 377–85.

Thurlby, P.L. and Trayhurn, P. (1980) Regional blood flow in genetically obese (ob/ob) mice. The importance of brown adipose tissue to the reduced energy expenditure on non-shivering thermogenesis. *Pflügers Arch.*, **385**, 193–201.

Trayhurn, P. (1979) Thermoregulation in the diabetic-obese (db/db) mouse. The role of non-shivering thermogenesis in energy balance. *Pflügers Arch.*, **380**, 277–32.

Trayhurn, P. and Fuller, L. (1980) The development of obesity in genetically diabetic-obese (db/db) mice pair-fed with lean siblings. The importance of thermoregulatory thermogenesis. *Diabetologia*, **19**, 148–53.

Trayhurn, P. and James, W.P.T. (1978) Thermoregulation and non-shivering thermogenesis in the genetically obese (ob/ob) mouse. *Pflügers Arch.*, **373**, 189–93.

Trayhurn, P., Jones, P.M., McGuckin, M.M. and Goodbody, A.E. (1982) Effects of overfeeding on energy balance and brown fat thermogenesis in obese (ob/ob) mice. *Nature (London)*, **25**, 323–5.

Triandafillou, J., Gwilliam, C. and Himms-Hagen, J. (1982) Role of thyroid hormone in cold-induced changes in rat brown adipose tissue mitochondria. *Can. J. Biochem.*, **60**, 530–7.

Tulp, O.L., Frink, R. and Danforth, E. jr. (1982) Effect of cafeteria feeding on brown and white adipose tissue cellularity, thermogenesis, and body composition in rats. *J. Nutr.*, **112**, 2250–60.

Tulp, O., Gambert, S. and Horton, E.S. (1979a) Adipose tissue development, growth, and food consumption in protein-malnourished rats. *J. Lipid Res.*, **20**, 47–54.

Tulp, O.L., Krupp, P.P., Danforth, E. jr. and Horton, E.S. (1979b) Characteristics of thyroid function in experimental protein malnutrition. *J. Nutr.*, **109**, 1321–2.

Tyzbir, R.S., Kunin, A.S., Sims, N.M. and Danforth, E. Jr. (1981) Influence of diet composition on serum triiodothyronine (T_3) concentration, hepatic mitochondrial metabolism and shuttle system activity in rats. *J. Nutr.*, **111**, 252–9.

Van Der Kroon, P.H.W. and Boldewijn, H. (1980) Trophoprivic hypothyroidism in the hereditary obese-hyperglycemic syndrome in mice (ob/ob). *IRCS Medical Science*, **8**, 859.

Van Der Kroon, P.H.W. and Speijers, G.J.A. (1979) Brain deviations in adult obese-hyperglycemic mice (ob/ob). *Metabolism*, **28**, 1–3.

Vander Tuig, J.G., Flynn, A.M. and Romsos, D.R. (1981) Ventromedial hypothalamic lesions reduce the number of Na^+, K^+-ATPase enzyme units in skeletal muscle of weanling rats. *Proc. Soc. Exp. Biol. Med.*, **167**, 475–9.

Vander Tuig, J.G., Trostler, N., Romsos, D.R. and Leveille, G.A. (1979) Heat production of lean and obese (ob/ob) mice in response to fasting, food restriction or thyroxine. *Proc. Soc. Exp. Biol. Med.*, **160**, 266–71.

Videla, L., Flattery, K.V., Sellers, E.A. and Israel, Y. Ethanol metabolism and liver oxidative capacity in cold acclimation. *J. Pharmacol. Exp. Ther.*, **192**, 575–82.

Williams, S.R. and Lefkowitz, R.J. (1979) Thyroid hormone regulation of alpha-adrenergic receptors: studies in rat myocardium. *J. Cardiovasc. Pharmacol.*, **1**, 181–9.

Wilson, J.H.P., Swart, G.R., Van Der Berg, J.W.O. and Lamberts, S.W.J. (1981) The effect of triiodothyronine on weight loss, nitrogen balance and muscle protein catabolism in obese patients on a very low calorie diet. *Nutrition Reports International,* **24,** 145–51.

Wimpfheimer, C., Saville, E., Voirol, M.J., Danforth, E. jr. and Burger, A.G. (1979) Starvation-induced decreased sensitivity of resting metabolic rate to triiodothyronine. *Science,* **205,** 1272–3.

Wood, R. and Carlson, L.D. (1956) Thyroxine metabolism in rats exposed to cold. *Endocrinology,* **59,** 323–30.

Wooten, W.L. and Cascarano, J. (1980) The effect of thyroid hormone on mitochondrial biogenesis and cellular hyperplasia. *J. Bioenergetics Biomembranes,* **12,** 1–12.

York, D.A., Bray, G.A. and Yukimura, Y. (1978*a*) An enzymatic defect in the obese (ob/ob) mouse: loss of thyroid-induced sodium and potassium-dependent adenosinetriphosphatase. *Proc. Nat. Acad. Sci.,* **75,** 477–81.

York, D.A., Otto, W. and Taylor, T.G. (1978*b*) Thyroid status of obese (ob/ob) mice and its relationship to adipose tissue metabolism. *Comp. Biochem. Physiol.,* **59B,** 59–65.

Zaror-Behrens, G. and Himms-Hagen, J. (1983) Cold-stimulated sympathetic activity in brown adipose tissue of lean and genetically obese (ob/ob) mice. *Am. J. Physiol.,* (in press).

Chapter Six

Energetics of Maintenance and Growth

A.J.F. Webster

This chapter is concerned with the factors that determine the utilization of food energy for the maintenance of life and the efficiency with which food energy surplus to that required for maintenance can be used to promote growth. The subject, of course, dates back to Lavoisier and de la Place (1784) and many of the basic concepts that I shall use in this chapter are the same now as then. There have been a number of classic treatises on energy metabolism of which the most famous five must be, in chronological order, Kellner (1900), Rubner (1902), Brody (1945), Kleiber (1961) and Blaxter (1962). All these contain both information and ideas which are still valuable to the modern student of energy metabolism. Indeed, most of the controversies in energy metabolism, such as that concerning diet-induced thermogenesis, have arisen several times during its history and each time have been resolved reasonably satisfactorily in the light of the information available at the time. Bernard Shaw said, 'Every good scientific discovery should be made once every ten years,' which is fair comment since each time a theory is repeated it is usually in the light of new evidence. It should not, however, be necessary each time to start from square one. There are giants in the field of energy metabolism. We should have the humility and good sense to stand on their shoulders.

In this chapter, food energy intake will be described throughout in terms of metabolizable energy (ME), which is gross energy minus the sum of combustible energy losses in faeces, urine and gases emitted from the digestive tract.* Maintenance, in this chapter, is defined as the situation that exists when intake of ME equals metabolic heat production (H). Energy retention (RE) is $ME - H$. It will be assumed that energy is retained in the growing animal only as protein (RE_p) and as fat (RE_f). The small amount of energy (about 2%) retained in other organic compounds, principally glycogen, is included for convenience within the non-nitrogenous body gains (i.e. fat).

*See Chapter 1 for discussion of terms and concepts used by various authorities in animal energetics.

6.1 MEASUREMENT OF ENERGY EXCHANGE

In order to solve the simple equation $ME - H = RE$, it is, of course, necessary to measure only two of these elements although it is better to measure all three, since if they agree they must be right. I cannot over emphasize the need for high precision in measurement of energy exchange. Imagine, for example, one wished to compare energy exchanges in two groups of adult humans of 40 years of age. The first group are lean individuals, weighing 70 kg, the same weight that they were at 20. The second group are obese individuals weighing 100 kg, having gained in the last 20 years 30 kg of pure fat. The former group will have maintained $ME = H$, having turned over in that time about 80×10^3MJ. The average individual from the fat group has retained 1.17×10^3MJ. Assuming at this stage that his energy turnover was also about 80×10^3MJ, his imbalance between ME and H has been less than 1.5%. In short, the difference of energy balance between a lean human and an absolute slob is very small indeed and likely to be within the limits of uncertainty attached to most measurements of energy exchange.

6.1.1 Energy retention

The most reliable measurements of energy exchange are those made over the longest possible period. In laboratory and meat animals the method of choice, if time and money permit, is that of comparative slaughter which estimates cumulative RE from analysis of the energy content of animals serially killed and analysed over the total period of interest. In man and large valuable animals for which the comparative slaughter method is, for a variety of reasons, not appropriate, there is a variety of methods for estimating the energy content of body gains. The use of specific gravity (SG) to estimate body composition has the theoretical merit that SG continually reflects the relative proportions of fat (SG = 0.92), muscle (SG = 1.06) and bone (SG = 1.50) in the body (Kraybill, Bitter and Hankins, 1952). Markers can also be used to estimate the volume of the major body compartments. Body water has been estimated from dilution of D_2O or THO (Foot and Greenhalgh, 1970). Lean body cell mass has been estimated using ^{42}K (Talso *et al.*, 1960) and body water and fat have been estimated simultaneously using respectively the markers THO and the radioactive gases krypton or zenon. All these estimates have a rather large element of uncertainty but they do have the merit of being applicable to long-term experiments which can integrate the large cumulative effects of small imbalances between ME and H.

6.1.2 Heat production

Short-term trials involve the measurement of H, i.e. animal calorimetry. Much has been written on methods of calorimetry and on ways and means of improving precision (see Blaxter, 1962, 1967). It is, however, a very good calorimeter that can measure heat production with an absolute error of less than 1.5%. Moreover, even if one can achieve absolute accuracy of measurement of H for an animal while in a calorimeter or attached to a respiration apparatus, problems arise immediately one wishes to extrapolate such values to predict H in a more normal environment. In the most simple analysis H is affected by body size, food intake, activity and the thermal environment. One may reasonably safely assume that the effect on H of size and food intake are the same in the calorimeter and out. The sensitivity of H to small changes in the thermal environment varies according to species. Cattle, sheep, horses and animals of comparable size and coat cover exhibit a wide thermoneutral range in which H is unaffected by variations in the warmness or coldness of the environment and body temperature is regulated at negligible metabolic cost by physiological control of evaporative heat loss by sweating and thermal panting. In such species effects on H of minor variations in the thermal environment can usually be ignored. Small laboratory animals like rats, rabbits and mice belong to that class of mammals which, having little capacity to regulate evaporative heat loss, depend on the regulation of H (and inevitably therefore on the regulation of ME intake) to maintain homeothermy. In such animals H is enormously sensitive to subtle environmental variations. Table 6.1 presents values for H ($kJ\,kg^{-0.75}\,d^{-1}$) in mice kept singly or in pairs at 24 or 28°C in a closed-circuit respiration chamber. The difference in H between 24 and 28°C was about 27%; that between singles and pairs was 9% and the interaction term was significant. The reduction in H due to keeping mice in pairs was 57 at 28°C and 80 at 24°C. These differences in H can be attributed both to the mice huddling together for warmth and to a reduction in activity when mice which were habitually reared together were kept together rather than isolated for the purposes of calorimetry. Table 6.1 also

Table 6.1 Influence of air temperature and social interactions on heat production of adult mice. (Mean values $kJ\,kg^{-0.75}\,d^{-1}$ with standard deviations)

	Air temperature (°C)		
	24	28	$H_{(24-28)}$
Heat production in calorimeter			
Single mice	918±16	721±29	197
Two mice	838±25	664±18	174
H (singles−pairs)	80	57	
ME for maintenance in cages	674±42		

Values taken from unpublished work at University of Bristol by A.A. Achmed.

presents estimates of H for these mice in their normal cages at 24°C derived from long-term measurements of ME. The difference between H in the cages and H for a single mouse in the calorimeter at the same air temperature is enormous (36%). Neither figure is seriously in error. They are simply different.

The figures presented in Table 6.1 illustrate very clearly just how much caution should be exercised in the design and the interpretation of calorimetric experiments with small animals. Since both ME intake and H are so exquisitely sensitive to small changes of environment it is essential to take these into account when investigating, for example, thermogenic consequences of variations in ME intake. For instance, it is not in the least surprising that cafeteria fed rats should both eat more and show a greater diet-induced thermogenesis at 22 than at 30°C. I shall return to this point later. Hairless man, who does his cultural best to avoid getting cold at all costs, appears to fall into the category of animals for which H varies with quite small variations in air temperature within what is commonly considered the thermoneutral range (Dauncey, 1981).

6.1.3 ME intake

One of the major sources of error in the measurement of energy balance arises from what is, in theory, the easiest of measurements, namely ME intake. One attempts to overcome these problems in animal experiments by devising metabolism cages which carefully separate and preserve faeces and urine for analysis. It is also important to restrict spillage of food as far as possible or at least to ensure that it does not get mixed with excreta. It is possible to consider a rotting mixture of faeces, urine and spilled food as unmetabolized energy but it would again be unrealistic to expect an accuracy of better than plus or minus 1.5% using this approach. With man and the large meat animals the main source of uncertainty derives from the fact that defaecation and urination are intermittent rather than continuous. In order to minimize these end errors energy balance trials with livestock designed for the very precise purpose of feed evaluation usually require collections of faeces and urine to be made for 8 days.

6.2 THE ENERGY COST OF MAINTENANCE AND FATTENING

The conventional approach adopted for the analysis of mammalian thermogenesis by Kellner (1900) and the schools of animal nutrition that have descended from him has been to assume that for an animal at rest in a thermoneutral environment, H is a function of animal size (itself a function of body weight, W) and ME intake.

$$H = f_1 W + f_2 \, \text{ME} \qquad (6.1)$$

This chapter will not hereafter deal with effects of physical activity or thermoregulations on H. All the energy costs of maintenance and growth can, in the simplest analysis, be reduced to this equation. Figure 6.1 illustrates the relationship between H, W and ME in a 50 kg sheep. Basal or fasting metabolism (F) is, by definition, H when ME intake is zero.

As ME increases so too does H in a curvilinear fashion such that each successive increment of ME produces a greater increment in H. Maintenance intake of ME (E_m) is achieved when ME $= H$ (in the example in Figure 6.1 at 8 MJ d^{-1}). In feed evaluation for ruminants it is conventional to

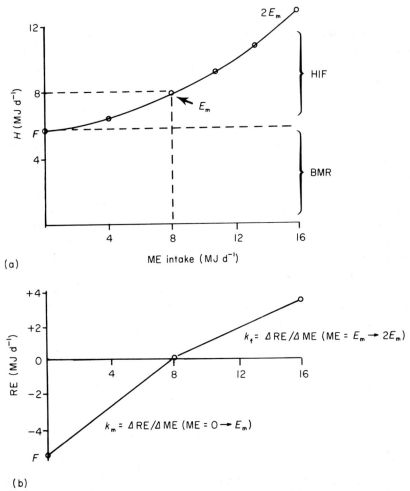

(a)

(b)

Fig. 6.1 The effect of increasing ME intake on (a) H and (b) RE for a 50 kg sheep. $F =$ fasting metabolism; $E_m =$ ME requirement for maintenance (when ME $= H$); BMR = basal metabolic rate. For further explanation see text.

describe the ratio of increments of energy retention (where RE = ME − *H*) to ΔME by two linear slopes:

The net availability of ME for maintenance,
$$k_m = \Delta RE / \Delta ME \text{ between } ME = 0 \rightarrow E_m$$
The net availability of ME for fattening,
$$k_f = \Delta RE / \Delta ME \text{ between } ME = E_m \rightarrow 2E_m$$

These terms are described in more detail by Blaxter (1962) and Blaxter and Boyne (1978).

6.3 FOOD INTAKE AND HEAT PRODUCTION

The interpretation of the effect of food intake on thermogenesis in mamals is a subject of perennial controversy. Figure 6.1 illustrates the marked effect of increasing ME intake for a conventional diet on *H* in an adult sheep (e.g. when ME = $2E_m$, *H* = 2*F*). The effect is more extreme in ruminants than in simple-stomached animals like the rat and man but it is a general truth that in no mammalian species can increments of ME be stored in the body nor substitute for body tissues in metabolism at an efficiency of 1.0. In other words, the more ME an animal eats the more heat it produces. The ruminant, or Kellner, school of nutritionists first called this the heat increment of feeding (HIF) and has for years measured heat increments produced over periods of 24 h or longer by feeds of different quality in order to determine their net energy content (defined as kJ RE kg^{-1}). The phrase 'specific dynamic effect' (SDE) emerged from Rubner's (1902) work with monogastric animals (dogs). A major criticism of Rubner's work, and of much subsequent calorimetry, is that conclusions have been based on short-term measurements of *H* made for a few hours in the post-prandial period and, as such, cannot describe properly the full effects of ME intake on *H*. Nevertheless, HIF and SDE are in theory synonymous both being operational definitions relating ΔH to ΔME. The most convenient units for man and meat animals are kJ *H* (MJ ME^{-1}). Neither describes any specific physiological consequence of the ingestion of food and for that reason the more general sounding term, HIF, is to be preferred. The fashionable new term 'diet induced thermogenesis' (DIT) sounds like a synonym for HIF. If this were so it would be a harmless enough term, albeit unnecessary. However, in current literature DIT has come to refer to the very large increase in *H* recorded in rats persuaded to overeat grossly by cafeteria feeding (Rothwell and Stock, 1979). In these special circumstances it appears that HIF is far greater than that normally associated with the processes of digestion and metabolism of ME and that this extra *H* is being produced for the sole purpose of wasting an excess of ME and thereby minimizing the rate at which the animals get fat. The similarity between this mechanism and regulatory, non-shivering cold thermogenesis in rats (in both cases *H* appears to be produced in brown adipose tissue) (Jansky, 1970; Smith and Horwitz, 1969) suggests that they may both be manifestations of the same mechanism. This point too will be discussed later. For the meantime suffice it to say that this specific phenomenon should perhaps be referred to as regulatory dietary induced thermogenesis (RDIT) to distinguish it from other specific elements which contribute to the totality of HIF.

Table 6.2 Some estimates of the heat increment of feeding (kJ (MJ ME)$^{-1}$) in adult or near adult animals retaining energy almost entirely as fat

Species	Dietary regime	HIF (kJ (MJ ME)$^{-1}$)	Reference
Rat	Synthetic, force-fed	160–190	McCracken and Gray (1976)
	Synthetic, eating *ad lib.*	260	Pullar and Webster (1977)
	Cafeteria ('overeating')	885	Stock (1982)
Pig	Mixed (principally carbohydrate)	260	Kielanowski (1976)
	Mixed (principally carbohydrate)	260–320	A.R.C. (1982)
	High protein v. low protein	285	Gurr *et al.* (1980)
Man	50 fat: 34 carbohydrate: 16 protein	125	Dauncey (1981)
	Alcohol	100 ⎫	Rosenberg and
	'Food'	250 ⎭	Durnin (1978)
Sheep	Intragastric infusion	370	Ørskov *et al.* (1979)
	Forage: cereal mixture	450 ⎫	Blaxter and Boyne (1978)
	All forage	620 ⎭	

Table 6.2 summarizes some recent determinations of HIF in rats, pig, man and sheep. The preferred values for pig and sheep represent best estimates derived from many thousands of measurements made of energy balance in these species. Other values are from individual publications and cannot, inevitably, be said to carry so much weight. However, several important generalizations can be drawn from this table. Firstly, in mature individuals of simple-stomached species (rat, pig and man) eating normal foods in excess of maintenance and depositing energy almost entirely as fat, HIF is about 200–300 kJ (MJ ME)$^{-1}$. This range can be accounted for relatively simply in terms of the proportions of carbohydrate, protein and fat in the diet. The heat increments associated with the synthesis of fats (triacylglycerols) are about 190, 400 and 15 kJ (MJ fat)$^{-1}$ from glucose, amino acids and long-chain fatty acids respectively (McGilvey, 1970). Gurr *et al.* (1980) recently claimed that pigs persuaded to overeat by being offered high fat, low protein diets *ad lib* showed an abnormally high HIF designed to 'waste' excess energy – a claim made originally by Miller and Payne (1962). In fact, as Table 6.2 shows, HIF in the experiments of Gurr *et al.* (1980) was very similar to that observed for pigs in more normal circumstances. For rats fed by intragastric infusion, HIF above maintenance was 160–190 kJ (MJ ME)$^{-1}$ (McCracken and Gray, 1976).

For adult ruminants like the sheep (which can eat a much wider range of feeds than simple-stomached species) HIF is both higher and more variable. Blaxter and Boyne (1978) have derived a series of experimental equations to describe the efficiency of utilization of different feeds by adult sheep below and above maintenance. In effect these equations describe a curvilinear (accelerating) increase in *H* with successive increments of ME to the limit of appetite. Table 6.2 linearizes these equations for two typical diets between maintenance and twice maintenance. In a recent experiment where sheep were sustained entirely by nutrients (mainly volatile fatty acids and casein) infused into the rumen and abomasum, HIF above maintenance was, on average, 370. (Ørskov *et al.*, 1979).

The observations in Table 6.2 provoke three main conclusions.

1. In both simple-stomached and ruminant animals HIF is considerably higher when food is eaten than when nutrients are infused into the gut.

2 When nutrients are infused, HIF is greater in ruminants than in simple-stomached animals. The reason for this can be attributed to the calorimetric efficiency of utilization of the end products of digestion, volatile fatty acids (VFAs) being incorporated into fat at a lower efficiency than glucose or long-chain fatty acids (Blaxter, 1962; MacRae and Lobley, 1982).

3. Each successive increment of ME produces a large increment of heat. This has been clearly documented for ruminants but is almost certainly true for all mammals. The extremely high HIF observed in rats persuaded to overeat grossly by cafeteria feeding (Rothwell and Stock, 1979, 1982) could be considered as an extreme manifestation of this general phenomenon (Table 6.2) but the difference of degree is so large that it is difficult to escape their conclusion that in this special case overeating has triggered an additional response, namely regulatory diet-induced thermogenesis (RDIT).

If at this stage one attemps without prejudice to analyse the contributors to HIF it seems that they must include the energy costs of ingestion and digestion of food, the work of active absorption and secretion of substances across the gut wall, increments in *H* in the food processing tissues such as the gut wall and the liver associated with increasing inflow of substrates and the costs of synthesis (turnover and net accretion) in body tissues, muscle, fat, mammary gland, etc. All these things *must* be considered as inevitable contributors to HIF. The concept of RDIT can be invoked *a posteriori* only if these essential elements fail to account for all of measured HIF.

The HIF in ruminants has been analysed into the following four elements which are capable of direct measurement or calculable with reasonable precision from things which can be measured directly (Webster, 1980).

1. The increase in heat production directly associated with (a) eating, and (b) rumination.

2. Heat production by microbial fermentation in the gut.

3. Heat arising from metabolic activities in the gut (the portal-drained viscera).

4. The increase in heat production associated with food intake but taking place in tissues other than those included in the activities described above. This reflects for the most part the efficiency of utilization of absorbed nutrients.

6.4 EATING AND RUMINATION

Continuous measurements of respiratory exchange consistently show that H in sheep increased rapidly by 40–80% during the course of a meal. This increase persists even through meals lasting up to 2 h but declines thereafter equally rapidly to rates not more than 15–20% greater than those recorded before the meal. The increase is not due to excitement since it, and the accompanying increase in heart rate, are not abolished by β-adrenergic blockade. Moreover, it is not significantly reduced if the animal is sham-fed, i.e. if the food is removed through an oesophageal fistula as it is swallowed and no comparable increase in H is invoked by putting food directly into the rumen. This elevation in H may thus be attributed largely to the energy cost of eating *per se* rather than to any consequence of the arrival of food in the gut. The main determinant of the energy cost of eating is the time spent in the act rather than the amount of food consumed, although this does vary from about 25 $J\,kgW^{-1}min^{-1}$ for pelleted feeds to 45 for hay. The latter feed requires a considerable amount of chewing and ensalivation but the former can probably be compared with most diets for monogastric animals. The energy cost of rumination is, curiously, only about 5 $J\,kgW^{-1}min^{-1}$, i.e. only about 10–20% of the energy cost of eating. Obviously the processes of chewing and ensalivation are common to both activities although the pattern and rate of both differ between eating and rumination. Wherein therefore lies the difference in H? Christopherson and Webster (1972) showed that when a sheep began to eat a meal plasma volume fell sharply and then stabilized but the volume of the extracellular fluid continued to fall steadily by as much as 1–1.5 litres through the period of eating. We suggested that much of the increase of H during eating might be due to the energy cost of this redistribution of body fluids by active secretion from the extra-vascular and vascular compartments into the lumen of the gut. No comparable shifts were observed during rumination.

6.4.1 Fermentation heat

Fermentation is a major contributor to HIF in all herbivores that rely on action by the microbes in the gut to break down structural carbohydrate of cell walls. If it is assumed that fermentation in the gut is strictly anaerobic, then the amount of heat produced can be calculated, in theory, from the heats of combustion of the substrates and end products of fermentation. For carbohydrates, HIF from fermentation accounts for about 6.0–6.5% fermented energy, for proteins the value is somewhat smaller (3.4%) (see Webster, 1980). Measurements made *in vivo* or *in vitro* of the heat produced during fermentation by rumen micro-organisms yield an average value of about 80 kJ. MJ fermented energy^{-1} which is slightly higher than the value predicted on the assumption of complete anaerobiosis.

6.4.2 Heat production in the tissues of the gut

Since nearly all the blood leaving the tissues of the gut wall drains into the liver by way of the hepatic portal vein, it is possible to estimate with reasonable precision the heat production of these portal drained tissues from blood flow and the arterio-venous difference in oxygen content or blood temperature. Estimates from different laboratories suggest with reasonable consistency that oxidative metabolism in the walls of the portal drained viscera contributes about 20% to the total heat production of a sheep on a maintenance ration. Data from other species are probably less reliable but Duratoye and Grayson (1971) obtained comparable values for the heat production of the portal-drained viscera of dogs. The tissues of the gut are clearly very metabolically active. Fell and Weekes (1975) showed that the metabolic activity of sheep rumen epithelium *in vitro* was relatively constant and deduced that metabolic rate was primarily a function of tissue mass; however, a threefold increase in food intake doubled the mass of rumen epithelium within about 7 days. Webster *et al.* (1975) observed that a twofold increase in food intake increased the heat production of the portal drained viscera by two-thirds. The ratio of the increase in, respectively, the structure and function of the gut to increasing food intake was exactly the same in both cases. This indicates clearly that, in ruminants at least, the presence of food has a direct affect upon the metabolic activity of the gut tissues and resurrects, in a modified way, the original suggestion of Kellner that HIF is due, in part at least, to what he called 'the work of digestion'. Table 6.3 summarizes the contributions made to HIF by eating and the work of digestion in a ruminant animal (the sheep). In the sheep the energy costs of eating and digestion of a chopped hay diet accounted for about 42% of total HIF which in this example was 340 below maintenance rising to 600 between maintenance and twice maintenance. A comparable analysis is not possible for simple-stomached animals but the difference in HIF between that induced by eating and gastric intubation tends to suggest that, although overall HIF in simple-stomached species is smaller than in ruminants, the proportions attributable to ingestion, digestion and metabolism in the gut wall may be similar.

Table 6.3 Analysis of the heat increment of feeding (HIF) in sheep fed chopped hay or a barley-based pellet below and above maintenance (E_m) (from Webster, 1980)

	Chopped hay		Barley pellets	
	Below E_m	Above E_m	Below E_m	Above E_m
Total HIF (kJ.MJME^{-1})	340	600	380	520
HI (kJ.MJME^{-1}) due to				
Eating	22	22	3	3
Rumination	5	5	nil	nil
Fermentation	82	82	33	33
Digestive tract	32	156	32	156
Residual	199	335	212	328

6.5 EFFICIENCY OF UTILIZATION OF ABSORBED NUTRIENTS

In the adult animal retaining energy almost entirely as fat, the efficiency of utilization of absorbed nutrients can be compared, in the first instance, with the calorimetric efficiency of fat synthesis from the various substrates made available by digestion in different species given different feeds. This comparison has been made for simple-stomached species by Reeds, Wahle and Haggarty (1982) and for ruminants by Blaxter (1962) and Milligan (1971). Calorimetric efficiency is defined as the ratio of the enthalpy of the synthesized fats to the enthalpy of their precursors (e.g. glucose, free fatty acids, short-chain volatile fatty acids). The predicted calorimetric efficiencies of fat synthesis are little lower than the HIF observed when, for example, rats or sheep are sustained by intragastric infusions. This provokes two important conclusions; firstly the turnover of fats in well-fed adult animals is relatively low (in effect most fats are only synthesized once) and secondly there is no suggestion in rats or sheep sustained by intragastric infusion of any form of RDIT invoked to 'waste' excess energy.

It is also necessary to consider why the ME requirement for maintenance of all species is also greater than basal (or fasting) metabolic rate (BMR). In some circumstances this can be attributed to the method of measurement. If maintenance requirement is measured over 24 h or longer it will obviously be greater than BMR measured say, in man, lying at rest for 30 min before the first meal of the day, for reasons that can simply be attributed to activity. Nevertheless, when both maintenance H and fasting H are measured over periods of 24 h, the former value exceeds the latter by an amount which is greater than that which can directly be attributed to the energy costs of ingestion and digestion (Tables 6.2 and 6.3). Moreover, HIF below maintenance is, once again, related to the nature of the substrates made available by digestion. In other words, the energetic efficiency with which the substrates made available by digestion substitute for energy reserves (principally fat) in meeting the catabolic processes necessary for maintenance is less than 1.0. When glucose is the major substrate, the efficiency of utilization is greater than 0.9. When volatile fatty acids form the major substrates the efficiency of utilization drops to about 0.8 (Table 6.3) (Blaxter, 1962; MacRae and Lobley, 1982).

In some experiments with ruminants fed, or infused with nutrients, in excess of maintenance requirement, HIF has been far greater than the average value of 370 kJ (MJ ME)$^{-1}$ observed by Ørskov *et al.* (1979). Armstrong and Blaxter (1957) reported higher values for HIF (670 kJ (MJ ME)$^{-1}$) when acetate alone was infused into the rumen of sheep otherwise receiving a maintenance ration. The metabolism of acetate was associated with a fall in blood sugar and an increased gluconeogenesis from protein. A reasonable interpretation for the high HI of acetate was that, in these special circumstances, there was an insufficient supply of glucose as a precursor for NADPH for fatty acid synthesis. The high HI may be attributed either to the energy costs of mobilizing body tissues for NADPH synthesis and/or to a 'luxus consumption' mechanism burning off acetate unable to be incorporated into fat. The essential difference between the experiments of Armstrong and Blaxter (1957) and those of Ørskov *et al.* (1979) is that in the latter case the sheep were never short of gluconeogenic precursors. These observations provoke the attractive hypothesis that when certain substrates are present in excess the animal body can dispose of them by combustion. However, a few unpublished observations made during the study by Ørskov *et al.* (1979) suggested that when very unbalanced molar proportions of VFAs were infused into sheep they did not produce excess heat. In fact, the sheep became ill and the experiments had to be terminated. In more normal circumstances, mammals tend to reduce their intake of unbalanced diets rather than burn off excesses of specific substrates (see Radcliffe and Webster, 1979). The possibility that high values of HIF can be associated with an unbalance of substrates remains unproven.

6.6 REGULATORY DIETARY-INDUCED THERMOGENESIS

The extremely high values for HIF recorded by Rothwell and Stock (1979, 1982) in rats persuaded to overeat grossly by 'cafeteria' feeding (though not the results of Gurr *et al.* (1980) for pigs) are so clearly in excess of those recorded for mammals in other circumstances that they cannot be explained adequately in terms of the mechanisms described so far and it becomes essential *a posteriori* to look for a further thermogenic process. This specific process, *regulatory diet induced thermogenesis,* involving the synthesis of increased amounts of brown adipose tissue (BAT) and an increase in (probably uncoupled) oxidation in this tissue, will be discussed at length in Chapter 7. It is important, however, to put this mechanism in perspective. In the first place, it has only been demonstrated in species which possess the ability to mobilize significant amounts of BAT in adult life. Secondly, it can only be triggered by spontaneous overeating (e.g. by cafeteria feeding) not by force feeding (Table 6.2). Thirdly, the magnitude of overeating and RDIT appear to be related to environmental temperature, the lower the temperature the greater being the potential for increase in both. Heroux (1969) observed that regulatory, non-shivering thermogenesis mediated by the action of noradrenaline on BAT in rats was not an inevitable consequence of cold exposure, but could only be induced in circumstances where rats could be persuaded to overeat certain diets.

It may be that for those small mammals which possess the ability to mobilize BAT, the stimulus of cold not only triggers BAT synthesis but overrides the normal satiety mechanisms thereby enabling the animals both to eat more and to dissipate a high proportion of the extra food energy as heat. The evolutionary advantages of such a mechanism for a variety of small mammals such as the common rat or extreme examples like the little brown bat *(Eptesicus fucus)* (Hayward and Lyman, 1967) in cool, temperate or boreal climates are obvious. The unusually attractive (and confusing) nature of the cafeteria diet may induce exactly the same mechanism in rats either in the absence of cold stress or (as suggested above) to a degree that is proportional to the cooling power of the environment.

The question remains: 'Can RDIT be induced in animals (such as man) which do not appear to have significant amounts of BAT?' Results to date from man are, to my mind, equivocal and cannot be said to prove anything. I have, however, been impressed by the calorimetric experiments done by Kirkwood (1981) with the kestrel *(Falco tinnunculus)*, a carnivorous bird which contains no obvious BAT. Adult kestrels, in captivity, can readily be persuaded to overeat. In the first week after their diet of day-old chicks was increased from maintenance to twice maintenance, HIF was 150 kJ MJ^{-1}, a value close to that predicted for the energy costs of fat synthesis from a diet of protein and fat. Two weeks later however, food intake had become erratic but HIF had increased to 600 kJ (MJ ME)$^{-1}$. It is difficult to avoid the conclusion that these adult birds were exhibiting a form of RDIT, one that certainly did not involve BAT. If BAT is not therefore an essential prerequisite for RDIT there is no reason to restrict consideration of it to small mammals such as the rat. It may occur too in man, in circumstances as yet unknown and for reasons as yet unexplained.

6.7 BODY SIZE AND HEAT PRODUCTION

6.7.1 Fasting metabolism in adults

The conventional approach to the measurement of the effect of body size on H has been to measure H in an animal at rest in a thermoneutral environment and when ME intake is zero. Such an approach assumes equation (6.1) is a satisfactory model for the prediction of H and that simply by removing food for a short period of time one can describe the effect of W on H in a way that is independent of prior nutrition and its interactions with the general metabolic state of the animal. In practice, life is not so simple. There is clear evidence for a variety of species that fasting metabolism (F) is influenced by the level of nutrition prior to the period of measurement.

The convention adopted by Blaxter (1962) for measuring F in ruminants involved successive 24 h measurements of H made on the third and fourth day of a period of starvation. Even such carefully designed trials do not yield measurements of H that can be related to size alone. Sheep fed on ME intake of twice maintenance prior to four days of starvation had a value for F 33% higher than those previously fed at maintenance (Marston, 1948). Recorded differences in basal metabolic rate (BMR) of man, measured over short periods before the first meal of the day, are inconsistent. Rose and Williams (1961) observed a difference of only 5% between habitual big and small eaters amongst Europeans, whereas Edmundson (1979) reported a 97% difference in BMR between big and small eaters amongst the native people of East Java (Fig. 6.2).

It is, however, reasonable to assume that for an adult or animal eating a reasonably normal amount of food (i.e. *ME* intake close to maintenance) the effects of prior nutrition on F will be fairly constant and in such circumstances F (or BMR) should be determined primarily by body size. The relationship between F and body weight *(W)* in adults of different species was explored in a series of classic studies by Brody (1945) and Kleiber (1961). In Brody's experiments, the slope of the line relating log F to log W was 0.734 and the expression $W^{0.73}$ was taken for many years as that which conferred proportionality on measurements made of F in adult animals of different species. It was also used to describe the effect of W on the energy requirements of growing animals although neither Brody nor Kleiber had assumed that the same exponent would necessarily fit interspecies comparisons between adults and interspecies comparisons between individuals at different stages of growth. Kleiber (1965) claimed, sensibly enough, that the use of the exponent 0.734 (or 0.73) implied an unwarranted degree of precision and suggested that the exponent 0.75, which is the decimal form of the 'more approximate' term three-quarters, would be more appropriate. This recommendation has been generally accepted and $W^{0.75}$ has become known as 'metabolic body size'. As a first approximation, one may estimate F in adult mammals to be 300 kJ $kg^{-0.75} d^{-1}$.

Simple rules exist therefore for predicting the effect of size on H for any adult animal. The question now arises, 'to what extent do different animals depart from these rules and why?' Figure 6.2 attempts to summarize estimates of F in adults or near adults of different species. The data have been drawn from a variety of sources and are grouped in columns according to the numbers of reliable estimates available. The preferred figures of the Agricultural Research Council (1981, 1982) for pigs, sheep and European cattle have to be the most precise, partly because they are based on thousands of measurements made under standard conditions and partly because commercial selection pressures in meat animals tend to reduce genetic variation. On the whole, however, the more measurements there are available for any species, the more nearly the average tends towards the interspecies mean which implies that reports of large deviations from this mean should be treated with some suspicion, particularly if they are based on relatively few measurements. There are some real exceptions. Values for sheep are consistently below 300 kJ kg$^{-0.75}$ d^{-1}. This point is discussed further below. In cattle there is a tendency for F to be related to the potential metabolic performance of the animal. One can rank European cattle *(Bos taurus)* in descending order of F, dairy cows, beef bulls, beef steers and

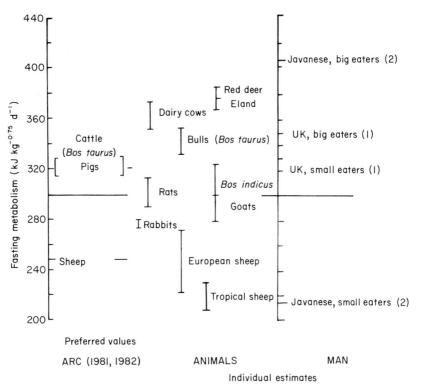

Fig. 6.2 Preferred mean values for fasting metabolism in different species, together with a selection made from individual estimates. The preferred mean values for cattle, pigs and sheep are from the Agricultural Research Council (1981, 1982). The values for big and small eaters among men are from (1) Rose and Williams (1961) and (2) Edmundson (1979). The sources of the data used for animals are too numerous to list here.

heifers. Tropical cattle *(Bos indicus)* overlap the bottom end of the range but tend, on average, to show lower values (Fig. 6.2). The values for man are extremely variable. It is difficult to see why these differences should be due to genetic variability when mean values for animals as disparate as rats, pigs and cattle are so similar. A lot of the variation can be attributed to differences in food intake prior to measurement of BMR but it is difficult to escape the conclusion that some of the techniques may have been at fault.

What factors, other than those obviously related to prior nutrition, might account for differences in F in mature animals? Table 6.4 compares a series of calorimetric measurements of E_m for a range of animals for which measurements of accurate estimates of body composition were also available (Webster, 1981). Since variations in k_m tend to be small, measurements of E_m are, obviously, closely correlated with F, though tending to exceed them by about 30%. Table 6.4 confirms the observations that sheep have lower values for F or E_m per unit of metabolic size ($W^{0.75}$) than most other species. Column 2 in Table 6.4 also shows that values were lower in old, fat than in old, thin sheep and both sets of values were lower than for yearlings (Toutain *et al.*, 1977). Graham, Searle and Griffiths (1974) have also reported a decline with age in F ($kJ\,kg^{-0.75}\,d^{-1}$) in sheep. The effect of body fat content on E_m was even greater when comparison was made between lean and congenitally obese Zucker rats.

Table 6.4 The maintenance requirement of different animals for metabolizable energy (ME) expressed per kg body weight$^{0.75}$ and per kg body protein$^{0.75}$ (from Webster, 1981)

	Body weight (kg)	ME for maintenance ($kJ\,d^{-1}$)	
		kg weight$^{0.75}$	kg body protein$^{0.75}$
Interspecies mean	—	420	—
Sheep			
yearling	56	385	1975
'old', fat	73	260	1295
'old', thin	58	310	1290
Steers			
yearling, Friesian	375	585	2305
Hereford × Friesian	375	500	2145
Zucker rats			
lean	0.35	425	1550
fatty	0.35	275	1550
Kestrel *(Falco tinnunculus)*	0.24	610	1750

Before attempting to postulate any subtle physiological explanation for these differences in E_m ($kJ\,kg^{-0.75}\,d^{-1}$) it is necessary to examine the extent to which they can be related simply to differences in body composition. For example, the low E_m values for sheep could be attributed simply to the fact that, compared with most species, they contain relatively large proportions of metabolically inactive mass in the form of gut contents, fat and wool. With some exceptions (e.g. BAT) the most metabolically active tissues are those which are rich in protein. Column 3 of Table 6.4 expresses E_m as kJ (kg body protein)$^{-0.75}$ d^{-1}. The difference between fat and thin individuals now disappears for both rats and sheep which suggests (i) that E_m is better expressed as a function of lean mass than of W and (ii) that fat metabolism (excluding BAT) contributes little to H in these circumstances. The very high values for E_m for the lean carnivorous kestrel also approximate more closely to the interspecies mean value when expressed in terms of protein mass. McCracken and Gray (1976) have shown that when mature rats are forcibly overfed, E_m increases in proportion to $W^{0.75}$ not to lean mass (which hardly alters). These, however, are special circumstances and it is reasonable to assume that the increase in E_m which parallels the increase in W is a response to prior nutrition.

Differences in E_m (kJ (kg protein)$^{-0.75}$ d^{-1}) between classes of cattle and between old and young sheep remain. These differences, like those of McCracken and Gray (1976) can tentatively (at this stage) be attributed to differences in rates of turnover of the major body constituents.

6.7.2 Predicted basal metabolism

It is clear from what has emerged so far that the effect of size on H in the adult animal cannot be separated from the effect of food intake simply by starving the animal for a little while. This problem is more acute in the growing animal, which has, one must assume, a greater capacity for cell synthesis and turnover than the adult and which is more likely to be disturbed, in a metabolic sense, by starvation. It is significant that the exponent now preferred by the Agricultural Research Council (1981, 1982) for predicting the effect of body weight on growing cattle and pigs is $W^{0.67}$ (or two-thirds). What this means, in effect, is that F per unit of body weight is higher in animals at the stage of peak growth than it is in adults (Webster, 1978). Again, this supports the conclusion that there are long-term effects of physiological state on H that cannot simply be expressed in terms of fasting metabolism or of responses to short-term changes in food intake.

If this is so, is there any alternative interpretation of measurements of H that would account not only for size but for physiological state?

The conventional approach to the prediction of energy retention in meat animals, illustrated in Fig. 6.1, can be described simply by the following equation

$$RE = k_f (ME - F/k_m)$$
(6.2)

Here F is determined directly as described already; k_m and k_f are, in fact, only expressions of the change that has occurred in H in a mature animal in the fourth week after a change in ration. None of these three values, F, k_m and k_f has any absolute physiological meaning. They are only systematically acquired observations whose value depends on the conventions adopted for their measurement. This does not render them invalid; k_m and k_f are particularly useful and repeatable ways of describing short-term effects of changes in ME intake on H in an adult. The problem arises when attempts are made to predict RE in growing animals by combining these factors with conventional measurements of F which have, by definition, to be made in an animal *that has stopped growing* (Webster, 1978).

A simple way out of this dilemma is to rearrange equation (6.2) so that the effect of body size can be predicted from successive measurements made of H (and thus of RE) in animals during continuous growth (Webster, Brockway and Smith, 1974).

$$F' = k_m (ME - RE/k_f) \qquad (6.3)$$

In this case F' is called *predicted basal metabolism* and it is not intended to describe any absolute property of the animal. It is simply defined as the intercept term, empirically derived, which when combined with the slope terms k_m and k_f, equally empirically derived, accurately predicts RE. Even more simply, it is the intercept term that gives the right answer within the experiments used to derive it. Values for F' based on over 200 calorimetric measurements made with growing cattle are presented in Table 6.5 which is taken from Webster (1978). All these measurements were derived from animals during rapid, uninterrupted growth. There was a consistent difference between entire (bulls) and castrate (steers) males and beef-type Hereford × Friesian cattle gave consistently lower values than dairy-type (pure Friesians). All these values for F' are higher than values of F

Table 6.5 Predicted basal metabolism (F') of young cattle at different live weights during continuous growth (from Webster, 1978)

Type/sex	Average daily weight gain (kg)	Predicted basal metabolism (F', MJ d^{-1}) at liveweights				
		100 kg	200 kg	300 kg	400 kg	500 kg
Dairy						
castrate male	0.9	14.3	23.7	31.9	39.3	46.3
entire male	1.4	16.0	26.9	—	—	—
Beef						
castrate male	0.9	14.1	21.2	27.6	33.3	38.4
entire male	1.3	16.1	25.0	32.7	39.3	45.1

summarized by Agricultural Research Council (1981) from conventional measurements made of fasting metabolism in animals grown slower then restricted to maintenance for about four weeks, then starved for four days. The following conclusions become inescapable.

1. In addition to the short-term effects of changes in ME intake on H (conventionally described by HIF) there are long-term effects which appear as variations in the relationship between F (or F') and $W^{0.75}$.

2. There are marked differences between individuals in H at any given W and intake of ME. In the adult these differences reduce when expressed per unit of lean mass. In growing animals they are most conveniently expressed in terms of F', derived by extrapolation from observations made during continuous growth.

6.8 THE ENERGETICS OF GROWTH

So far the discussion of mammalian thermogenesis has considered only the simple model described by equation (6.1), namely H as a function of W and ME. No attempt has been made to consider in any detail effects on H of the nature of the synthetic processes taking place in the body. In the adult animal altering its body composition almost entirely by gaining or losing fat, such an approach is valid. In the growing animal however, it is necessary to consider the extent to which H is affected by the balance between, in the first instance, energy deposition as protein and as fat and ultimately, by the true energy costs of protein and fat synthesis, deposition of these things being merely the balance between synthesis and degradation.

6.8.1 Protein and fat deposition

During normal growth, the partition of retained energy between protein and fat is determined overwhelmingly by two things, stage of maturity and plane of nutrition relative to maintenance. A normal rat eating *ad libitum* will retain over 40% of energy as protein at a liveweight of about 100 g; by the time liveweight has reached 400 g, less than 20% of energy is retained as protein, nearly all the rest as fat. In more general terms, the more mature the animal the greater the proportion of energy retained as fat. This generalization applies to all mammals in most normal circumstances. Secondly, the growing animal whose intake of a balanced diet is restricted by food availability or digestibility will tend to maintain protein deposition at the expense of fat. Again, in other words, the more slowly an animal is allowed to grow the greater the ratio of protein to fat deposition. There are exceptions to these rules for genetic oddities and for mammals given diets abnormally unbalanced with respect to energy and protein but these are outside the scope of this chapter (see Radcliffe and Webster, 1978, 1979).

Kielanowski (1976) first attempted to partition, by multiple regression analysis, the utilization of ME by growing pigs into that associated with protein deposition (RE_p), fat deposition (RE_f) and maintenance (aW^n), itself assumed to be a simple function of body weight. The pigs in his experiments were between 20 and 90 kg and body composition was varied by adjustments to the composition and quantity of the diet.

His approach is described by equation (6.4).

$$ME = aW^n + bRE_p + cRE_f \qquad (6.4)$$

It follows that,

$$H = aW^n + (b-1)RE_p + (c-1)RE_f \qquad (6.5)$$

Kielanowski recognized at the outset the theoretical limitations of this approach which are that during growth in normal animals the proportion of energy retained as protein starts small and gets smaller while that retained as fat increases, so the variables are not independent. Moreover, the coefficient that one obtains for both these functions is dominated by the exponent that one selects *a priori* to relate the largest element of H to body weight. Normally the exponent used has been 0.75 although, as indicated already, 0.67 may be a better exponent to describe the effect of increasing W or H in growing animals. Not surprisingly, therefore, the first attempts to assess the energy costs of protein and fat synthesis in this way were wildly variable. Nevertheless, by the time that Kielanowski (1976) reviewed all the available data for the rat and pig a consensus was emerging that the net efficiencies of utilization of ME for protein and fat deposition in simple-stomached animals were 45 and 75% respectively.

A net efficiency of utilization of increments of ME for fat deposition of 0.75 corresponds to a HIF of 250 kJ $(MJ ME)^{-1}$, a value which compares to those observed in practice for pigs and rats in the late stages of growth (Table 6.2). It is hardly surprising that equation (6.4) predicts the efficiency of fat deposition when protein deposition is negligible. The more important question is how precisely does it predict the energetic efficiency of protein deposition in the young animal?

In an attempt to circumvent the serious theoretical objections to the use of multiple regression analysis (equation 6.4) Pullar and Webster (1977) measured energy balance during growth in lean and congenitally obese Zucker rats which differ markedly in the way that they partition ME between H, RE_p and RE_f at the same age (or W) and ME intake; i.e. they permit the exploration of variations between individuals in H that are reasonably independent of the major determinants ME and W considered by equation (6.1). These observations are summarized in Table 6.6 which shows that the experimental design realized its objective in that RE_p and RE_f varied in a way that was, to a considerable degree, independent of variations in ME and W.

If one makes the single assumption that the apparent energy costs of protein and fat deposition are independent of genotype or stage of maturity then for each phenotype at each body weight one can solve equations (6.4) or (6.5) deriving a different intercept term aW for each group at each body weight. From the results listed in Table 6.6, the solution to equation (6.4) becomes

$$ME = aW + 2.25RE_p + 1.36RE_f \qquad (6.6)$$

Table 6.6 Aspects of growth and energy exchanges during growth in lean and congenitally obese (fatty) Zucker rats (for further explanation see text)

	Lean rats				Fatty rats			
	$W = 200$ g		$W = 350$ g		$W = 200$ g		$W = 300$ g	
	Low	High	Low	High	Low	High	Low	High
Energy metabolism (kJ d^{-1})								
ME intake	240	315	240	315	240	315	240	315
Heat production	174	203	214	245	135	160	156	185
RE, protein	24.5	40.5	11.2	23.3	15.8	23.9	5.6	15.4
RE, fat	41.7	71.7	15.0	47.1	89.1	131.6	78.3	113.6
Maintenance H (a)								
per kg $W^{0.75}$ d^{-1}	410		422		280		269	
per kg $P^{0.75}$ d^{-1}	1587		1550		1562		1550	
Protein metabolism (g d^{-1})								
Synthesis*	7.1	14.0	8.6	11.9	5.6	9.9	6.1	9.8
Deposition	1.0	1.7	0.48	0.99	0.67	1.0	0.24	0.66
Synthesis/deposition	7.1	8.2	17.9	12.0	8.3	9.9	25.4	14.8
Energy costs of protein synthesis								
$H_{(p)}$, kJ d^{-1}†	32	63	39	53	25	44	27	44
$H_{(p)}$H^{-1}	0.18	0.31	0.18	0.22	0.18	0.27	0.17	0.24
HI (high–low) kJ d^{-1}	29		31		25		29	
HI$_{(p)}$ (high–low) kJ d^{-1}	31		14		19		17	

*Estimated from the difference between leucine flux and leucine catabolism (Reeds and Lobley, 1980).
†Assuming 4.5 kJ (g protein)$^{-1}$.

The amounts of ME required to deposit 1 kJ of protein and fat are therefore 2.25 and 1.36 kJ respectively. Since the energy value of protein and fat are 23.5 and 39.3 kJ g^{-1}, the ME requirements for deposition of 1 g protein or 1 g fat are 52.9 and 53.4 kJ, or more simply, one can say that the apparent energy costs of protein and fat deposition in the rat are both about 53 kJ ME g^{-1}. These figures correspond closely to preferred values for pigs

(Kielanowski, 1976; A.R.C., 1982) and may, I think, be used with confidence for predicting the energy costs of protein and fat deposition in simply-stomached animals given balanced diets based largely on protein and carbohydrate. The efficiency of fat digestion from dietary fat is, of course, greater.

The reciprocals of the ME costs of RE_p and RE_f are, of course, the net efficiencies of deposition, namely 0.44 for RE_p and 0.75 for RE_f. It also follows that the heat increments apparently associated with protein and fat deposition are 560 and 260 kJ (MJ ME)$^{-1}$ respectively. This suggests superficially that HIF could be very high in young animals. In fact, predicted differences in HIF due to the changing partition of RE between protein and fat during normal growth are quite small because even the very young animal retains a high proportion of energy as fat. In the very young rat retaining 40% of energy as protein, HIF is 430 kJ (MJ ME)$^{-1}$, in late growth where 20% energy is retained as protein HIF reduces to 310 kJ (MJ ME)$^{-1}$.

Table 6.6 also expresses the intercept terms for these animals per kg $W^{0.75}$ ($aW^{0.75}$) and per kg body protein$^{0.75}$ ($aP^{0.75}$). As indicated earlier (Table 6.4) the two Zucker rat phenotypes differed markedly when expressed in terms of whole body weight but showed an almost exact similarity when expressed in terms of $aP^{0.75}$. It is not fruitful to speculate too deeply on the meaning of the apparent maintenance requirement of the growing animal since the phrase has no physiological meaning; like predicted basal metabolism (F'), it is simply an intercept term that gives the right answer. Nevertheless the constancy of $aP^{0.75}$ between phenotypes suggests that the component of H not obviously related to RE_p and RE_f in the growing animal is intimately related to lean tissue mass and, by implication, to the work it has to do.

6.8.2 Protein and fat synthesis

Although the values presented above for the energy costs of protein and fat deposition form a useful and reliable basis for predicting the energy costs of growth, they have no real physiological meaning; it would be far more useful to know how much thermogenesis is associated with the total synthesis of protein and fat in the body. Protein synthesis, in particular, is an essential aspect of both maintenance and growth and it is obvious that as an animal approaches maturity the ratio of protein synthesis to protein deposition must increase towards infinity (Table 6.6).

The amount of heat liberated during the synthesis of proteins and fatty acids can be calculated stoichiometrically, taking into account such factors as the energy cost of formation of the bonds in the macromolecule and the cost of synthesis of ATP and reduced pyridine nucleotides required for the process of polymerization (see Millward, Garlick and Reeds, 1976; Reeds *et al.*, 1982). The amount of heat liberated during the synthesis of a mole of fat (tripalmitylglycerol) depends on the precursors, being 480 kJ mol^{-1} when the precursor is fat, 6104 kJ mol^{-1} from carbohydrate and 12767 kJ mol^{-1}

from amino acids (Reeds *et al.*, 1982). The energy cost of protein synthesis cannot be calculated with such certainty but there is general agreement that it is about 4.5 kJ (g protein)$^{-1}$. (Buttery and Annison, 1973; Millward *et al.*, 1976). These values must be considered as minimal estimates since they do not take into account the possibility that there are other thermogenic processes which are inextricably linked to the synthesis of protein or fat in the live animal.

The biggest single obstacle to the estimation of the contribution made by protein (and fat) synthesis to mammalian thermogenesis is the degree of uncertainty attached to estimates of protein (or fat) synthesis. The development of methods for measuring protein turnover in mammalian tissues and in the whole body has been reviewed at length in a book by Waterlow, Garlick and Millward (1978). The simplest expression of their general conclusions is that at the time of their book there was no best method: all were subject to an unsatisfactory element of uncertainty. The estimates of protein synthesis in Zucker rats (Table 6.6) were based on the difference between leucine flux and the rate of leucine oxidation assuming that body protein contains 6.7% leucine by weight (Reeds and Lobley, 1980). It is probable that this approach underestimates total protein synthesis.

Notwithstanding the limitations of the values for protein synthesis in Table 6.6 they can, I think, be used to make some important statements concerning the contribution of protein synthesis to thermogenesis and to HIF in particular. There was a remarkably consistent relationship between the estimated energy cost of protein synthesis $H_{(p)}$ and H for rats on the low plane of nutrition; $H_{(p)}/H$ being 0.18 irrespective of phenotype or body size. On the high plane of nutrition, $H_{(p)}/H$ was more variable, ranging from 0.22 to 0.31, but the ratio increased in each with increasing food intake. The increase in $H_{(p)}$ with increasing ME was, on average, 20 kJ; the increase in H, on average, 28 kJ. These results suggest that $H_{(p)}$ makes a minor but significant contribution to H for rats growing slowly on restricted amounts of food (i.e. about 20%). However, $H_{(p)}$ does appear to make a major contribution to HIF above maintenance (i.e. 20/28 or 72%). The recent observations by Reeds *et al.* (1982) suggest that for growing Zucker rats at least, rates of fat synthesis are very close to those of fat deposition. Using stoichiometry, one can predict that the contribution of the energy costs of fat synthesis to H is trivial, being about 1% in lean rats and 4% in fatties. This observation is consistent with the fact that maintenance H in these (Table 6.6) and most other experiments (Table 6.2) is related to lean mass rather than to total body weight. In these experiments therefore the contribution of fat to thermogenesis was very small. However, protein turnover, although contributing only about 20–25% to total thermogenesis, was the major contributor to HIF.

The main difference between these experiments and those involving either 'cafeteria' feeding (Rothwell and Stock, 1979) or force-feeding (McCracken and Gray, 1976; McCracken and McNiven, 1982) is that these rats were not overeating. Even the congenitally obese Zucker rat does not 'overeat' during growth in the strictest sense; it regulates food intake very precisely so as to achieve as nearly as possible a normal rate of growth of the lean body mass (although it gets very fat in the process, Radcliffe and Webster, 1978).

In these circumstances, there is no suggestion that regulatory diet-induced thermogenesis is being invoked. Thus RDIT appears (in rats at least) to be a phenomenon only in animals seduced or forced to overeat. If they are seduced (e.g. by cafeteria feeding) RDIT is very marked and almost certainly involves brown adipose tissue. If they are force fed, RDIT is smaller, but McCracken and McNiven's (1982) data did show that in these circumstances the maintenance component of H (or the intercept term) was related more closely to total body mass than to lean mass. This suggests that force-feeding may also trigger increased thermogenesis in fat, although not to the same degree as cafeteria feeding. It may equally increase protein turnover, thus increasing $H \, kJ \, g^{-1} \, P^{-0.75}$.

6.8.3 Sites of heat increment

In order to confirm the hypothesis that in most normal circumstances most of the heat increment of feeding occurs in the lean body mass, and that much of it is associated with increased protein turnover, it is necessary not only to consider the association between H and different metabolic functions (as has been done already) but to measure H in different organs, or at least regions of the body, in different circumstances. The classic technique for estimating organ thermogenesis is to measure blood flow through that organ and the difference in O_2 content between arterial and venous blood, $(a-v)_{O_2}$. This depends on the presence of single vessels entering or leaving the organ which are large enough to catheterize. One can, however, obtain a more approximate estimate of H simply from measurement of blood flow since there are finite limits to $(a-v)_{O_2}$. Foster and Frydman (1978) have made comprehensive measurements of blood flow to different organs of rats using radioactive microspheres and augmented these measurements where possible with measurements of $(a-v)_{O_2}$. Their observations on warm-acclimated rats at rest are shown in Table 6.7, together with my estimates of organ thermogenesis assuming that $(a-v)_{O_2}$ was 6.9 ml O_2 (100 ml blood)$^{-1}$ for the body in general but only 3.8 ml O_2 (100 ml blood)$^{-1}$ for the viscera draining into the hepatic portal vein (Webster, 1981). In this example, muscle contributed 20% to H and liver and gut a further 20%. It would be interesting to see the extent to which these tissues contributed to HIF in rats kept at two planes of nutrition.

Table 6.7 Distribution of blood flow and estimated O_2 consumption for rats at rest

	Blood flow (ml min^{-1})	Heat production (J min^{-1})	Heat production (% of total)
Whole rat	89.0	125	—
Muscle (including heart)	18.3	26	20
Brain, skeleton, spinal cord	12.7	18	14
Liver	3.5($\pm$12.9)	15	12
Gut	12.9	10	8
Skin	10.0	14	11
Kidneys	12.5	18	14
BAT	0.8	1	1

Data taken from Foster and Frydman (1978) and Webster (1981).

Table 6.8 Fractional distribution of blood flow, protein synthesis and heat production in ruminants

	% total Cardiac output	% total Protein synthesis	% total Heat production
Abdominal organs	48	50	40
Muscle	15	20	
Skin	11	17	60
Others	26	13	

Data from a variety of sources reviewed by Webster (1981).

Table 6.8 presents a less discrete breakdown of blood flow, protein synthesis and heat production in the organs of sheep (Webster, 1981). In this case the abdominal organs (gut and liver) contributed 40–50% of estimated total cardiac output, heat production and protein synthesis. Moreover, it is known that a threefold increase in food consumption in sheep doubles not only the protein mass of the rumen epithelium (Fell and Weekes, 1975) but also H in the portal-drained viscera (Webster, 1980). This is further evidence in support of the case that a major part of HIF can be attributed to increased protein turnover.

6.8.4 Protein synthesis and heat: an interspecies comparison

The existence of a relationship between protein synthesis (Ps) and thermogenesis (H) in mammals is well known (see Waterlow *et al.*, 1978);

indeed some degree of relationship is inevitable. However, if Ps accounts for only 20% of H then one would not expect the relationship to be very close. Figure 6.3 plots available data for Ps and H on a log–log basis for a variety of species. Values for Ps are taken from various sources referred to in the review by Reeds and Lobley (1980). Values for H for the same species at comparable sizes and ME intakes were taken largely from the observations of my colleagues and I at the Rowett Institute, Aberdeen.

The fact that the points lie close to a straight line is not an unexpected result of a log–log plot. What is of extreme interest is that the slope of the line does not differ significantly from 1.0. In its simplest form therefore the interspecies relationship between H and Ps is

$$H = 20 \text{ kJ} \, (\text{g Ps}^{1.0})^{-1} \tag{6.7}$$

Since the theoretical minimal energy cost of Ps is about 4.5 kJ g^{-1}, equation (6.7) tends to confirm that in animals at rest Ps contributes about

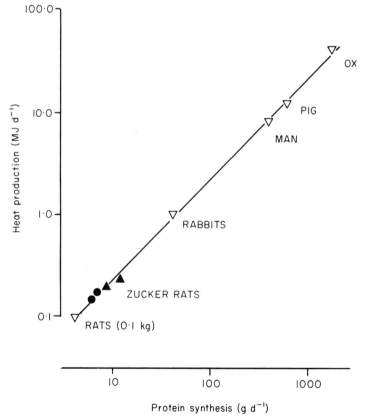

Fig. 6.3 The relationship between heat production (MJ d^{-1}) and estimated protein synthesis (g d^{-1}) in animals of different species expressed on a log:log basis. For sources see text.

20–25% to *H*. However, the closeness of the correlation between *H* and Ps on an interspecies basis (Fig. 6.3) and of that between HI and $HI_{(p)}$ (Table 6.6) are both remarkable.

6.9 CONCLUSIONS

Much of this chapter, and much of this book, are concerned with why animals differ in their heat production in a way that is unrelated to size. This problem is of major practical importance both to our understanding of energy balance in adult man and to the efficient feeding of the meat animals. The material presented elsewhere in the book shows unequivocally that in certain, rather exceptional circumstances, species such as the laboratory rat, which possess brown adipose tissue, vary *H* by invoking a mechanism (RDIT) which invokes increased thermogenesis in brown fat. Perhaps the most important conclusion to be drawn from *this* chapter is that in a wide range of normal circumstances and for a wide range of species, differences in *H*, particularly those differences in *H* associated with differences in food intake, can be attributed to normal metabolic processes such as protein synthesis in the lean body mass, and particularly in those organs such as the liver and gut which have to process the food. There has been no direct evidence that these conclusions apply to man, but, on first principles, it makes sense to assume that they do, since they are common to all mammals. It is only after such obvious mechanisms have been taken into account that it should be necessary to invoke 'futile cycles' or other mechanisms such as RDIT which, in man at least, serve no other apparent purpose than to waste energy.

REFERENCES

Agricultural Research Council (1981) *The Nutrient Requirements of Farm Livestock* No. 2. Ruminants, 2nd ed. Agricultural Research Council, London.

Agricultural Research Council (1982) *The Nutrient Requirements of Farm Livestock* No. 3. Pigs, 2nd ed. Agricultural Research Council, London.

Armstrong, D.G. and Blaxter, K.L. (1957) The utilisation of acetic, propionic and butyric acids by fattening sheep. *Br. J. Nutr.*, **11**, 413–27.

Blaxter, K.L. (1962) *The Energy Metabolism of Ruminants*. Hutchinson, London.

Blaxter, K.L. (1967) Techniques in energy metabolism and their limitations. *Proc. Nutr. Soc.*, **26**, 86–96.

Blaxter, K.L. and Boyne, A.W. (1978) The estimation of the nutritive value of feeds as energy sources for ruminants and the derivation of feeding systems. *J. Agric. Sci., Camb.*, **90**, 47–68.

Brody, S. (1945) *Bioenergetics and Growth*. Reinhold Publs., New York.

Buttery, P.J. and Annison, E.F. (1973) in *The Biological Efficiency of Protein Production* (ed. J.G.W. Jones), University Press, Cambridge, p. 141.

Christopherson, R.J. and Webster, A.J.F. (1972) Changes during eating in oxygen consumption, cardiac function and body fluids of sheep. *J. Physiol.*, **221**, 441–57.

Dauncey, M.J. (1981) Influence of mild cold on 24 h energy expenditure, resting metabolism and diet-induced thermogenesis. *Br. J. Nutr.*, **45**, 257–67.

Duratoye, A.W. and Grayson, J. (1971) Heat production in the gastrointestinal tract of the dog. *J. Physiol.*, **214**, 417–26.

Edmundson, W. (1979) Individual variations in basal metabolic rate and work efficiency. *Ecol. Food Nutr.*, **8**, 189–95.

Fell, B.F. and Weekes, T.E.C. (1975) in *Digestion and Metabolism in the Ruminant* (eds I.W. McDonald and A.C.I. Warner), Univ. New Eng. Publ. Unit., pp. 101–118.

Foot, Janet Z. and Grenhalgh, J.F.D. (1970) The use of deuterium oxide space to determine the amount of body fat in pregnant Blackface ewes. *Br. J. Nutr.*, **24**, 815–25.

Foster, D.O. and Frydman, M.L. (1978) Non-shivering thermogenesis in the rat. 2. Measurements of blood flow with microspheres point to brown adipose tissue as the dominant site of the calorigenesis induced by noradrenaline. *Can. J. Physiol. Pharmacol.*, **56**, 110–22.

Graham, N.McC., Searle, T.W. and Griffiths, D.A. (1974) Basic metabolic rate in lambs and young sheep. *Aust. J. Agric. Res.*, **25**, 957–72.

Gurr, M.I., Mawson, R., Rothwell, N.J. and Stock, M.J. (1980) Effects of manipulating dietary protein and energy intake on energy balance and thermogenesis in the pig. *J. Nutr.*, **110**, 532–42.

Hayward, J.S. and Lyman, C.P. (1967) Non-shivering thermogenesis during arousal from hibernation and evidence for the contributing of brown fat. in *Mammalian Hibernation* III (eds K.C. Fisher *et al.*), American Elsevier, New York, pp. 346–55.

Heroux, O. (1969) Catecholamines, corticosteroids and thyroid hormones in non-shivering thermogenesis under different environmental conditions. in *Physiology and Pathology of Adaptation Mechanisms* (ed. E. Bajusz), Pergamon Press, Oxford.

Jansky, L. (1970) *Non-shivering Thermogenesis*. Swets & N.B. Zeitlinger, Amsterdam.

Kellner, O. (1900) *The Scientific Feeding of Animals*. English edition Translated (1926) by W. Goodwin. Duckworth, London.

Kielanowski, J. (1976) Energy cost of protein deposition. in *Protein Metabolism and Nutrition* (ed. D.J.A. Cole *et al.*), EAAP Publ. No. 16, Butterworths, London, pp. 207–216.

Kirkwood, J.K. (1981) *Bioenergetics and Growth in the Kestrel (Falco tinnunculus)*. Ph.D. Thesis. University of Bristol.

Kleiber, M. (1961) *The Fire of Life*. Wiley, New York.

Kleiber, M. (1965) Metabolic body size. in *Energy Metabolism* (ed. K.L. Blaxter), EAAP Publ. No. 11, Butterworths, London.

Kraybill, H.F., Bitter, H.L. and Hankins, O.G. (1952) Body composition of cattle 2. Determination of fat and water content from measurement of body specific gravity. *J. Appl. Physiol.*, **4**, 575–83.

Lavoisier, A.L. and de la Place, P.S. (1784) Memoire sur la chaleur. *Memoires de l'Academie Royale*, p. 355.

Marston, H.R. (1948) Energy transactions in the sheep. *Aust. J. Sci. Res.* Ser. B **1**, 93–112.

McCracken, K.J. and Gray, R. (1976) Energy cost of digestion and maintenance of obese tissue. in *Energy Metabolism of Farm Animals* (ed. M. Vermorel), EAAP Publ. No. 19, Butterworths, London, pp. 114–145.

McCracken, K.J. and McNiven, M.A. (1982) Energy cost of deposition and maintenance of 'obese tissue' in the adult rat. *Proc. Nutr. Soc.*, **41**, 31A.

McGilvey, R.W. (1970) *Biochemistry, a Functional Approach.* W.B. Saunders, Philadelphia & London.

MacRae, J.C. and Lobley, G.E. (1982) Factors which influence energy losses of ruminants. *Livestock Prod. Sci.*, **9**, 447–56.

Miller, P.S. and Payne, P.R. (1962) Weight maintenance and food intake. *J. Nutr.*, **78**, 255–62.

Milligan, L.P. (1971) Energetic efficiency and metabolic transformations. *Fed. Proc.*, **30**, 1454–8.

Millward, D.J., Garlick, P.J. and Reeds, P.J. (1976) The energy cost of growth. *Proc. Nutr. Soc.*, **35**, 339–49.

Ørskov, E.R., Grubb, D.A., Smith, J.S., Webster, A.J.F. and Corrigall, W. (1979) Efficiency of utilisation of volatile fatty acids for maintenance and energy retention in sheep. *Br. J. Nutr.*, **41**, 541–51.

Pullar, J.D. and Webster, A.J.F. (1977) The energy costs of protein and fat deposition in the rat. *Br. J. Nutr.*, **37**, 355–63.

Radcliffe, J.D. and Webster, A.J.F. (1978) Sex, body composition and regulation of food intake during growth in the Zucker rat. *Br. J. Nutr.*, **39**, 483–92.

Radcliffe, J.D. and Webster, A.J.F. (1979) The effect of varying the quality of dietary protein on food intake and growth in the Zucker rat. *Br. J. Nutr.*, **41**, 111–24.

Reeds, P.J. and Lobley, G.E. (1980) Protein synthesis, are there real species differences? *Proc. Nutr. Soc.*, **39**, 43–51.

Reeds, P.J., Wahle, K.W.J. and Haggarty, P. (1982) Energy costs of protein and fatty acid synthesis. *Proc. Nutr. Soc.*, **41**, 155–61.

Rose, G.A. and Williams, R.J. (1961) Metabolic studies on large and small eaters. *Br. J. Nutr.*, **15**, 1–15.

Rosenburg, K. and Durnin, J.V.G.A. (1978) The effect of alcohol on resting metabolic rate. *Br. J. Nutr.*, **40**, 293–8.

Rothwell, N.J. and Stock, M.J. (1979) A role of brown adipose tissue in diet-induced thermogenesis. *Nature*, **281**, 31–34.

Rubner, M. (1902) *Die Gesetze des Energie verbrauchs bei de Ernahung Leipzig*, p. 109.

Smith, R.E. and Horwitz, B.A. (1969) Brown fat and thermogenesis. *Physiol. Rev.*, **49**, 330–425.

Talso, P.J., Miller, C.E., Carballo, A.J. and Vasquez, I. (1960) Exchangeable potassium as a parameter of body composition. *Metabolism*, **9**, 456–71.

Toutain, P.L., Toutain, Claire, Webster, A.J.F. and McDonald, J.D. (1977) Sleep and activity, age and fatness and the energy expenditure of confined sheep. *Br. J. Nutr.*, **38**, 445–54.

Waterlow, J.C., Garlick, P.J. and Millward, D.J. (1978) *Protein Turnover in Mammalian Tissues and in the Whole Body.* North-Holland Publ. Co., Amsterdam, N.Y., Oxford.

Webster, A.J.F. (1978) Prediction of the energy requirements for growth in beef cattle. *World Rev. Nutr. Dietet.*, **30**, 189–227.

Webster, A.J.F. (1980) The energy costs of digestion and metabolism in the gut. in *Digestive Physiology and Metabolism in Ruminants* (eds Y. Ruckebusch and P. Thivend), MTP Press, Lancaster, pp. 423–38.

Webster, A.J.F. (1981) The energetic efficiency of metabolism. *Proc. Nutr. Soc.*, **40**, 121–8.

Webster, A.J.F., Brockway, J.M. and Smith, J.S. (1974) Prediction of the energy requirements for growth in beef cattle. 1. The irrelevance of fasting metabolism. *Anim. Prod.*, **19**, 127–39.

Webster, A.J.F., Osuji, P.O., White, F. and Ingram, J.F. (1975) The influence of food intake on portal blood flow and heat production in the digestive tract of sheep. *Br. J. Nutr.*, **34**, 125–39.

Webster, A.J.F., Smith, J.S. and Mollison, G.S. (1977) Prediction of the energy requirements for growth in beef cattle. 3. Body weight and heat production in Hereford×British Friesian bulls and steers. *Anim. Prod.*, **24**, 237–44.

Chapter Seven

Diet-Induced Thermogenesis

Nancy J. Rothwell and Michael J. Stock

7.1 INTRODUCTION

Diet-induced thermogenesis (DIT) refers to the increase in metabolic rate that follows the ingestion of food, as well as changes associated with chronic alterations in the overall level of energy intake (i.e. the plane of nutrition). The term DIT therefore includes phenomena such as the specific dynamic action (SDA) or thermic effect (TE) of food, and is synonymous with the heat increment of feeding (HIF) used by agricultural nutritionists (see previous chapter for dicusssion of terminology).

Apart from differences in terminology, two schools of thought can be distinguished when considering the metabolic origins and significance of DIT. One school believes that DIT represents the energy cost of digestion, absorption and assimilation of nutrients as well as the cost of synthesizing body fat and protein. The other maintains that, in addition to this 'obligatory' DIT, there is an 'adaptive' component that serves to dissipate energy consumed in excess of requirements. This adaptive DIT is due to energetically non-conservative mechanisms that provide a form of output control in the regulation of energy balance. However, the distinction between these two forms of DIT – obligatory and adaptive, is not always clear, since the energy cost of some obligatory processes vary, and might participate in the adaptive responses to changes in the plane of nutrition. Further complications result from semantic, methodological and interpretational differences between various workers. In this chapter, an attempt will be made to clarify some of these differences. The evidence for an adaptive role for DIT will be considered, and its physiological and biochemical origins discussed.

It was probably Rubner (1902) who first noticed that the consumption of food stimulated heat production, thus resulting in a loss of potentially useful food energy which might otherwise be retained by the body. Rubner fed single nutrients and demonstrated that protein had a much greater effect than either fat or carbohydrate. However, subsequent studies (e.g. Forbes and Swift, 1944; Hamilton, 1939) with complete diets, revealed that the effects of food on heat production were not directly related to the protein content of the diet but to the ratio of nutrients. This led Mitchell (1962) to propose that a balanced diet was one that produced the least DIT (or HIF), and Kleiber (1945) described the stimulatory effects of unbalanced diets on heat production as an example of 'homeostatic waste'.

One of the most dramatic examples of this homeostatic waste was described by Miller and Payne (1962), who compared the energy requirements for weight maintenance of two pigs fed either a normal diet in restricted amounts (i.e. energy restricted) or a very low (2%) protein diet fed *ad libitum* (i.e. protein restricted). Weight maintenance was achieved in spite of the fact that the protein-restricted pig ate five times more energy than its partner. This experiment has since been repeated with full measurements of energy balance (Gurr *et al.*, 1980) and some of the data are shown in Table 7.1 where it can be seen that the energy intake and expenditure of the protein-restricted pigs was twice that of the energy-restricted pigs. The protein-restricted group did gain more body energy, but the energy cost of this gain (i.e. cost of fat synthesis) accounted for only 20% of their raised energy expenditure. Thus, the remaining 80% has to be ascribed to adaptive DIT. Similar effects of low-protein diets have been reported in rats (McCracken and Gray, 1976; Stirling and Stock, 1968) and overfed human subjects (Miller and Mumford, 1967).

Table 7.1 Energy balance in weanling pigs

	Final body weight (kg)	Metabolizable energy intake (MJ)	Body energy gain (MJ)	Energy expenditure (MJ)
Energy restricted				
Pig 1	6.2	34.0	− 9.9	43.9
Pig 2	6.0	34.3	− 7.7	42.0
Protein restricted				
Pig 1	5.6	89.8	5.2	84.6
Pig 2	5.7	88.4	7.6	80.8

Values are for individual pigs over a 21-day experiment (Adapted from Gurr *et al.*, 1980).

The work on low-protein diets led to the idea that mechanisms of energy dissipation could play an important role in regulating energy balance and body composition. This was not a new idea because at the turn of this century Voit and Neumann had suggested that a form of heat production, which they called 'luxuskonsumption', enabled the body to dispose of energy consumed in excess of requirements. Neumann (1902) and later Gulick (1922) used themselves as subjects and observed that they could maintain body weight in spite of large and long-term differences in energy intake. However, it was not until the 1960s, when controlled overfeeding studies in man were undertaken, that this idea started to receive more attention. Many of these studies (e.g. Miller, Mumford and Stock, 1967; Apfelbaum, Botscarron and Lascatis, 1971; Sims *et al.*, 1973; Goldman *et al.*, 1975) produced evidence for adaptive increases in energy expenditure during overnutrition, although in some of the other studies the evidence was equivocal (see Garrow, 1978 for a review).

The notion that man can compensate for excessive intakes of energy via luxuskonsumption (i.e. DIT) is still met with some resistance and the problem will not be finally resolved until complete, continuous and precise energy balance measurements have been made on chronically overfed lean subjects. Nevertheless, the wide (two-fold) range in habitual energy intake seen in individuals maintaining a similar body weight (Walker, 1965; Widdowson, 1947), and the large differences in the TE of a standard meal in large and small eaters (Morgan *et al.*, 1982) suggests that DIT could be a major variable in human energy metabolism.

7.2 FORCE-FEEDING AND DIET-INDUCED THERMOGENESIS

The problems associated with estimating changes in energy balance in overfed human subjects can be avoided in animals, particularly laboratory rodents, because it is possible to make precise measurements of carcass energy content. However, the researcher is faced with a different problem, that of inducing hyperphagia in these animals. Ever since the studies by Adolph (1947), the rat has been considered as the perfect example of an animal that exhibits precise control over its energy intake, i.e. the rat merely eats for calories. This is generally true for rats fed conventional laboratory stock diets and many workers have therefore resorted to force-feeding to induce high energy intakes. However, tube-feeding disrupts the normal nibbling meal pattern of rats, and the effect of large and less frequent meals is to produce dramatic increases in metabolic efficiency. Cohn and Joseph (1959), for example, found that tube-feeding rats 100% or 80% of normal *ad libitum* energy intake resulted in greater fat gains than those of free-feeding controls. Similarly, Rothwell and Stock (1978) reported greater body weight and fat gains in rats receiving varying fractions of normal intake by stomach tube, even though these animals adjusted voluntary intake to compensate precisely for the energy delivered by stomach tube (see Fig. 7.1).

The greater energy retention of rats whose intake is similar to, or even less than, free-feeding controls illustrates the potent effects of varying meal frequency on metabolic efficiency. It is not surprising, therefore, to find reports that rats fed by gastric intubation fail to exhibit adaptive changes in DIT during overfeeding (Armitage, Hervey and Tobin, 1981). Nevertheless, even in rats force-fed by gastric intubation, it is still possible to demonstrate the effects of low protein diets on DIT (McCracken and Gray, 1976; and Table 7.2). The results shown in Table 7.2 provide another

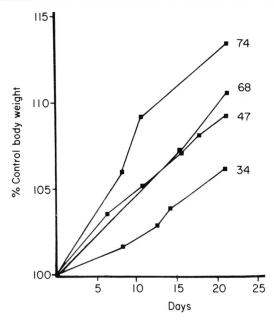

Fig. 7.1 Body weight (as percentage of control body weight) of rats tube-fed at four levels (34, 47, 68 and 74%) of *ad libitum* intake. Rats were also allowed free access to the diet. Total energy intake (voluntary plus that tube-fed) was not significantly different from that of *ad libitum* fed controls at any level of tube-feeding, but body weight (and fat) gain was significantly greater in the tube-fed groups. (Data from Rothwell and Stock, 1978)

example of the low-protein effect seen in the pig experiments (Table 7.1), only this time the intake of the high- and low-protein diets was fixed at the same level. Included in Table 7.2 are the calculated net efficiencies (i.e. energy gained per unit of energy consumed above maintenance), which are higher than one would normally obtain with *ad libitum* fed animals (see

Table 7.2 Effect of dietary protein concentration on energy balance in force-fed rats

Protein energy concentration (%)	Metabolizable energy intake (kJ d^{-1})	Body energy gain (kJ d^{-1})	Energy expenditure (kJ d^{-1})	Net efficiency (%)
10	322 ± 1	109 ± 6	223 ± 6	67 ± 3
2.5	337 ± 2	73 ± 7*	264 ± 8*	41 ± 3*

Adult, female rats were tube fed (three times daily) at twice maintenance levels for 35 days. The greater energy expenditure, and hence lower net efficiency, on the low protein diet cannot be ascribed to the cost of nutrient processing or synthesis of fat and protein since these were the same or lower than on the high protein diet. Mean values ± SEM. $n = 6$. *$P < 0.01$ (Unpublished data of M. Lotfi and M.J. Stock).

values in Table 7.4, for example). These higher efficiencies are presumably due to the effects of force-feeding but, nevertheless, the differential effects of protein concentration on efficiency are retained.

It appears, therefore, that adaptive responses to changes in nutrient content are not affected by the method of feeding, but the effects of changing the level of intake are attenuated by force-feeding. At present, no explanation can be given for this. There have been many studies that demonstrate increases in lipogenic capacity following a reduction in meal frequency (e.g. Tepperman and Tepperman, 1958; Leveille and Hanson, 1965), but these do not help to explain the phenomonon in energetic terms, i.e. from where is the extra energy for lipogenesis derived. In view of the recent developments concerning the role of the sympathetic nervous system and brown adipose tissue in DIT (see below), perhaps the effects of tube-feeding on these systems should now be investigated.

7.3 VOLUNTARY HYPERPHAGIA AND DIET-INDUCED THERMOGENESIS

The seemingly precise control of appetite in rats eating a laboratory stock diet contrasts with the variable and often excessive energy intake of man. However, this does not appear to be due to inherent species differences in feeding behaviour, but rather to their different diets. If rats are offered a varied and palatable diet, similar to that of Western man, they will also eat excessive amounts of food (Sclafani and Springer, 1976). Presenting rats with a large variety of human food items and constantly changing the choice of foods offered (the 'cafeteria' diet) results in increases in voluntary energy intake of over 80% and similar increases in DIT (Rothwell and Stock, 1979).

Many of the foods offered have a high energy density and this, together with increases of up to 40–60% in the weight of food eaten, accounts for the hyperphagia. The bulk of food may in fact be the main limitation on intake since further increases (over 126%, equivalent to 3.7 times maintenance requirements) can be achieved by offering very high fat, energy-dense foods (N.J. Rothwell, M.J. Stock and B.P. Warwick, unpublished data). A high protein (20%) stock diet is normally available and the final nutrient ratio selected by cafeteria rats indicates that the diet is well-balanced and more than adequate for growth, since fat, carbohydrate and protein normally represent about 50, 35 and 15% of total energy intake respectively. Thus, stimulation of DIT by the cafeteria diet cannot be ascribed to a low protein intake, as in the experiments described earlier.

The level of DIT produced in hyperphagic cafeteria fed rats depends on many factors, but age and genetic background appear to be very important. The genetic influence on the response to hyperphagia is profound and, even within the same strain, differences can arise in rats derived from different colonies (Rothwell and Stock, 1980a). The cafeteria experiment shown in

Table 7.3 demonstrates this effect in adult Sprague–Dawley rats. Marked increases in energy expenditure are seen in both cafeteria groups, but the intra-strain variation in DIT results in a two-fold difference in energy retention for the same degree of hyperphagia. The genetic differences between two colonies of Sprague–Dawley rat are subtle by comparison with the extremes in body fat seen in genetically obese rodents and their lean littermates. Thus, it is not surprising to find very noticeable differences in energy gain and DIT when lean and obese (ob/ob) mice are presented with the cafeteria diet (see Trayhurn and James, Chapter 8).

Table 7.3 Intra-strain differences in energy balance

	Metabolizable energy intake $(kJ\,d^{-1})$	Body energy gain $(kJ\,d^{-1})$	Energy expenditure $(kJ\,d^{-1})$
Control diet			
Colony A	231 ± 3	9 ± 1	223 ± 3
Colony B	229 ± 8	10 ± 4	219 ± 10
	N.S.	N.S.	N.S.
Cafeteria diet			
Colony A	438 ± 5	95 ± 10	343 ± 10
Colony B	429 ± 5	40 ± 5	389 ± 6
	N.S.	$P<0.01$	$P<0.001$

Rats from Colony A and B were Sprague–Dawley males of the same age (75 days). Mean values±SEM, $n = 8$ (Data from Rothwell and Stock, 1980*a*).

If young, weanling Sprague–Dawley rats are fed the cafeteria diet, heat production is often raised sufficiently to prevent any excess weight gain (Rothwell and Stock, 1980*b*), but as the animals get older the level of DIT declines and the amount of excess energy gained increases. These age-related changes are illustrated in Table 7.4, which shows recent data obtained in another strain of rat (lean Zucker). The degree of hyperphagia on the cafeteria diet was similar for both young and old rats (73% increase in energy intake) but only in the old rats was body energy gain significantly elevated. However, energy expenditure in both young and old cafeteria rats was increased, by 77 and 57% respectively, and this confirms earlier work (Rothwell and Stock, 1979) on the quantitative importance of DIT. Thus, even in old hyperphagic rats, the degree of obesity was attenuated considerably by the increase in heat production, and if these animals had not increased DIT, their fat gain would have been three times greater.

In the experiment described above, the fraction of DIT due to the obligatory energy cost of fat synthesis accounted for only 4 and 14% of the extra heat production in young and old rats respectively. This is probably an overestimate, since it assumes that fat deposited in the body was derived from *de novo* synthesis. In fact, measurements of *in vivo* lipogenesis (N.J. Rothwell, M.J. Stock and P. Trayhurn, unpublished data) indicate that, because of the high intake of dietary fat, lipogenesis is reduced in cafeteria fed rats. Liver lipogenesis is reduced by 83%, and in epididymal fat the reduction is 53%. The energy cost of depositing lipid derived directly from the diet is very low (6 kJ g^{-1}, compared with 14 kJ g^{-1} for *de novo* synthesis), and would only account for 1.8 and 6.0% of the increased heat production of the young and old cafeteria rats in the study described in Table 7.4.

Table 7.4 Energy balance in young (5.5 week) and old (5.5 month) lean male Zucker (+ /?) rats

	Young control	Young cafeteria	Old control	Old cafeteria
Metabolizable energy intake (kJ rat^{-1} d^{-1})	210±10	360±20*	245±15	420±15*
Body energy gain (kJ rat^{-1} d^{-1})	33±6	49±8*	23±5	75±12*
Energy expenditure (kJ rat^{-1} d^{-1})	175±5	311±10*	222±10	347±10*
Net efficiency (%)	34±4	20±1†	37±10	33±4

Metabolizable energy intake and body energy gain were determined by bomb calorimetry of all foods, faeces, urine and carcasses. Energy expenditure was calculated from the differences between metabolizable energy intake and body energy gain. Mean values (± SEM) for 6–8 rats per group over 24 days. *$P < 0.001$ †$P < 0.01$, compared to respective control group (Data from Rothwell and Stock, 1982a).

Thus, obligatory DIT accounts, at most, for only about 10% of the raised heat production, and the remainder must be due to adaptive changes in energy dissipation. The magnitude of this adaptive component can be considerable (equivalent to 2.5 times the fasting heat production in young hyperphagic animals) and, not surprisingly, is associated with high rates of oxygen consumption (Andrews and Donne, 1982; Rothwell and Stock, 1982b). Fig. 7.2 shows a 24-hour record of VO_2 in control and cafeteria rats and illustrates the effect of nocturnal food consumption on metabolic rate in the cafeteria group. Further confirmation of these findings has come from a study by Trayhurn *et al.*, 1982a) on cafeteria feeding lean and genetically obese (ob/ob) mice. All animals increased food intake on the cafeteria diet, but lean mice deposited only 4% of their excess in the carcass, whereas the obese mutants retained 54% of the extra energy. This study not only provides a dramatic demonstration of the effect of genotype on the capacity for DIT and energetic efficiency, but also invalidates any criticisms

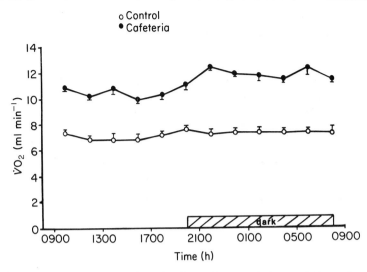

Fig. 7.2 Oxygen consumption (ml min⁻¹) for 2-hour periods over 24 hours in control and cafeteria-fed rats. Mean values ($n = 4$), bar denotes SEM. All values for cafeteria rats are significantly greater ($P < 0.001$) than control values. (From Rothwell and Stock, 1982*b*)

of the methods employed to measure energy balance in rats or mice on cafeteria diets.

7.4 MECHANISMS OF DIET-INDUCED THERMOGENESIS

7.4.1 Similarities to non-shivering thermogenesis

Several mechanisms have been proposed to explain the high metabolic rates of hyperphagic animals, such as substrate cycles, protein turnover and active sodium transport, but as long ago as 1968 it was suggested that DIT might involve similar mechanisms to non-shivering thermogenesis (Stirling and Stock, 1968).

It is now well established that cold exposure results in a marked increase in sympathetic activity, which is largely responsible for the elevated metabolic rates of cold-adapted animals (Himms-Hagen, 1967, 1976; Jansky, 1973). Hyperphagic cafeteria-fed rats also show enhanced thermogenic responses to noradrenaline (Fig. 7.3) and their high levels of $\dot{V}O_2$ can be completely abolished by β-adrenergic blockade with propranolol (Rothwell and Stock, 1979). These findings support the proposal that cold exposure and overfeeding elicit the same thermogenic mechanisms, and imply that the two stimuli may be interactive. Exposure to cold causes an increase in food intake which may in itself contribute to the thermogenesis, while fasting completely abolishes NST (Bignall, Heggeness and Palmer, 1977). Cold-adapted animals brought into a warm environment (24°C) also show a greater thermogenic response to the cafeteria diet or to a single meal than animals with no previous experience of cold (Rothwell, Saville and Stock, 1982a). Conversely, cafeteria fed rats maintained at 24°C can maintain a high core temperature when acutely exposed to 5°C, without shivering, whereas stock fed controls shiver violently. Injection of these animals with propranolol caused the cafeteria group to shiver violently, but rectal temperature then fell rapidly to the level of controls, demonstrating that cafeteria fed rats maintained in the warm are apparently adapted to cold (Rothwell and Stock, 1980c). Thus, DIT and NST can substitute for one another, and combining cafeteria feeding with cold exposure results in dramatic increases in energy expenditure and thermogenic capacity (Rothwell and Stock, 1980c).

The metabolic origins of NST were in doubt for many years. In hibernators and neonates, it was accepted that brown adipose tissue (BAT) was the major source of the increased heat production, but in other, adult mammals it was considered that the total mass of tissue was too small to account for the very large increases in metabolic rate. However, Foster and Frydman (1978) have now dispelled many of these doubts by clearly demonstrating the remarkable thermogenic capacity of BAT in the adult rat. These workers measured blood flow, with radioactively-labelled microspheres, and oxygen extraction of BAT *in vivo* in warm- and cold-adapted rats, and found that this tissue could account for at least 60% of

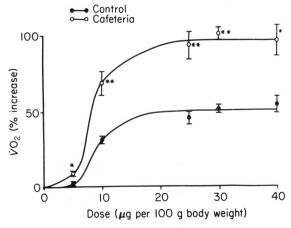

Fig. 7.3 Effect of noradrenaline on oxygen consumption in cafeteria and control animals. Oxygen consumption ($\dot{V}_2$) was measured for 2 h before and 2 h after subcutaneous injection of different doses of noradrenaline. Results are presented as oxygen onsumption after noradrenaline as a percentage of pre-injection values. Mean values ($n = 6$), bar denotes SEM. *$P < 0.05$; **$P < 0.01$ compared to controls. (From Rothwell and Stock, 1979)

NST. These experiments were initially performed in anaesthetized animals but have now been confirmed in conscious rats exposed to varying ambient temperatures (Foster and Frydman, 1979).

Cafeteria feeding results in two- to threefold increases in BAT mass which are not simply due to deposition of triglyceride, but represent a significant increase in protein content. In older animals, this response is due to hypertrophy of the tissue but in young rats there is an initial increase in brown adipocyte size which is then followed by hyperplasia (Tulp *et al.*, 1980; Brooks, Rothwell and Stock, 1982). The measurements of blood flow and oxygen extraction, performed by Foster and Frydman (1978), have now been repeated in cafeteria fed rats with similar results. During noradrenaline infusion it was found that BAT accounts for all of the twofold increase in thermogenic capacity of the cafeteria fed rats (Fig. 7.4), thus confirming BAT as the major effector tissue for adaptive DIT as well as NST (Rothwell and Stock, 1981*a*).

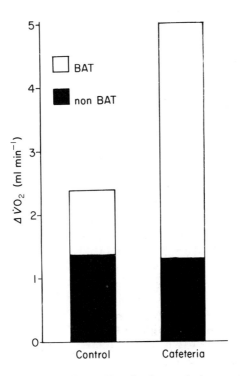

Fig. 7.4 The contribution made by BAT to the thermogenic response of control and cafeteria rats infused with noradrenaline. The thermogenic response ($\Delta \dot{V}O_2$) represents the difference between rats infused with saline or noradrenaline. (From Rothwell and Stock, 1981*a*)

7.4.2 Biochemical mechanisms

There are two mechanisms which may be responsible for the high metabolic rates of BAT, one of which involves uncoupling of oxidative phosphorylation and the other, increased utilization of ATP.

Nicholls (1979, see also Chapter 2) has suggested that BAT mitochondria possess a proton conductance pathway which allows protons to 'leak' back across the inner mitochondrial membrane without obligatory synthesis of ATP. Thus, respiration is effectively uncoupled from phosphorylation, with a resultant increase in heat production. This pathway can be inhibited by purine nucleotides which bind to a protein of molecular weight 32 000, and the activity of the pathway can be assessed from the effects of purine nucleotides (particularly GDP) on mitochondrial respiration, or from the binding of radioactively labelled nucleotides to mitochondria. Several groups have reported that the activity of the proton conductance pathway is elevated in cold-adapted animals (Nicholls, 1979; Desautels and Himms-Hagen, 1979; Sundin and Cannon, 1980) and now there are also data to show that it is increased in animals exhibiting DIT (Brooks *et al.*, 1980). Mitochondrial respiration in BAT from cafeteria-fed rats is elevated by about 30% compared to stock-fed controls but addition of GDP to the medium *in vitro* causes a reduction in respiratory rate, which is greatest in mitochondria from cafeteria-fed rats, and abolishes the differences between the two groups. Subsequent addition of the uncoupling agent FCCP stimulates maximal rates of oxygen consumption, which are identical in mitochondria from control and cafeteria-fed rats (Brooks *et al.*, 1980).

The specific binding of [^{3}H]GDP to BAT mitochondria is elevated by about 150% in the cafeteria group, but since the mass of mitochondria is also doubled, the total binding in the interscapular depot is increased by about fourfold (Brooks *et al.*, 1980). Full kinetic analysis of the purine nucleotide binding to BAT mitochondria from control and cafeteria-fed rats has now been performed, and Scatchard plots reveal an increase in the maximum number of binding sites in the cafeteria group but no change in affinity (Brooks *et al.*, 1982). However, unlike cold-adapted animals, no change in the concentration of the 32 000 peptide has been found in BAT mitochondria from cafeteria rats (Himms-Hagen, Triandafillou and Gwilliam, 1981; S.L. Brooks, D. Ricquier, N.J. Rothwell and M.J. Stock, unpublished data), although the total mitochondrial protein is markedly elevated in these animals. There are concurrent increases in the activity of mitochondrial respiratory enzymes (e.g. cytochrome oxidase, α-glycerophosphate dehydrogenase), but no change in their specific activity. In this respect, the mitochondrial changes are similar to those seen in cold-adapted animals (Barnard, Skala and Lindberg, 1970; Heick *et al.*, 1973).

The information presently available suggests that the only noticeable difference between rats exhibiting DIT and NST is the unaltered mitochondrial concentration of the 32000 protein in the former. It has been argued (Himms-Hagen *et al.*, 1981) that hyperphagia produces changes similar to those seen in warm-adapted rats chronically treated with twice-daily injections of noradrenaline. Animals treated in this way exhibit increased thermogenic responses to noradrenaline, hypertrophy of BAT, (LeBlanc and Villemaire, 1970) greater mitochondrial enzyme activity, but no change in the concentration of the GDP-binding protein (Desautels and Himms-Hagen, 1979). It is suggested that adrenergic activation of BAT unmasks pre-existing GDP-binding sites. There is some evidence for an unmasking effect of noradrenaline, but this is an acute response. A single injection of noradrenaline, given 1 hour before sacrifice, results in a 67% increase in the maximum number of binding sites in BAT mitochondria from cafeteria-fed rats without affecting the affinity constant (Brooks *et al.*, 1982). In stock-fed control animals, however, the absolute level of binding is about 30–50% lower and is not significantly affected by noradrenaline injection. This implies that the effects of cafeteria feeding on binding capacity cannot be due entirely to this acute unmasking effect of adrenergic activation. If it were, a noradrenaline injection should raise GDP binding in control animals to the same as that in cafeteria animals. In the absence of any change in the mitochondrial content of the 32000 binding protein or its binding constant, one can only presume that either chronic adrenergic activation during hyperphagia exposes further sites or that another unmasking factor is involved. The former explanation is unlikely since Desautels and Himms-Hagen (1979) failed to produce an increase in GDP-binding by chronic treatment with twice-daily injection or noradrenaline.

The arguments for a chronic unmasking effect of hyperphagia all rely on the failure to observe changes in the concentration of the mitochondrial binding protein. However, the methods available for quantifying this are not precise and it is worth noting that the 10–12-fold increase in GDP binding capacity usually observed in the cold-adapted rat is associated with only a 50% increase the amount of protein (Ricquier and Kader, 1976). Cafeteria feeding produces only a 2–3-fold increase in binding capacity and so it is perhaps not surprising that any increase in the 32000 protein goes undetected.

Horwitz (1973) has proposed that increased activity of the Na^+K^+-ATPase enzyme may contribute to BAT thermogenesis in cold-adapted animals, and the activity of this enzyme is also elevated in BAT homogenates from cafeteria rats (Rothwell, Stock and Wyllie, 1981). This increase may be due to endogenous noradrenaline present in the homogenate because no difference in activity is observed between membrane fractions of BAT from control and cafeteria rats, though addition of noradrenaline (Fig. 7.5) causes a much greater stimulation of the enzyme in cafeteria rats (Rothwell *et al.*, 1981). Preliminary studies indicate that stimulation of the Na^+K^+-ATPase by noradrenaline is not necessarily associated with increased ion pumping so that the enzyme may be acting inefficiently (M.G. Wyllie, unpublished data). A remarkable correlation has been observed between the *in vitro* Na^+K^+-ATPase activity in BAT homogenates and *in vivo* oxygen consumption in control and cafeteria rats (correlation coefficient = 0.9) which, although not necessarily causal, does suggest that this enzyme is involved in DIT (Rothwell *et al.*, 1981).

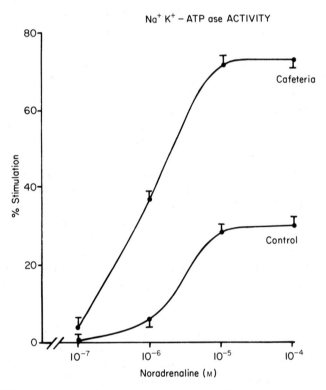

Fig. 7.5 Effect of concentration on the noradrenaline-stimulated increase in NA^+K^+-ATPase activity of BAT membrane fractions from control ($n = 4$) and cafeteria-fed ($n = 6$) rats. Mean values, bar denotes SEM. (From Rothwell, Stock and Wyllie, 1981)

The relative importance of the proton conductance pathway or Na^+K^+-ATPase, and their contribution to *in vivo* metabolic rates, are unknown. It might be considered that these two mechanisms are mutually exclusive since one involves reduced production of ATP and the other increased utilization. However, it is possible that both mechanisms could act simultaneously because even in active BAT, mitochondrial respiration may be partially coupled so that some ATP is synthesized and ATP could also be derived from substrate level phosphorylation.

7.4.3 Neuroendocrine control

The data discussed earlier clearly demonstrated that the sympathetic nervous system is the primary activator of NST and DIT. Both these forms of heat production can be mimicked by noradrenaline, inhibited by β-blockade and are associated with increased urinary excretion of catecholamines and elevated rates of noradrenaline turnover (Landsberg *et al.*, 1981). Recently, the pharmacology of DIT has been investigated in more detail, and the use of selective agonists and antagonists has revealed that the thermogenic response to diet is mainly associated with the β_1 receptor However, complete inhibition of DIT can be achieved only by combined treatment with β_1 and β_2 antagonists, indicating either a mixed receptor population or possibly a separate, as yet unidentified, β-receptor (N.J. Rothwell, M.J. Stock and D. Stribling, unpublished data).

It has recently been suggested that the parasympathetic nervous system also influences DIT. The rise in metabolic rate and BAT temperature after a single meal are potentiated by treatment with the muscarinic antagonist atropine sulphate (Rothwell, Saville and Stock, 1981). Genetically obese Zucker rats fail to increase oxygen consumption after food, but this response can be completely restored to normal by atropine treatment, indicating that high parasympathetic activity may be partly responsible for the defective thermogenesis and increased fat deposition in these mutants (Rothwell *et al.*, 1981). Furthermore there is evidence for elevated activity of the parasympathetic in many obese animals (e.g. ob/ob and db/db mice and rats with lesions of the ventromedial hypothalamus) and weight gains can often be inhibited by vagotomy (see Jeanrenaud, 1978 for review).

The long-term effects of the parasympathetic on DIT are unknown, but it now seems that chronic increases in sympathetic activity alone cannot account for all of the adaptive changes seen in cold-adapted or cafeteria-fed rats (see previous section on biochemical mechanisms). It therefore appears that some other trophic factor must be involved in DIT and NST, and the most likely is probably the thyroid hormones, since T_3 levels are elevated in cold-adapted (Reichlin *et al.*, 1973) and cafeteria-fed rats (Rothwell and Stock, 1979). It has been reported that experimental hyperthyroidism in rats inhibits purine nucleotide binding to BAT mitochondria (Sundin, 1981), but we have found no change in binding assessed from Scatchard analysis (N.J. Rothwell, M.E. Saville and M.J. Stock, unpublished data). Moderate hyperthyroidism does, however, stimulate BAT, Na^+K^+-ATPase activity in the rat, and addition of T_3 or T_4 to the incubation medium *in vitro* causes sensitization of the enzyme to noradrenaline (Rothwell *et al.*, 1982b). These varied responses of BAT may reflect the levels of thyroid hormones achieved, so that moderate increases stimulate BAT thermogenesis whereas severe hyperthyroidism could be inhibitory.

Another hormone thought to play an important role in DIT is insulin.

Very large doses of exogenous insulin induce hypoglycaemic hyperphagia and obesity, but absence of this hormone inhibits DIT and NST. Streptozotocin-induced diabetic rats maintained on a cafeteria diet, fail to show the normal increases in $\dot{V}O_2$ and thermogenic responses to noradrenaline seen in hyperphagic rats, unless injected with insulin shortly before the measurements (Rothwell and Stock, 1981b). Diabetic rats also fail to exhibit NST and cannot survive cold environments unless replaced with insulin (Drury, 1957). Previous work has shown that insulin can stimulate $\dot{V}O_2$ in BAT (Dawkins and Hull, 1963) and potentiate the response to noradrenaline (Shackney and Joel, 1966), but the absence of thermogenesis in diabetic animals may also represent a central requirement for insulin. Diabetic rats increase metabolic rate after injection with noradrenaline (Rothwell and Stock, 1981b), indicating that the thermogenic effector mechanisms can function normally and that the impaired DIT of diabetic cafeteria-fed rats described above may partly result from an inability to activate these mechanisms.

7.4.4 Central control

As yet, very little is known about the central control of DIT, but there is some evidence to suggest that this may be closely related to the control of food intake. The dual centre hypothesis proposed that the ventromedial hypothalamus (VMH) acts as a satiety centre and the lateral hypothalamus (LH) as a hunger or feeding centre, and that these two areas of the brain exert the major control over food intake. This view is now considered rather simplistic but, nevertheless, the profound effects of the VMH and LH on food intake are very well known.

Electrolytic destruction of the LH causes aphagia, whereas lesions of the VMH induce hyperphagia and obesity. However, VMH-lesioned animals will become obese even when intake is restricted to the level of lean controls (Han, 1967), suggesting that increased efficiency of energy retention (i.e. reduced expenditure) is a major cause of obesity. Seydoux *et al.*, (1981) have reported that BAT from these animals is atrophied and insensitive to noradrenaline or nerve stimulation. Also, electrical stimulation of the VMH causes a marked rise in temperature of the interscapular BAT depot which is very similar to the effects of sympathetic nerve stimulation, and can be inhibited by β-adrenergic blockade (Perkins *et al.*, 1981). These findings indicate that the VMH is concerned with the control of energy expenditure as well as food intake and could also explain the insulin requirement for DIT. This area of the hypothalamus contains cells which are sensitive to glucose and may be dependent on insulin for glucose entry (Oomura *et al.*, 1978).

Glucose deprivation in the hypothalamus induces hyperphagia, but Shiraishi and Mager (1980) have reported that glucoprivation after hypothalamic injections of the glucose analogue 2-deoxy-D-glucose also causes a marked reduction in metabolic rate, which can be inhibited by vagotomy or atropine. It is also possible that the lateral hypothalamus inhibits thermogenesis, because stimulation of this area increases parasympathetic activity whereas electrolytic lesions result in large increases in metabolic rate (R. Keesey, personal communication). Other areas of the brain, apart from the hypothalamus, may also play a role in controlling DIT and NST and, in particular, there is some evidence to suggest that the lower brain stem may provide a thermogenic drive. Bignall *et al.*, (1975) found that infant rat pups (5 and 10 days old) developed high body temperatures and maximal rates of oxygen consumption following midpontine brain stem transection. The same effect of decerebration can be produced in adult animals and the increases in metabolic rate were found to be inhibited by β-adrenergic blockade (N.J. Rothwell, M.J. Stock and A. Thexton, unpublished data). Furthermore, decerebration results in large increases in the temperature of the interscapular BAT depot (Fig. 7.6) suggesting that the rise in oxygen consumption and body temperature is due to sympathetic activation of BAT thermogenesis. If confirmed, thse findings would indicate that the hypothalamus and other higher brain areas modulate thermogenesis via inhibitory and disinhibitory influences on this efferent thermogenic drive from the lower brain stem.

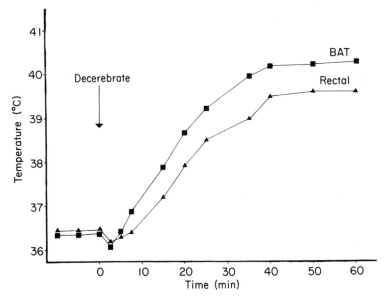

Fig. 7.6 Interscapular BAT and rectal temperatures in an anaesthetized rat following midpontine brain stem transection. (Unpublished data of N.J. Rothwell, M.J. Stock and A. Thexton)

7.5 OTHER FACTORS AFFECTING DIET-INDUCED THERMOGENESIS

The most important influences on DIT are environmental temperature, age and genetic background, all of which have been discussed earlier. However, several other factors can have quite marked effects on thermogenic capacity such as dietary composition, season and behavioural conditioning.

The traditional view has been that high fat diets always lead to an increased efficiency of energy utilization and excess weight gains, whereas diets containing a large proportion of carbohydrate tend to reduce weight gains (Peckham *et al.*, 1962; Schemmel and Mickelsen, 1973). High carbohydrate diets usually result in elevated plasma thyroid hormone levels (Davidson and Chopra, 1979) and in human overfeeding studies, weight gains were much lower when the excess intake was comprised mainly of carbohydrate rather than fat (Danforth *et al.*, 1979). Rats allowed a high-carbohydrate, low-fat cafeteria diet gain less weight than animals maintained on a balanced cafeteria diet, but this is partly due to lower levels of food intake in the former group. Animals presented with a high-fat, low-carbohydrate cafeteria diet exhibit marked hyperphagia but excess weight gains are relatively small because of very large increases in energy expenditure (N.J. Rothwell, M.J. Stock and B.P. Warwick, unpublished data). This suggests that differences between the two nutrients may be partly due to their effects on energy intake but may also be dependent on changes in the thermogenic capacity of the animal. A single carbohydrate meal for example produced similar increases in metabolic rate in control and cafeteria-fed rats, where fat induces a much larger response in the latter group, indicating that fat may have a greater thermogenic effect in animals where DIT is already high (N.J. Rothwell, M.E. Saville and M.J. Stock, unpublished data).

Seasonal variations in energy intake and expenditure are obvious in hibernators and many other wild animals, and are largely dependent on food availability and environmental temperature. However, seasonal changes in the level of DIT in cafeteria-fed rats maintained under constant conditions of lighting and temperature have been observed. Energy expenditure, noradrenaline responses and BAT mass are all significantly greater in cafeteria-fed rats during the winter months (Oct–Feb), in spite of similar levels of hyperphagia throughout the year, but these parameters remain relatively constant in stock-fed controls (Fig. 7.7).

It has also been noticed that DIT, like feeding behaviour, may be susceptible to conditioning. The energy expenditure of cafeteria-fed rats is much higher at night when food intake is greatest, and the level of purine nucleotide binding in BAT also increases within three hours of darkness. However, a similar increase in binding can be observed in the cafeteria group when no food is available, but this effect does not occur in the

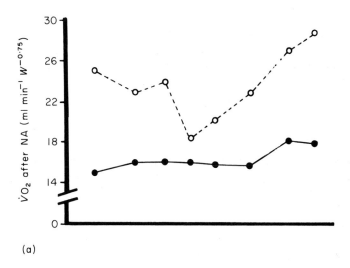

(a)

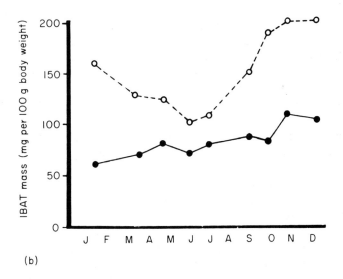

(b)

Fig. 7.7 (a) Resting oxygen consumption after injection of noradrenaline (25 μg per 100 g body weight) and (b) interscapular BAT mass of separate groups of male rats (approx. 6 weeks old) maintained on either stock (●) or cafeteria (○) diets for 2–3 weeks at different times of the year. Each point represents a mean of six to ten animals and all values for cafeteria rats are significantly ($P < 0.01$) higher than control values. (Unpublished data of N. J. Rothwell and M. J. Stock)

stock-fed animals (S.L. Brooks, N.J. Rothwell and M.J. Stock, unpublished data).

7.6 REDUCED THERMOGENESIS

Undernutrition or complete starvation produced reductions in energy expenditure in man and experimental animals (Benedict *et al.*, 1919; Dontcheff, 1973). The reduction is partly due to decreases in body mass (i.e. maintenance requirements) and spontaneous activity, but there is also evidence for a simultaneous increase in metabolic efficiency. The effect of these changes is to allow the body to conserve its energy stores and thus increase survival time.

The effects of fasting or food restriction on metabolic efficiency are seen most clearly during rehabilitation, when the increased efficiency produces greater increases in weight gain and energy retention than would be normally expected. Levitsky, Faust and Glassman (1976), for example, found that the increase in food intake of rats previously fasted for 1–4 days was independent of the length of the fast or the amount of weight lost, and a return to normal body weight was achieved even when intake was controlled at pre-fast levels. Similarly Oscai (1980) found that rats consumed normal amounts of food after 30 days of food restriction, but if given sufficient time, could still return spontaneously to normal body weight. These observations are consistent with the reports of increased efficiency of growth during the rehabilitation of undernourished young rats (Mahendra and Miller, 1969; Miller and Wise, 1973) and lower metabolic rates in response to food in energy-restricted rats (Forsum, Hillman and Nesheim, 1981; Boyle *et al.*, 1981). In this respect, however, it should be noted that exaggerated thermic responses to a single meal have been observed in babies recovering from protein-energy malnutrition (Brooke and Ashworth, 1972). This effect may be due to the obligatory DIT associated with restoring body protein.

The extent of the metabolic adaptions to chronic low energy intakes in man can be seen when comparing intakes of impoverished groups from developing countries with conventional Western standards for energy requirements. Norgan, Ferro-Luzzi and Durnin (1974), for example, reported exceptionally low energy intakes for New Guinean adults, and Miller *et al.* (1976) found that energy intakes of Ethiopian highlanders were equivalent to about 70% of FAO/WHO recommendations, even after taking account of the low body weights of the subjects. Thus, even though a reduced body mass represents part of the adaption to limited availabiliy of food, increased metabolic efficiency clearly makes an important contribution to the survival of these poor communities. This metabolic strategy becomes even more important when the physiological requirement for additional energy increases as a result of pregnancy and lactation. A remarkable demonstration of this is described in a recent report on child-bearing Gambian women (Prentice *et al.*, 1981). During the season when food was most abundant, energy intakes were only 62–64% of international recommendations and were theoretically incompatible with successful child-nurturing. However, not only was maternal nutritional status maintained, irrespective of the number of children born, but breast milk output and quality and early infant growth were close to values from well-nourished communities.

These examples suggest that the ability to inactivate DIT is as important to the regulation of energy balance during food restriction as the ability to activate it in response to overnutrition.

The changes in sympathetic activity seen during under- and overfeeding (see Landsberg and Young, Chapter 4) indicate that changes in metabolic efficiency are, at least partly, under autonomic control, and Trayhurn and co-workers (1982b) have found marked reductions in BAT thermogenesis in lactating mice which could make significant amounts of energy available to meet the cost of milk production. The ability to activate and inactive thermogenesis to meet fluctuations in energy supply and demand has obvious evolutionary and selective advantages and is perhaps best examplified by seasonal hibernators who switch from hyperphagia, increased efficiency and fat deposition in the autumn, to high rates of thermogenesis and fat depletion in the spring. However, chronic exposure over successive generations to limited food supplies appears to favour the evolution of 'thrifty' genotypes that have lost the capacity for thermogenesis. Example of this metabolic adaptation are the desert animals (e.g. spiny mouse, tuco tuco) and primitive human tribes (Pima Indian, Aborigine) that have evolved and can survive in exceptionally arid environments. Nevertheless, this loss of metabolic flexibility can prove a disadvantage when abundant food is available since all these examples of thrifty genotypes become markedly obese (and often diabetic) when brought into the laboratory or, in the human examples, exposed to the Western urban environment (see Rothwell and Stock, 1981c for discussion of the relative advantages of 'thrifty' versus 'wasteful' genotypes).

7.7 SUMMARY

There is little doubt that a variety of tissues and metabolic pathways can influence the overall efficiency with which energy is utilized in the body. Their relative importance in determining the level of DIT will depend on circumstances, and factors such as nutrient balance, meal frequency, level of intake, etc. could influence DIT in several ways. However, it is now obvious that, in terms of adaptive DIT, the mechanism involves the same autonomic, endocrine and biochemical processes that are involved in NST. The similarities between these two forms of adaptive thermogenesis are summarized in Table 7.5. The contribution made by BAT thermogenesis to

Table 7.5 Similarities between diet-induced and non-shivering thermogenesis

1. Increased energy intake and expenditure.
2. Reduced efficiency of energy utilization.
3. Improved cold tolerance and lower threshold temperature for shivering.
4. Increased sympathetic activity.
5. Activation by β-adrenergic agonists.
6. Inhibition by β-adrenergic antagonists, ganglionic blocking agents and hypoxia.
7. Raised plasma T_3, reduced plasma insulin and/or improved glucose homeostasis.
8. Hypertrophy and hyperplasia of BAT.
9. Increased BAT mitochondrial respiration and proton conductance.
10. Increased BAT Na^+- K^+-ATPase activity.

the regulation of energy balance has yet to be fully assessed but, because of its large capacity for dissipating significant amounts of energy as heat, its potential value should not be underestimated. For example, in cold-adapted rats exhibiting NST and hyperphagic rats exhibiting DIT, the interscapular BAT depot can utilize 1.7 ml of oxygen g^{-1} min^{-1} when stimulated with noradrenaline. If this level of thermogenesis applied to the other BAT depots, it would allow a rat to eat twice its maintenance requirements (i.e. 840 kJ $W^{-0.75}$ d^{-1}) without going into positive energy balance. In the young control and cafeteria rats shown in Table 7.4, the level of DIT in the cafeteria groups was estimated to be 513 kJ $W^{-0.75}$ d^{-1} greater than in the control group, and could therefore be dissipated by BAT.

The role of BAT in the regulation of energy balance in man has yet to be established, and it will prove very difficult to do so. For example, a 20–25% increase in metabolic rate could be accomplished by as little as 40–50 g of active BAT (i.e. less than 0.1% of body weight). Thus, an apparently trivial amount of tissue could have a profound influence on energy balance, since 20% of daily energy expenditure could make the difference between maintaining body weight or gaining at the rate of 20 kg per year.

REFERENCES

Adolph, E.F. (1947) Urges to eat and drink in rats. *Am. J. Physiol.*, **151**, 110–25.

Andrews, J.F. and Donne, B. (1982) 24h oxygen consumption of rats stimulated by cafeteria feeding. *Proc. Nutr. Soc.* **41**, 36A.

Apfelbaum, M., Botscarron, J. and Lacatis, D. (1971) Effects of calorie restriction and excessive calorie intake on energy expenditure. *Am. J. Clin. Nutr.*, **24**, 1405–9.

Armitage, G., Hervey, G.R. and Tobin, G. (1981) Energy expenditure in rats over-fed by tube-feeding. *J. Physiol.*, **312**, 58P.

Barnard, T., Skala, J. and Lindberg, O. (1970) Changes in interscapular brown adipose tissue of the rat during perinatal and early postnatal development and after cold acclimation. Activities of some respiratory enzymes. *Comp. Biochem. Physiol.*, **33**, 499–508.

Benedict, F.G., Miles, W.R., Roth, P. and Smith, M. (1919) *Human Vitality and Efficiency under Prolonged Restricted Diet.* Carnegie Institute of Washington, Publication 280.

Bignall, K.E., Heggeness, F.W. and Palmer, J.E. (1975) Effect of neonatal decerebration on thermogenesis during starvation and cold exposure in the rat. *Exp. Neurol.,* **49,** 174–88.

Bignall, K.E., Heggeness, F.W. and Palmer, J.E. (1977) Sympathetic inhibition of thermogenesis in the infant rat: possible glucostatic control. *Am. J. Physiol.,* **233,** R23–9.

Boyle, P.C., Storlein, L.H., Harper, A.E. and Keesey, R.E. (1981) Oxygen consumption and locomotor activity during restricted feeding and realimentation. *Am. J. Physiol.,* **241,** R392–7.

Brooke, O.G. and Ashworth, A. (1972) The influence of malnutrition on the postprandial metabolic rate and respiratory quotient. *Br. J. Nutr.,* **27,** 407–15.

Brooks, S.L., Rothwell, N.J. and Stock, M.J. (1982) Effects of diet and acute noradrenaline treatment on brown adipose tissue development and mitochondrial purine-nucleotide binding. *Q.J. Exp. Physiol.,* **67,** 259–68.

Brooks, S.L., Rothwell, N.J., Stock, M.J., Goodbody, A.E. and Trayhurn, P. (1980) Increased proton conductance pathway in brown adipose tissue mitochondria of rats exhibiting diet-induced thermogenesis. *Nature,* **286,** 274–6.

Cohn, C. and Joseph, D. (1959) Changes in body composition with force feeding. *Am. J. Physiol.,* **196,** 965–8.

Danforth, E., Horton, E.S., O'Connell, M., Sims, E.A.H., Burger, A.G., Ingbar, S.H., Braverman, L. and Vagenakis, A.G. (1979) Dietary-induced alterations in thyroid hormone metabolism during overnutrition. *J. Clin. Invest.,* **64,** 1336–47.

Davidson, M.B. and Chopra, I.J. (1979) Effect of carbohydrate and non-carbohydrate sources of calories on plasma 3, 5, 3′-triiodothyronine concentrations in man. *J. Clin. Endocr. Metab.,* **48,** 577–81.

Dawkins, M.J.R. and Hull, D. (1963) Brown adipose tissue and the response of newborn rabbits to cold. *J. Physiol.,* **172,** 216–38.

Desautels, M. and Himms-Hagen, J. (1979) Roles of noradrenaline and protein synthesis in the cold-induced increase in purine nucleotide binding by rat brown adipose tissue mitochondria. *Can. J. Biochem.,* **57,** 968–76.

Dontcheff, L. (1973) Effects of prolonged starvation on energy expenditure. in *Energy Balance in Man* (ed. M. Apfelbaum), Masson, Paris, pp. 113–8.

Drury, D.R. (1957) Storage and other disposal of fed carbohydrate. *Artic Aeromed. Lab. Tech. Rep.,* **46,** 1–10.

Forbes, E.B. and Swift, R.W. (1944) Associative dynamic effects of protein carbohydrate and fat. *J. Nutr.,* **27,** 453–68.

Forsum, E., Hillman, P.E. and Nesheim, M.C. (1981) Effect of energy restriction on total heat production, basal metabolic rate and specific dynamic action of food in rats. *J. Nutr.,* **111,** 1691–7.

Foster, D.O. and Frydman, M.L. (1978) Nonshivering thermogenesis in the rat. II Measurements of blood flow with microspheres point to brown adipose tissue as the dominant site of the calorigenesis induced by noradrenaline. *Can. J. Physiol. Pharmacol.,* **56,** 110–22.

Foster, D.O. and Frydman, M.L. (1979) Tissue distribution of cold-induced thermogenesis in conscious warm- or cold-acclimated rats reevaluated from changes in tissue blood flow. The dominant role of brown adipose tissue in the replacement of shivering by nonshivering thermogenesis. *Can. J. Physiol. Pharmacol.*, **57**, 257–70.

Garrow, J.S. (1978) The regulation of energy expenditure in man. in *Recent Advances in Obesity Research* II. (ed. G.A. Bray), London, Newman, pp. 200–10.

Goldman, R.F., Haisman, M.F., Cynum, G., Danforth, E., Horton, E.S. and Sims, E.A.H. (1975) Experimental obesity in man: Metabolic rate in relation to dietary intake. in *Obesity in Perspective* (ed. G.A. Bray), DHEW Publ. No. (NIH) 75–708, Washington DC, Nat. Inst. Health. pp. 165–86.

Gulick, A. (1922) A study of weight regulation in the adult human body during overnutrition. *Am. J. Physiol.*, **60**, 371–95.

Gurr, M.I., Mawson, R., Rothwell, N.J. and Stock, M.J. (1980) Effects of manipulating dietary protein and energy intake on energy balance and thermogenesis in pigs. *J. Nutr.*, **110**, 532–42.

Hamilton, T.S. (1939) The heat increments of diets balanced and unbalanced with respect to protein. *J. Nutr.*, **17**, 583–92.

Han, P.W. (1967) Hypothalmic obesity in rats without hyperphagia. *Trans. N.Y. Acad. Sci.*, **30**, 229–43.

Heick, H.M.C., Vachon, C., Kallai, M.A., Begin-Heick, N. and LeBlanc, J. (1973) The effects of thyroxine and isopropylnoradrenaline on cytochrome oxidase activity in brown adipose tissue. *Can. J. Physiol. Pharmacol.*, **51**, 751–8.

Himms-Hagen, J. (1967) Sympathetic regulation of metabolism. *Pharmacol. Rev.*, **19**, 367–461.

Himms-Hagen, J. (1976) Cellular thermogenesis. *Ann. Rev. Physiol.*, **38**, 315–51.

Himms-Hagen, J., Triandafillou, J. and Gwilliam, C. (1981) Brown adipose tissue of cafeteria-fed rats. *Am. J. Physiol.*, **241**, E116–20.

Horwitz, B.A. (1973) Oubain-sensitive component of brown fat thermogenesis. *Am. J. Physiol.*, **224**, 352–5.

Jansky, L. (1973) Non-shivering thermogenesis and its thermoregulatory significance. *Biol. Rev.*, **48**, 85–132.

Jeanrenaud, B. (1978) An overview of experimental models of obesity. in *Recent Advances in Obesity Research*, II. (ed. G.A. Bray), London, Newman, pp. 111–22.

Kleiber, M. (1945) Dietary deficiencies and energy metabolism. *Nutr. Abstr. Rev.*, **15**, 207–22.

Landsberg, L., Saville, E., Young, J.B., Rothwell, N.J. and Stock, M.J. (1981) Chronic overfeeding stimulates sympathetic nervous system activity of brown adipose tissue *in vivo*. *Clin. Res.*, **29**, 542A.

LeBlanc, J. and Villemaire, A. (1970) Thyroxine and noradrenaline on noradrenaline sensitivity, cold resistance and brown fat. *Am. J. Physiol.*, **217**, 1742–5.

Leveille, G.A. and Hanson, R.W. (1965) The influence of periodicity of eating on adipose tissue metabolism in the rat. *Can. J. Physiol. Pharmacol.*, **43**, 857–68.

Levitsky, D.A., Faust, I. and Glassman, M. (1976) The ingestion of food and the recovery of body weight following fasting in the naive rat. *Physiol. Behav.*, **17**, 575–80.

Mahendra, C.C. and Miller, D.S. (1969) The effect of depletion – repletion on calorie utilisation. *Proc. Nutr. Soc.*, **28**, 72A.

McCraken, K.J. and Gray, R. (1976) A futile energy cycle in adult rats given a low-protein diet at high levels of energy intake. *Proc. Nutr. Soc.*, **35**, 59A–60A.

Miller, D.S. and Mumford, P. (1967) Gluttony 1. An experimental study of over-eating on high or low protein diets. *Am. J. Clin. Nutr.*, **20**, 1212–22.

Miller, D.S., Mumford, P. and Stock, M.J. (1967) Gluttony 2. Thermogenesis in overeating man. *Am. J. Clin. Nutr.*, **20**, 1223–9.

Miller, D.S. and Payne, P.R. (1962) Weight-maintenance and food intake. *J. Nutr.*, **78**, 255–62.

Miller, D.S. and Wise, A. (1973) The effect of age on the thermic energy of 'catch-up' growth. *Proc. Nutr. Soc.*, **32**, 41A.

Miller, D.S. *et al.* (1976) The Ethiopia applied nutrition project. *Proc. Roy. Soc. B*, **194**, 23–48.

Mitchell, H.H. (1962) *Comparative Nutrition of Man and Domestic Animals.* Academic Press, New York.

Morgan, J.B., York, D.A., Wasilewska, A. and Portman, J. (1982) A study on the thermic responses to a meal and to a sympathomimetic drug (ephedrine) in relation to energy balance in man. *Br. J. Nutr.*, **47**, 21–32.

Neumann, R.O. (1902) Experimentelle Beitrage Zur Lehre von dem taglichen Nahrungsbedarf des Menschen nuter besonderer Berucksichtigung der not wendigen Eiweissmenge. *Arch. Hyg.*, **45**, 1–87.

Nicholls, D.G. (1979) Brown adipose tissue mitochondria. *Biochim. Biophys. Acta*, **549**, 1–29.

Norgan, N.G., Ferro-Luzzi, A. and Durnin, J.V.G.A. (1974) The energy and nutrient intake and energy expenditure of 204 New Guinean adults. *Phil. Trans. Roy. Soc. B.*, **268**, 309–48.

Oomura, Y., Ohta, M., Ishibashi, S., Kita, H. Okajima, T. and Ohno, T. (1978) Activity of chemosensitive neurons related to the neurophysiological mechanisms of feeding. in *Recent Advances in Obesity Research*, II. (ed. G.A. Bray), London, Newman, pp. 17–26.

Oscai, L.B. (1980) Evidence that size does not determine voluntary food intake in the rat. *Am. J. Physiol.*, **238**, E318–21.

Peckham, S.C., Centerman, H.W. and Carroll, J. (1962) The influence of a hyper-caloric caloric diet on gross body and adipose tissue composition in the rat. *J. Nutr.*, **77**, 187–97.

Perkins, M.N., Rothwell, N.J., Stock, M.J. and Stone, T.W. (1981) Activation of brown adipose tissue thermogenesis by the ventromedial hypothalamus. *Nature*, **289**, 401–2.

Prentice, A.M., Whitehead, R.G., Roberts, S.B. and Paul, A.A. (1981) Long-term energy balance in child-bearing Gambian women. *Am. J. Clin. Nutr.*, **34**, 2790–9.

Reichlin, S., Bollinger, J., Nejad, I. and Sullivan, P. (1973) Tissue thyroid concentration of rat and man determined by radioimmunoassay: biological significance *Sinai J. Med.*, **40**, 502–10.

Ricquier, D. and Kader, J.C. (1976) Mitochondrial protein alteration in active brown fat. A sodium dodecyl sulfate-polyacrylamide gel electrophoretic study. *Biochem. Biophys. Res. Comm.*, **73**, 577–83.

Rothwell, N.J. and Stock, M.J. (1978) A paradox in the control of energy intake in the rat. *Nature*, **273**, 146–7.

Rothwell, N.J. and Stock, M.J. (1979) A role for brown adipose tissue in diet-induced thermogenesis. *Nature*, **281**, 31–5.

Rothwell, N.J. and Stock, M.J. (1980*a*) Intra-strain difference in the response to overfeeding in the rat. *Proc. Nutr. Soc.*, **39**, 20A.

Rothwell, N.J. and Stock, M.J. (1980*b*) Thermogenesis induced by cafeteria feeding in young growing rats. *Proc. Nutr. Soc.*, **39**, 5A.

Rothwell, N.J. and Stock, M.J. (1980*c*) Similarities between cold and diet-induced thermogenesis in the rat. *Can. J. Physiol. Pharmacol.*, **58**, 842–8.

Rothwell, N.J. and Stock, M.J. (1981*a*) Influence of noradrenaline on blood flow to brown adipose tissue in rats exhibiting diet-induced thermogenesis. *Pflugers Archiv.*, **389**, 237–42.

Rothwell, N.J. and Stock, M.J. (1981*b*) A role for insulin in the diet-induced thermogenesis of cafeteria-fed rats. *Metabolism*, **30**, 673–8.

Rothwell, N.J. and Stock, M.J. (1981*c*) Thermogenesis: comparative and evolutionary considerations. in *The Body Weight Regulatory System: Normal and Disturbed Mechanisms* (eds L.A. Cioffi, W.P.T. James and T.B. Van Itallie), Raven Press, New York, pp. 335–43.

Rothwell, N.J. and Stock, M.J. (1982*a*) Effects of feeding a palatable 'cafeteria' diet on energy balance in young and adult lean (+/?) Zucker rats. *Br. J. Nutr.* **47**, 461–71.

Rothwell, N.J. and Stock, M.J. (1982*b*) Energy expenditure of 'cafeteria fed' rats determined from measurements of energy balance and 24-hour oxygen consumption. *J. Physiol.*, **324**, 59–60P.

Rothwell, N.J., Saville, M.E. and Stock, M.J. (1981) Acute effects of food, 2-deoxy-D-glucose and noradrenaline on metabolic rate and brown adipose tissue in normal and atropinised lean and obese (fa/fa) Zucker rats. *Pflugers Archiv.*, **392**, 172–7.

Rothwell, N.J., Saville, M.E. and Stock, M.J. (1982*a*) Factors influencing the acute effect of food on oxygen consumption in the rat. *Int. J. Obesity*, **6**, 53–9.

Rothwell, N.J., Saville, M.E., Stock, M.J. and Wyllie, M.G. (1982*b*) Catecholamine and thyroid hormone influence on brown fat Na^+, K^+-ATPase activity and thermogenesis in the rat. *Hormone Metab. Res.*, **14**, 261–5.

Rothwell, N.J., Stock, M.J. and Wyllie, M.G. (1981) Na^+, K^+-ATPase activity and noradrenaline turnover in brown adipose tissue of rats exhibiting diet-induced thermogenesis. *Biochem. Pharmacol.*, **30**, 1709–12.

Rubner, M. (1902) *Die Gesetze des Energieverbrauchs bei der Ernahrung*. Deutiche, Leipzig.

Schemmel, R. and Mickelsen, O. (1973) Influence of diet, stain, age and sex on fat depot mass and body composition of the nutritionally obese rat. in *The Regulation of Adipose Tissue Mass* (eds J. Vague and J. Boyer), American Elsevier, New York, pp. 238–53.

Sclafani, A. and Springer, D. (1976) Dietary obesity in adult rats: Similarities to hypothalamic and human obesity syndromes. *Physiol. Behav.*, **17**, 461–71.

Seydoux, J., Rohner-Jeanrenaud, F., Assimacopoulous-Jeannet, F., Jeanrenaud, B. and Girardier, L. (1981) Functional disconnection of brown adipose tissue in hypothalamic obesity in rats. *Pflugers Archiv.*, **390**, 1–4.

Shackney, S.E. and Joel, C.D. (1966) Stimulation of glucose metabolism in brown adipose tissue by addition of insulin *in vitro*. *J. Biol. Chem.*, **241**, 4004–10.

Shiraishi, T. and Mager, M. (1980) Hypothermia following injection of 2-deoxy-D-glucose into selected hypothalmic sites. *Am. J. Physiol.*, **239**, R265–9.

Sims, E.A.H., Danforth, E., Horton, E.S., Bray, G.A., Glennon, J.A. and Salans, L.B. (1973) Endocrine and metabolic effects of experimental obesity in man. *Recent Prog. Horm. Res.*, **29**, 457–96.

Stirling, J.L. and Stock, M.J. (1968) Metabolic origins of thermogenesis induced by diet. *Nature*, **220**, 801–2.

Sundin, U. (1981) Thyroxine and brown fat thermogenesis. in *Contributions to Thermal Physiology* (eds Z. Szelenyi and M. Szekely), Pergamon, 499–501.

Sundin, U. and Cannon, B. (1980) GDP-binding to the brown fat mitochondria of developing and cold adapted rats. *Comp. Biochem. Physiol.*, **65B**, 463–71.

Tepperman, J. and Tepperman, H.M. (1958) Effects of antecedent food intake pattern on hepatic lipogenesis. *Am. J. Physiol.*, **193**, 55–64.

Trayhurn, P., Jones, P.M., McGuckin, M.M. and Goodbody, A.E. (1982*a*) Effects of overfeeding on energy balance and brown fat thermogenesis in obese (ob/ob) mice. *Nature*, **295**, 323–5.

Trayhurn, P., Douglas, J.B. and McGuckin, M.M. (1982*b*) Brown adipose tissue thermogenesis is suppressed during lactation in mice. *Nature*, **298**, 59–60.

Tulp, O., Frink, R., Sims, E.A.H. and Danforth, E. (1980) Overnutrition induces hyperplasia of brown fat and diet-induced thermogenesis in the rat. *Clin. Res.*, **28**, 621A.

Walker, S.E. (1965) A 5-year study of the daily food consumption of South African University students. *Br. J. Nutr.*, **19**, 1–12.

Widdowson, E.M. (1947) *A Study of Individual Childrens Diets*. London, MRC Spec. Rep. Ser. No. 257.

Chapter Eight

Thermogenesis and Obesity

P. Trayhurn and W.P.T. James

8.1 INTRODUCTION

Obesity is generally considered to be the major nutritional disorder in the Western world, affecting, for example, some 10% of the middle-aged population in the UK, with up to 50% of the same group being overweight and at an increased risk of disease and early death. Given the increased incidence of important diseases such as coronary artery disease, hypertension, diabetes, gallstones and osteoarthritis in the overweight adult, it is clearly necessary to identify the factors which contribute to weight gain. Excess weight is usually taken to result from a pathological increase in body fat. In practice both lean tissue and fat are gained in obese individuals, and the distinction between a pathological and physiological change in fat stores is necessarily somewhat arbitrary. Examples of physiologically programmed increases in body fat include the selective deposition of fat during adolescence in girls and during pregnancy. In some animals marked increases in fat occur in the period before hibernation, and it may well be that by understanding the basis for 'physiological' changes in body fat we can then define the abnormalities which lead to 'excessive' fat deposition in both animals and in man.

This chapter is concerned with linking the knowledge gained from studies on animal obesity to the continuing attempts to understand the metabolic basis of obesity in man. We have not set out to provide a comprehensive review of obesity, but have instead focused on the current evidence and views which link the aetiology of the obese state to the central theme of this book.

8.1.1 Selective advantage of obesity

Although obesity may be *disadvantageous* to an animal since it reduces speed and manoeuverability, as well as being associated with increased morbidity and mortality, it has been widely argued that the 'tendency to obesity' has a selective *advantage* in an evolutionary sense (Neel, 1962;

James and Trayhurn, 1976; Coleman, 1978). The concept is of a 'thrifty genotype' associated with a high metabolic efficiency rather than genetic selection for excessive food intake. A high metabolic efficiency will mean a rapid accumulation of energy when food is available, but a lower energy need for maintaining body function during periods of food shortage.

Some evidence in support of the concept of the thrifty genotype comes particularly from observations made on desert rodents and on human populations such as the Australian aborigines. The spiny mouse (*Acomys cahirinus*), although lean in its native environment, becomes obese when allowed to live in laboratory conditions where food is freely available (Wise, 1977*a,b*). If the food is particularly palatable and energy dense, then severe obesity ensues. In man, societies subject to severe selection pressures with poor food availability for generations show the same response to the provision of a readily available energy dense diet. Thus aborigines on becoming 'acculturated' into the environment of urban Australia – with the ready availability of high-fat Western foods – become very obese and also exhibit a high incidence of diabetes.

8.1.2 Genetic component of obesity

The susceptibility of particular strains of animals or races of man to become obese is matched by increasing evidence to support the view that within a human population the recognized large variation in genetic attributes is accompanied by a genetically determined susceptibility of some individuals to weight gain. In a recent review Bray (1981) has summarized the present position by stating that genetic factors 'are primarily operative in making an individual resistant to those factors in the environment which would tend to enhance obesity'. Much of the evidence for a genetic component to obesity in man comes from twin studies and studies on the clustering of obesity within families. There are also several well-recognized genetically inherited conditions in which obesity occurs, and the best known of these is the Prader-Willi syndrome. This syndrome occurs at a frequency of approximately 1 in 20 000, and has been attributed to hypothalamic dysfunction.

In farm animals differences between strains in the tendency to fatness are well recognized, for example the difference between the lean Pietrain and Landrace breeds of pig and the fatter Large White and Berkshire varieties. Until relatively recently, breeding for fatness was a prime goal in animal farming, but current emphasis is on the production of a lean carcass. Differences in the degree of fatness also occur in different strains of laboratory animals, and at the extreme end of the spectrum there are a number of animals in which there is frank obesity. This obesity may result either from a polygenic inheritance, in which case a number of interacting genes are involved, or from a single gene mutation (Table 8.1).

Table 8.1 The major forms of genetically transmitted obesity

Mode of inheritance		
Single gene		Polygenic
Dominant	Recessive	
Yellow obese mouse	Obese (ob/ob) mouse	NZO mouse
Adipose (Ad) mouse	Diabetic-obese (db/db) mouse	KK mouse
	'fat' mouse	PBB mouse
	Zucker, or fatty (fa/fa) rat	'Wellesley' mouse (C3Hf×1F₁)
		BHE rat
		B10 4.24 Syrian hamster
		Obese OS chicken
		*Spiny mouse (*Acomys cahirinus*)
		*Sand rat (*Psammomys obesus*)
		*Tuco-tuco (*Ctenomys talarum*)
		*Djungarian hamster (*Phodopus sungorus*)

*Obesity highly dependent in these species on environment.

8.1.3 Animal models

The great difficulty in identifying the factors responsible for the development of obesity in man has led to considerable interest in animal models, and a wide range of such models is now used in experimental studies on energy regulation (Bray and York, 1979). They can be divided into those in which obesity is induced experimentally and those in which it is genetically transmitted. The experimental manipulations include lesioning of the ventromedial hypothalamus, the administration of gold thioglucose or monosodium glutamate, and, in some strains, feeding high-fat diets (Table 8.2). Most obese mutations have been obtained in mice and the most widely studied animal is the obese ob/ob mouse (also called the obese-hyperglycaemic mouse), found originally in the late 1940s in the Jackson Laboratories, Bar Harbor, USA. The 'ob' gene has now been crossed on to a variety of genetic backgrounds, and in the UK it is generally used on the mixed 'Aston' background. The degree of interest in the ob/ob mouse is such that of all the recognized mutations in laboratory rodents (numbering hundreds), only the nude athymic mouse has been the subject of more investigations (Festing, 1979). Yet only a minority of studies have been concerned with the central problem, i.e. of energy balance. Many of the

Table 8.2 The major forms of experimentally induced obesity

Surgical	Chemical	Endocrinological	Dietary
Lesions of the ventromedial hypothalamus	Gold thioglucose	Insulin administration	High-fat feeding
Castration	Monosodium glutamate	Corticosteroid administration	'Cafeteria' feeding
	Bipiperidyl mustard		Tube feeding

Note: Some methods of inducing obesity (e.g. high-fat feeding) are specific to particular strains of laboratory animal.

documented abnormalities in obese animals, such as hyperinsulinaemia, insulin resistance and other endocrinological changes, are almost certainly related to the effects of obesity itself rather than being the cause of the condition (see Bray and York, 1979).

8.1.4 Hyperphagia and metabolic efficiency

The fundamental tenet of energy regulation is that for an animal to be in energy balance, energy output must be equal to energy intake. Obesity can only result, therefore, from energy intake being in excess of energy expenditure. This simple and obvious statement has generally led to the assumption that obesity must result from an abnormally high intake in both animals and man – and this view has until recently been widely held. In experimental studies on animals the distinction is much easier than in human investigations, but it is clearly possible for individuals who develop obesity to have a normal energy intake but a low expenditure. In this case obesity would not be primarily the result of hyperphagia, but of an increased metabolic efficiency.

A number of studies have been made on the food intake of lean and obese subjects, and the main conclusion that can be derived is that it is not possible to demonstrate that the obese, in general, eat more than the lean. There are, however, a number of problems associated with such studies. These problems include the fact that most measurements have been done, inevitably, with individuals who are already obese (i.e. during the so-called 'static' phase of obesity), rather than during the early stages of the development of the disorder ('dynamic' phase). There is also the general problem of the validity and accuracy of measurements of voluntary food intake in man. This may be a particular problem with the weight conscious adult in a culture where the widespread belief is that obesity is the result of over-indulgence. In the absence of foolproof methods of measuring energy turnover by monitoring food intake, it is difficult to refute the claim that while under investigation the obese consciously reduce their normal intake – or even under-record their true intake. Increasing emphasis is therefore being given to measuring energy expenditure and changes in body energy content as a method for assessing the basis of energy inbalance in man.

Studies on man have increasingly depended on the development of an understanding of the factors which modulate energy expenditure in animal models of obesity. These are therefore described first before the discussion is extended to human investigations.

8.2 ENERGY BALANCE IN GENETICALLY OBESE RODENTS

Early studies had indicated that obesity could develop in the ob/ob mouse in the absence of hyperphagia and this indicated that the ob/ob mutant was expending less energy than normal in order to accumulate the additional triacylglycerol (Alonso and Maren, 1955; Hollifield and Parson, 1958; Chlouverakis, 1970; Welton, Martin and Baumgardt, 1973; Dubuc, 1976). In keeping with this idea it was also shown that body weight could be maintained on a below-normal energy intake. Fair-feeding studies with several different obese mutants have confirmed that a common feature of genetically obese rodents is the ability to gain excess body fat on a normal intake (e.g. Cox and Powley, 1977).

In recent years full energy balance measurements have been conducted in pair-feeding experiments on young lean and ob/ob mice (Woodward, Trayhurn and James, 1977; Thurlby and Trayhurn, 1979), diabetic-obese (db/db) mice (Trayhurn and Fuller, 1980) and the fatty (fa/fa) rat (Zucker, 1975; Deb, Martin and Hershberger, 1976; Pullar and Webster, 1974, 1977). In all three mutants clear evidence for an increase in metabolic efficiency has been obtained, i.e. the energy gain per kJ food energy of the obese was greater than that of the lean. In some cases the difference in efficiency that was observed is very substantial. In ob/ob mice on the Aston background, for example, the obese were found to deposit 2.3 times more of the dietary energy than the lean, and gross efficiency (energy gain/digestible energy intake) was 32% for the obese but only 14% for the lean (Woodward *et al.*, 1977).

When obese mutants (aged four weeks and more) are allowed to feed *ad-libitum* their energy gain is even higher than in the pair-fed condition, indicating that hyperphagia can be of considerable significance in the development of the obesity. However, several studies have now indicated that during the suckling period and in the early post-weaning phase, when an excessive deposition of body fat has already begun and physically obvious obesity is established, energy intake is similar in lean and mutant siblings; this is the case for the ob/ob mouse (Lin, Romsos and Leveille, 1977; Rath and Thenen, 1979; Contaldo *et al.*, 1981), the diabetic-obese mouse

(Contaldo, 1981) and the fatty rat (Boulangé, Planche and de Gasquet, 1979). Thus, it now appears that in all those obese mutants for which information is available, the initiation and early development of the obese state is entirely due to a high metabolic efficiency, with hyperphagia only occurring at a later stage. Whether or not this means that the hyperphagia is merely a secondary factor (e.g. to hyperinsulinaemia) in obese mutants, or whether it is an aspect of the primary genetic defect which is only expressed once a number of other changes have occurred is not clear. At present, it seems more likely that it is a true secondary factor, in which case there may be valuable clues to the regulation of food intake emerging from studies on the development of hyperphagia in obese mutants.

8.2.1 Energy expenditure in ob/ob mice

The most extensive studies on energy expenditure in obese animals have been conducted on the ob/ob mouse. Figure 8.1 shows the main components of the energy expenditure of normal, non-growing, mice housed under standard environmental conditions. The figure only includes the components which contribute to the 'maintenance requirement', since, at least in the case of the ob/ob mutant, there is no difference in the energy cost of growth between lean and obese animals (Woodward *et al.*, 1977). The two main components of the energy expended in weight maintenance are the basal metabolic rate – a particularly elusive term when applied to animals – and the thermoregulatory thermogenesis needed to maintain homeothermy. In small mammals such as the mouse the energy cost of thermogenesis is high, particularly when they are housed, as is normally the case, well below their lower critical temperature. The high energy cost of thermoregulation in such animals is, of course, a consequence of their large surface area/volume ratio, which results in a high rate of heat loss.

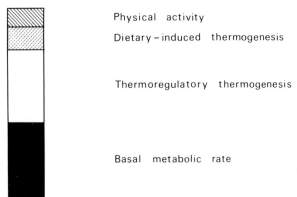

Physical activity

Dietary – induced thermogenesis

Thermoregulatory thermogenesis

Basal metabolic rate

Fig. 8.1 Partition (approximate) of the maintenance energy expenditure of mice housed at 20°C.

8.2.2 Thermogenesis in the ob/ob mouse

One of the earliest abnormalities observed in the ob/ob mouse was the failure to maintain body temperature on exposure to low environmental temperatures (Mayer and Barnett, 1953; Davis and Mayer, 1954). This abnormality is surprising since superficially the obese animal appears well insulated by virtue of its large subcutaneous fat depots. Measurements of resting oxygen consumption showed, however, that on cold exposure (4°C) the obese ob/ob mouse has a smaller increase in metabolic rate than lean animals (Davis and Mayer, 1954). Thus the cold-induced hypothermia of the ob/ob mutant is the result of a reduced capacity for thermogenesis. When the non-shivering portion of cold-induced thermogenesis was measured by monitoring the increase in resting metabolic rate at thermoneutrality after large doses of noradrenaline (up to 1000 μg (kg body weight)$^{-1}$), a specific reduction in capacity was found in the ob/ob mouse (Trayhurn and James, 1978). Whether or not the ability to produce heat through shivering is also impaired in these animals is unknown, although on casual observation they are clearly able to shiver in the cold.

Studies on pre-weanling animals have shown that the impairment in the response to cold occurs well before the development of obesity so it is not a secondary phenomenon. It has in fact been noted at as early as 10–12 days of age (Trayhurn, Thurlby and James, 1977) – the age by which mice first develop an appreciable capacity for thermoregulation. Cold-induced hypothermia has been used as the basis of a simple test for identifying animals bearing the ob/ob genotype before they can be distinguished visually from their normal siblings (Trayhurn *et al.*, 1977).

In addition to their susceptibility to severe cold, ob/ob mice also have a lower than normal temperature at any temperature below the thermoneutral zone (Yen, Fuller and Pearson, 1974; Kaplan and Leveille, 1974; Ohtake, Bray and Azukizawa, 1977; Trayhurn and James, 1978). At thermoneutrality (32–33°C) the resting metabolic rate of ob/ob animals is the same as, or higher than, that of lean siblings when expressed on a whole-body basis (the most appropriate baseline in energy balance experiments), but at lower environmental temperatures metabolic rate is increased less in the obese than in the lean. Consequently, at ambient temperatures between about 10 and 30°C, the resting metabolic rate of the obese is consistently some 20% less than that of the lean (Trayhurn and James, 1978). If the metabolic rate of obese mice is expressed in relation to some index of body size (e.g. per g, or $g^{0.75}$), then the abnormal response to changes in environmental temperature is partly obscured and the obese appear to have a much lower energy expenditure than the lean at all temperatures (Kaplan and Leveille, 1974), largely because the energy expenditure of white adipose tissue is very much lower than that of lean tissues.

These data on metabolic rates at different temperatures, together with the reduced capacity for non-shivering thermogenesis, led us to the view that under normal environmental conditions the ob/ob mouse expends less energy on non-shivering thermogenesis than lean mice, and that this is the primary reason for the high metabolic efficiency of the mutant (Trayhurn and James, 1978; Trayhurn *et al.*, 1979). This view has been substantiated by subsequent nutritional studies. In the first type of study the maintenance requirement of young lean and obese mice was found to be similar at thermoneutrality, while at a lower temperature it was significantly less in the obese than in the lean (Trayhurn *et al.*, 1979; Lin *et al.*, 1979). In the second type of nutritional study young ob/ob mice were pair-fed to the *ad-libitum* energy intake of lean siblings at several different environmental temperatures, and energy gain was measured (Thurlby and Trayhurn, 1979). Pair-feeding at thermoneutrality largely, although not entirely, abolished the extra energy gain of the obese over the lean seen at lower temperatures. The degree of 'excess' energy gained by the obese was directly related to the temperature of the study – the lower the temperature the greater the excess gain of the obese.

This intimate interaction between environmental temperature, metabolic rate and the excess energy gain clearly indicates that the high metabolic efficiency of the ob/ob mouse is dependent on the ability to display reduced non-shivering thermogenesis, and this is only possible at environmental temperatures below thermoneutrality.

8.2.3 Thermogenesis in other genetically obese rodents

The diabetic-obese mouse has similar thermoregulatory abnormalities to the ob/ob mouse – at least when the 'db' gene is on the C57Bl/Ks background. The db/db mouse dies rapidly of hypothermia when placed in a 4°C cold-room, and has a reduced capacity for non-shivering thermogenesis (Trayhurn, 1979). The mutant also has a low body temperature at normal environmental temperatures, and its resting metabolic rate (per animal) at thermoneutrality is similar to that of lean siblings, while at temperatures below the thermoneutral zone it is approximately 20–25% below that of the lean (Trayhurn, 1979). These observations are directly comparable with the data on the ob/ob mouse, and suggest that a reduced energy expenditure on non-shivering thermogenesis is also the mechanism by which the diabetic-obese mutant achieves its high efficiency. This view has again been supported by a pair-feeding study conducted at different environmental temperatures (Trayhurn and Fuller, 1980).

Detailed studies on thermoregulation and its relationship to energy balance have not been made on other genetically obese animals. The Zucker fatty rat does, however, appear to have a reduced body temperature although the reduction is less than that observed in ob/ob and db/db mice (Godbole, York and Bloxham, 1978; Levin, Triscari and Sullivan, 1980;

Armitage *et al.*, 1981). The reduced body temperature has again formed the basis of a test for the early identification of the fa/fa genotype (Godbole *et al.*, 1978) and one study has reported that the Zucker rat dies of hypothermia in the cold, although this was only following some 30 h exposure to 4°C (Trayhurn, Thurlby and James, 1976).

The capacity for non-shivering thermogenesis is not impaired in the Zucker rat (Rothwell, Saville and Stock, 1981), and measurements of energy expenditure over a range of environmental temperatures between 5 and 30°C do not suggest that expenditure on non-shivering thermogenesis is reduced in this mutant (Armitage *et al.*, 1981). Thus this genotype appears to maintain its core temperature below normal despite total heat output exceeding that of its lean siblings. However, the studies conducted hitherto have been made on fatty animals which are already in the hyperphagic state, in which case 24 h energy expenditure measurements, as made by Armitage *et al.* (1981), may be dominated by the excess inflow of food. This hyperphagia may also have a secondary effect on the capacity for non-shivering thermogenesis by a diet-induced hypertrophy of brown adipose tissue (see below). It would therefore be of interest to assess the capacity for non-shivering thermogenesis and total energy expenditure under normal conditions before the dominating effect of hyperphagia has developed. The mild hypothermia, despite the higher energy output, in the hyperphagic fatty rat suggests that the lowering of the body temperature is a controlled primary feature which is not dependent on an inability to generate heat.

A reduced body temperature or an impaired response to cold exposure have been noted with several of the less commonly studied genetically obese rodents, including the Yellow obese mouse (Turner, 1948), the Ad mouse (Trayhurn *et al.*, 1979) and the spiny mouse (Wise, 1977a,b). There are, however, insufficient metabolic rate or energy balance data to relate clearly any putative thermogenic defects in these animals to their metabolic efficiency and obesity.

8.2.4 Diet-induced thermogenesis in genetically obese rodents

There is currently a considerable renewal of interest in the role of diet-induced thermogenesis in the regulation of energy balance (see Chapter 7). The fact that the hyperphagia of genetically obese animals leads to the accelerated development of obesity clearly indicates that the capacity of these animals for 'regulatory' diet-induced thermogenesis is limited. Whether their capacity for this form of thermogenesis is abnormal, or defective, in the sense that it is less than that of their lean siblings is less obvious, particularly since rats of different ages and strains would appear to differ in their ability to use 'regulatory' diet-induced thermogenesis as a means of minimizing energy deposition during over-feeding (Rothwell and Stock, 1979a,b; Rolls, Rowe and Turner, 1980; Armitage *et al.*, 1981). The earlier studies on the energy efficiency of the ob/ob mouse at different temperatures suggest that in the situation where food intake is limited by pair-feeding then any defect in dietary-induced thermogenesis is small.

In order to assess the capacity for diet-induced thermogenesis in lean and ob/ob mice, higher food intakes are needed; this can be achieved by using the cafeteria feeding regimen to induce voluntary overfeeding (Sclafani and Springer, 1976; Rothwell and Stock, 1979*b*; Rolls *et al.*, 1980). In an experiment using a cafeteria diet lean mice ('Aston') were found to deposit only 4% of their excess energy intake during overfeeding, suggesting that they have a very substantial capacity to dissipate excess energy by dietary-induced thermogenesis, and thereby regulate energy balance (Trayhurn *et al.*, 1982). In contrast, the obese mice deposited some 55% of their extra energy intake on the cafeteria diet. It therefore appears that regulatory diet-induced thermogenesis is defective in the ob/ob mutant. This has led to the suggestion that obesity in the ob/ob mouse can be viewed as a direct consequence of a reduction in non-shivering and diet-induced thermogenesis – reduced non-shivering thermogenesis leads to a low maintenance requirement and excessive energy gain on a 'normal' energy intake, while a reduction in the capacity for diet-induced thermogenesis leads to an impairment in the ability to dissipate the excess energy intake once hyperphagia is established soon after weaning (Trayhurn *et al.*, 1982). The same situation is also likely to obtain in the case of the db/db mouse.

Studies with feeding low protein diets have indicated that the fatty rat also has an impairment in its capacity for diet-induced thermogenesis (Young, Tulp and Horton, 1980). The acute response to food (i.e. the thermic effect) is also considered to be reduced in this mutant (Rothwell *et al.*, 1981).

8.3 MECHANISMS OF HEAT PRODUCTION

In principle the reduced energy expenditure of the ob/ob and db/db mutants on non-shivering thermogenesis – the earlier of the two main thermogenic defects manifested in these animals – could be due to a limitation in the supply of substrate. However, studies *in vivo* have shown that the ob/ob mouse has a normal rise in the concentration of free fatty acids in plasma following either acute exposure to the cold (Carnie and Smith, 1978) or the administration of noradrenaline (Abraham *et al.*, 1971). In addition, although earlier *in vitro* studies on lipolysis suggested that white adipose tissue from ob/ob mice is resistant to lipolytic agents, recent experiments with adipocyte preparations have shown that this is not the case when the results are expressed in terms of the most appropriate reference point – the individual adipocyte (Carnie and Smith, 1978). The supply of substrate is therefore unlikely to be a limiting factor in non-shivering thermogenesis in the ob/ob mouse. The effect of isoproterenol, a more specific β-agonist than noradrenaline, on the plasma free fatty acid concentration in db/db mice suggests that substrate supply is also not impaired in this mutant (Allan and Yen, 1976).

These findings have led to a direct focus on investigating various putative biochemical mechanisms of heat generation in obese animals. Studies in this area have, necessarily, been greatly influenced by the prevailing views on the mechanistic basis of non-shivering thermogenesis (see Chapter 2). For example, it has been suggested that protein turnover is abnormally low in ob/ob mice (Miller *et al.*, 1977, 1979), but this has been disputed in subsequent work employing more appropriate techniques (James, Trayhurn and Garlick, 1981). It has also been argued that undefined changes in endocrinological status result in decreased substrate cycling of inter-mediates of glucose metabolism in ob/ob mice (Newsholme *et al.*, 1979), but kinetic studies have failed to demonstrate this in the Zucker rat (Wade, 1980).

Considerable emphasis has been placed on the thermogenic importance of the Na^+ pump, particularly by workers in the United States. Reduced Na^+K^+-ATPase activity has been reported, either by direct measurement of enzyme activity or by [^{3}H]ouabain binding studies, in skeletal muscle, kidney and liver of mature ob/ob mice (York, Bray and Yukimura, 1978; Lin *et al.*, 1978, 1979a; Guernsey and Morishige, 1979). A reduction in enzyme activity has also been found in the liver of the db/db mouse (Bray, York and Yukimura, 1978), and a decrease in the number of 'enzyme units' has been reported for skeletal muscle, although not the liver, of 14-day old ob/ob mice (Lin *et al.*, 1979b). These results and the conclusions deriving from them have recently been disputed by Clausen and Hansen (1982) who have reported that there is no reduction in the number of Na^+K^+-ATPase enzyme units in skeletal muscle of ob/ob mice, and that Na^+ transport is of little quantitative importance to energy expenditure.

The central criticisms of the mechanisms which have been discussed so far is that their energetic significance is in considerable doubt, and that they may be associated with tissues which have little or no role in non-shivering thermogenesis. However, the whole question of the biochemical basis of thermogenesis has been greatly simplified following measurements of regional blood flow with radioactively labelled microspheres by Foster and Frydman (1978, 1979). These authors convincingly demonstrated in some meticulous experiments that brown adipose tissue is the main site of non-shivering thermogenesis in adult rats with skeletal muscle playing at most only a very minor role. The implications of this work were readily appreciated by several groups involved in the regulation of energy balance, and it may be argued that the repercussions of Foster and Frydman's results have been much greater in obesity research than elsewhere.

8.3.1 Brown adipose tissue in the ob/ob mouse

Measurements of regional blood flow have been made using [46]Sc-labelled microspheres in young lean and obese ob/ob mice, both in the basal state and when non-shivering thermogenesis was stimulated by exogenous noradrenaline (Thurlby and Trayhurn, 1980). Blood flow to brown adipose tissue proved to be greater in lean mice than in the obese, with the tissue receiving 15.5% of the cardiac output after noradrenaline in the lean compared with only 8.1% of cardiac output in the obese. Noradrenaline increased the peak blood flow to interscapular brown adipose tissue of the lean nearly 40-fold while the increase in the obese was only 13-fold (Fig. 8.2). The total oxygen consumption of brown adipose tissue in the stimulated state in the lean was 2.5 times that of the obese.

From this blood flow study it has been suggested that differences in the metabolic activity of brown adipose tissue alone can account for the difference in the capacity for non-shivering thermogenesis, as well as the difference in the normal energy expenditure on this process, between lean and obese animals (Thurlby and Trayhurn, 1980). The blood flow measurements on mice, like those on rats (Foster and Frydman, 1978, 1979; Rothwell and Stock, 1981) also indicated that skeletal muscle is of little quantitative importance to non-shivering thermogenesis, at least in small rodents.

The principal mechanism for non-shivering thermogenesis in brown adipose tissue is considered to be the proton conductance pathway which operates across the inner mitochondrial membrane, and the details of this pathway are set out in Chapter 2. A reduced activity of the proton conductance pathway, inferred from a purine nucleotide binding assay, was originally found in ob/ob mice by Himms-Hagen and Desautels (1978), and

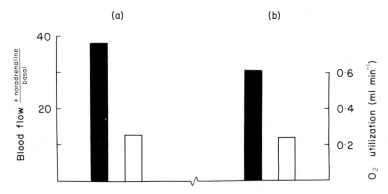

Fig. 8.2 The effect of noradrenaline on the activity of brown adipose tissue in lean (■) and obese ob/ob (□) mice. (a) blood flow (b) oxygen utilization (Taken from Thurlby and Trayhurn, 1980).

this was the first experimental evidence implicating brown adipose tissue in energy regulation and obesity. On acute cold exposure lean mice show an increase in purine nucleotide binding, but no increase occurs in the ob/ob mutant (Himms-Hagen and Desautels, 1978). Thus ob/ob mice appear to be unable to respond to acute cold exposure, although it is now clear that they can adapt to chronic exposure to the cold (Hogan and Himms-Hagen, 1980; A.E. Goodbody and P. Trayhurn, unpublished observations). The long-term adaptation includes increases in mitochondrial purine nucleotide binding, as well as increases in the total activity of cytochrome oxidase in brown adipose tissue.

Several other differences in brown adipose tissue have been observed between lean and obese ob/ob mice, some of which are mitochondrial whilst others are extra-mitochondrial. Brown adipose tissue is hypertrophied in ob/ob mice, because of a greatly increased amount of triacylglycerol. Measurements of tissue DNA content suggest, however, that there is no hyperplasia in the obese (A.E. Goodbody, unpublished observations). The other main changes in brown adipose tissue mitochondria of the obese mouse include an apparent reduction in the number of cristae (Hogan and Himms-Hagen, 1980), and a greater capacity to transport calcium than lean mice (Fraser and Trayhurn, 1981); the obese also show lower rates of uncoupled respiration (A.E. Goodbody and P. Trayhurn, unpublished observations). These changes are consistent with the mitochondria from the obese being more tightly 'coupled' with the former having a lower activity of the proton conductance pathway.

If a reduction in the activity of brown adipose tissue is to be invoked as a major factor in the development of obesity in the ob/ob mutant, then it is clearly necessary to determine the activity of the tissue in very young animals. A recent study has shown that uncoupled respiration is lower in mitochondria from 14-day old ob/ob mice than from normal mice, but that there is no difference between the two groups in the coupled respiration (i.e. +purine nucleotide). In addition, purine nucleotide binding is much less in 14-day old, and even in 10-day old, ob/ob animals than in normal mice (Goodbody and Trayhurn, 1982). Thus the thermogenic activity of brown adipose tissue is greatly reduced in ob/ob mice before obesity begins to develop, and this would support the view that this tissue plays an important role in the aetiology of the disorder.

Girardier (1981) and Assimacopoulos-Jeannet *et al.* (1982) have indicated that interscapular brown adipose tissue is 'functionally disconnected' in the adult ob/ob mouse, and that there is an increase in the number of β-receptors on the plasma membrane of the tissue. This raises the important question of whether it is the sympathetic drive which is defective in the ob/ob mutant, rather than brown adipose tissue itself. The recent finding of subnormal responses in brown adipose tissue to sympathetic nerve stimulation (Seydoux *et al.*, 1982) does not argue against the central importance of the sympathetic drive if the nerve supply also has a trophic as well as a direct stimulant effect.

8.3.2 Brown adipose tissue in the diabetic-obese mouse

The only other genetically obese animal in which the activity of brown adipose tissue has been reported is the diabetic-obese (db/db) mouse. In adult diabetic-obese mice the activity of cytochrome oxidase in brown adipose tissue from the interscapular site is lower than that of lean (db/+) siblings, and the activity of the proton conductance pathway, as gauged from the purine nucleotide binding assay, is also lower (Goodbody and Trayhurn, 1981). Since the reduced activity of the proton conductance pathway is apparent in 14-day old, as well as in adult animals (Goodbody and Trayhurn, 1981) these results again indicate that reduced thermogenesis in brown adipose tissue is likely to underlie the development of obesity in the diabetic-obese mutant as well as in the ob/ob mouse.

8.3.3 Ventromedial hypothalamic obesity

The classical observation derived from studies on rats lesioned in the ventromedial region of the hypothalamus is that of hyperphagia, which has generally been assumed to be the cause of the subsequent obesity. However, it is now clear that there is an increased metabolic efficiency in the lesioned animal, and in rats lesioned shortly after weaning obesity can develop in the absence of any hyperphagia (see Bernardis and Goldman, 1976; Vander Tuig, Knehans and Romsos, 1982).

Lesioning the ventromedial hypothalamus reduces fatty acid oxidation in brown adipose tissue (Seydoux *et al.*, 1981), a feature related to a 'functional disconnection' of the tissue. Preliminary studies have also indicated that the depletion of hypothalamic noradrenaline leads to a reduction in purine nucleotide binding to brown adipose tissue mitochondria (B.J. Sahakian and P. Trayhurn, unpublished observations). Both of these observations indicate that there is a reduction in brown adipose tissue thermogenesis in lesioned animals, and this is consistent with at least part of the sympathetic innervation to brown fat being derived from the ventromedial hypothalamus (Shimazu and Takahashi, 1980; Perkins *et al.*, 1981).

8.4 THERMOGENESIS IN MAN

The possibility that thermogenesis may play an important role in energy regulation in man, and that human obesity is related more to a subnormal thermogenic response than to overeating has long been advocated by Miller (1975). Interest in this view has widened during the past five years, and this is partly because thermogenic abnormalities have been so clearly documented in animal studies. It is important, however, to emphasize at this point that human obesity is a heterogeneous condition and that even at best an impairment in thermogenesis is unlikely to provide a comprehensive explanation of its aetiology.

In considering the relationship between thermogenesis and obesity in man four fundamental questions have to be confronted. First, is there an adaptive thermogenesis in man, in response either to cold or to food (or to other factors), similar to that in small animals? Secondly, what is the quantitative importance of any such thermogenesis to energy expenditure in man? Thirdly, are differences in thermogenesis responsible for the differences between individuals in the tendency to develop obesity? Lastly, is brown adipose tissue responsible for the major portion of any non-shivering or regulatory diet-induced thermogenesis in man?

Two factors make it unlikely that cold-induced thermogenesis would be a quantitatively important component of normal energy expenditure in adult man. Man's large body size, with a relatively small surface area/volume ratio, leads to a low rate of heat loss in comparison to that of small laboratory mammals. This in turn results, of course, in a low requirement for thermoregulatory heat production. Such considerations do not apply to the newborn infant, and the capacity of the human neonate for non-shivering thermogenesis is considerable (Hey, 1975). Behavioural thermoregulation, including altering the amount of clothing worn and the degree of external heating, operates almost exclusively on the control of heat loss and is a major factor in reducing the requirement for thermogenesis in man.

Shivering is a well-recognized mechanism for cold-induced thermogenesis in humans, but whether or not adult man also has some capacity for non-shivering thermogenesis has been less certain. Non-shivering thermogenesis does, however, occur in the human neonate although the capacity declines markedly over the first few months of life (Hey, 1975). Evidence for the occurrence of non-shivering thermogenesis in adult man comes primarily from the increase in resting metabolic rate (10–20%) which is observed on the infusion of noradrenaline, and this can almost certainly be taken to reflect non-shivering thermogenesis (Rennie *et al.*, 1962; Jung *et al.*, 1979; Joy, 1963: Doi *et al.*, 1979). The most convincing demonstration, however, of non-shivering thermogenesis in adult man is that of Jessen, Rabol and Winkler (1980) who measured the increase in resting metabolic rate induced by cold exposure in patients treated with chlorpromazine.

Metabolic adjustments to cold in man have been documented (Rennie *et al.*, 1962), but recently Dauncey (1981) has shown a response to relatively small changes in temperature. A twenty-four hour study of total energy expenditure using a whole-body calorimeter set at either 28°C or 22°C has shown that man can rapidly adjust thermogenesis without apparent shivering or discomfort (Dauncey, 1981). In obese subjects similar studies have failed to show the response of the lean group (Blaza and Garrow, 1980), and tolerance of acute exposure to severe cold has been reported to be reduced in obese subjects (Buskirk *et al.*, 1963; Andrews and Jackson, 1978). The acute fall in body temperature when the obese are stressed may not, however, have the same physiological significance in man as in small animals because there are unusual experimental conditions and the obese individuals may have adapted in vasomotor terms to their large insulating layer of fat.

Differences in the apparent capacity for non-shivering thermogenesis between lean and obese subjects have been investigated by measuring the increase in resting metabolic rate during the infusion of noradrenaline (Jung *et al.*, 1979). Although lean subjects exhibited an increase amounting to approximately 20% of their basal metabolic rate, the obese subjects showed only a 10% response. A similar reduced response was obtained in a group of 'post-obese' people who had the same type of familial obesity as the obese group, but who had slimmed to near normal weights. This persistent reduction in noradrenaline-stimulated thermogenesis suggests a constitutional difference rather than one which is secondary to the obese state. Such studies are inevitably difficult to perform, and there are particular difficulties in ensuring that a maximal or near maximal response is obtained in all groups. In the experiments of Jung *et al.* (1979), however, the concentration of noradrenaline in venous blood in the arm opposite to that which was being used for the infusion was similar in all three groups of subjects. In addition, the effect of noradrenaline on the plasma concentration of various substrates, particularly free fatty acids, was not significantly different between groups. This observation also suggests that the reduced thermogenic response of the obese is not due to any abnormality in substrate provision, and this parallels the findings in ob/ob mice (see above).

8.4.1 Diet-induced thermogenesis in man

A reduction in diet-induced thermogenesis has generally been considered to be the main thermogenic factor involved in determining weight gain in man. Man does show an immediate thermic response to food, i.e. the short-term component of diet-induced thermogenesis, but this is to a large extent inevitable since much of the increase in metabolic rate which follows the ingestion of nutrients may reflect the obligatory energy costs of handling and storing those nutrients.

Several studies have now demonstrated a reduced thermic response to a single meal in obese compared with lean subjects (Pittet *et al.*, 1976; Kaplan and Leveille, 1976; Shetty *et al.*, 1981). This has been reported with both a meal of mixed nutrient composition, and with a single glucose load (Pittet *et al.*, 1976). A reduction in the thermic response to a single meal has been observed in 'post-obese' as well as in obese subjects (Shetty *et al.*, 1981) and this has close parallels with the observations made on the metabolic response to the infusion of noradrenaline.

In detailed studies on the thermic effect of different nutrients we have been unable to observe differences between lean and obese subjects on feeding either a starch or a protein meal (James and Trayhurn, 1981a). This has led to the suggestion that the differences between the lean and obese in the thermic effect of a mixed meal are related primarily to the fat content of the diet (James and Trayhurn, 1981*a*, 1981*b*). This is, however, difficult to reconcile with the results of Pittet *et al.* (1976) on the effects of glucose alone, but the same authors in subsequent studies (Felber *et al.*, 1981) have shown that glucose oxidation is only less in obese than lean subjects if the obesity is accompanied by glucose intolerance.

The effects of long-term overfeeding on energy balance have been the subject of a number of investigations. The best known of such investigations is the now classic Vermont prison study conducted by Sims and his colleagues (see Sims *et al.*, 1973; Sims, 1976). Substantial overfeeding was obtained over periods of more than 30 weeks with remarkably small changes in weight. The key observations were the great individual variation in the extent to which weight was gained and the apparent ability of some individuals to increase substantially their maintenance energy requirements. Although like so many other studies on man the Vermont prison study can be criticized on the grounds that true energy balance measurements were not made, the prolonged nature of the study and the stability of body weights suggests that substantial adaptation in energy expenditure must have occurred.

Smaller-scale overfeeding studies have been widely undertaken, and they fall essentially into those which indicate that increases in energy expenditure have occurred in excess of that expected for simple nutrient storage, and those which report a degree of weight gain consistent with no adaptation having occurred. Garrow (1978) has recently analysed these various investigations and shown that adaptive increases in energy expenditure do occur but only with substantial overfeeding.

It is noteworthy that overfeeding studies generally involve a high dietary fat intake, and in a recent investigation using whole-body calorimetry to investigate the effect on 24 h energy expenditure of overfeeding with fat alone, a clear difference in the response of lean and obese subjects has been observed (Zed and James, 1982); the increase in energy expenditure of the obese was only half that of the lean. This study was, however, notable not so much for the difference in thermogenesis observed between lean and obese, but for the remarkably high efficiency at which dietary fat was deposited in all subjects – at the theoretical maximum for the obese, and close to the maximum in the case of the lean.

This observation serves to underline forcibly the central problem which has been continually encountered in studies on differences in thermogenesis between lean and obese subjects – the quantitative importance. Although, as we have indicated, a number of differences in thermogenesis can be observed between the lean and the obese, and although these differences generally occur in the right direction, to demonstrate that they are of real quantitative significance to energy balance in man and to the aetiology of obesity is not easy. The usual argument that obesity develops over a long time scale, and that small differences in energy balance on a day-to-day basis become very important is relevant, but there is increasing evidence that obese subjects when studied in whole-body calorimeters have a *higher*, not a lower, 24-hour rate of energy expenditure (James *et al.*, 1978). This implies that whatever the thermogenic defects which can be demonstrated in obesity, some increase in food intake must also occur and that it would be wrong to think in terms of thermogenesis alone as the exclusive basis for established metabolic obesity. This, of course, does not preclude thermogenic defects from playing a central role in *initiating* the development of obesity in man.

8.4.2 Brown adipose tissue in man

The current emphasis on brown adipose tissue as the main effector of both non-shivering and diet-induced thermogenesis in laboratory rodents has led to considerable discussion and speculation as to its possible role in human obesity. The human neonate is well endowed with brown adipose tissue (Aherne and Hull, 1966; Heim, 1971), and the tissue has been estimated to comprise 2–5% of body weight. It is generally considered to 'atrophy' to some extent during the first couple of years of life, possibly developing into white adipose tissue. Such changes parallel the apparent decline in the capacity for non-shivering thermogenesis which occurs over the same period.

There are several studies reporting the presence of brown adipose tissue in adult man, including its identification during the later decades of life (Heaton, 1972; Tanuma *et al.*, 1975, 1976). These studies, which have been at the level of light microscopy, suggest, however, that the tissue is not particularly active; it certainly does not have the appearance characteristic of the neonate or of cold-adapted rodents. Nevertheless, the potential for brown adipose tissue thermogenesis would seem to be present, although 'activity' as such has not been directly demonstrated. The recent observations by Huttenen and his colleagues (Huttenen, Hirvonen and Kinnula, 1981) that brown adipose tissue was more extensively distributed in the bodies of Finnish men who had lived an outdoor existence in cold matches the observations on cold-induced hyperplasia of brown adipose tissue in rodents. At present there is a need for a method of firmly identifying brown adipose tissue in man (and other species), and one approach may be to use an antibody to the 32 000 molecular weight 'uncoupling protein' since this protein appears to be characteristic of brown adipose tissue mitochondria.

Neither differences in the amount of brown adipose tissue nor differences in its activity have yet been demonstrated between lean and obese subjects. Investigations of this type, although central to substantiating the hypothesis that reduced activity in brown adipose tissue is important in the aetiology of obesity in man, represent a formidable and daunting task. Despite the absence of data on the activity of brown adipose tissue in human obesity, much interest has been aroused by the possibility that energy expenditure in the obese may be raised by pharmacological agents which stimulate brown adipose tissue.

8.5 CONCLUDING COMMENTS

We have attempted to summarize and review the various studies in both animals and man which have linked thermogenesis to the regulation of energy balance and the aetiology of obesity. We have also laid considerable emphasis on the recent developments involving brown adipose tissue, and such emphasis appears at present to be fully justified. Although in some respects the facts which are now available are considerable, it is clear that the problem is still in its infancy; at best we have only a simple outline of a link between thermogenesis and obesity. The central regulatory determinants of energy balance are still obscure and it is difficult to explain, for example, why there is not a compensatory reduction in food intake in an animal which is beginning to deposit excessive amounts of energy by virtue of a high metabolic efficiency.

There is also no satisfactory explanation for the apparent association of a low level of energy expenditure and hyperphagia in so many obese animals, particularly the genetically obese. Such an association may also occur in man, with hyperphagia developing in some instances once obesity has been established by virtue of a high metabolic efficiency (James *et al.,* 1981). Future work on energy balance is likely to concentrate on the central regulatory mechanisms, and the link between food intake and metabolic efficiency.

REFERENCES

Abraham, R.R., Dade, E., Elliott, J. and Hems, D.A. (1971) Hormonal control of intermediary metabolism in obese hyperglycemic mice. II. Levels of plasma free fatty acid and immuno-reactive insulin and liver glycogen. *Diabetes,* **20,** 535–41.

Aherne, W. and Hull, D. (1966) Brown adipose tissue and heat production in the newborn infant. *J. Pathol. Bacteriol.,* **91,** 223–34.

Allan, J.A. and Yen, T.T. (1976) Lipolytic response of 'diabetic' mice (db/db) to isoproterenol and propranolol *in vivo. Experientia,* **32,** 836–7.

Alonso, L.G. and Maren, T.H. (1955) Effect of food restriction on body composition of hereditary obese mice. *Am. J. Physiol.,* **183,** 284–90.

Andrews, F. and Jackson, F. (1978) Increasing fatness inversely related to increase in metabolic rate but directly related to decrease in deep body temperature in young men and women during cold exposure. *Irish J. Med. Sci.,* **147,** 329–30.

Armitage, G., Harris, R.B.S., Hervey, G.R. and Tobin, G. (1981) The part played by variation of energy expenditure in the regulation of energy balance. in *The*

Body Weight Regulatory System: Normal and Disturbed Mechanisms (eds L.A. Cioffi, W.P.T. James and T.B. Van Itallie), Raven Press, New York, pp. 137–41.

Assimacopoulos-Jeannet, F., Giacobino, J.-P., Seydoux, J., Girardier, L. and Jeanrenaud, B. (1982) Alterations of brown adipose tissue in genetically obese (ob/ob) mice. II. Studies of β-adrenergic receptors and fatty acid degradation. *Endocrinology*, **110**, 439–43.

Bernardis, L.L. and Goldman, J.K. (1976) Origin of endocrine-metabolic changes in the weanling rat ventromedial syndrome. *J. Neur. Sci. Res.*, **2**, 91–116.

Blaza, S.E. and Garrow, J.S. (1980) The thermogenic response to comfortable temperature extremes in lean and obese subjects. *Proc. Nutr. Soc.*, **39**, 85A.

Boulangé, A., Planche, E. and de Gasquet, P. (1979) Onset of genetic obesity in the absence of hyperphagia during the first week of life in the Zucker rat (fa/fa). *J. Lipid Res.*, **20**, 857–64.

Bray, G.A. (1981) The inheritance of corpulence. in *The Body Weight Regulatory System: Normal and Disturbed Mechanisms* (eds L.A. Cioffi, W.P.T. James and T.B. Van Itallie), Raven Press, New York, pp. 185–95.

Bray, G.A. and York, D.A. (1979) Hypothalamic and genetic obesity in experimental animals: An autonomic and endocrine hypothesis. *Physiol. Rev.*, **59**, 719–809.

Bray, G.A., York, D.A. and Yukimura, Y. (1978) Activity of $(Na^+ + K^+)$-ATPase in the liver of animals with experimental obesity. *Life Sci.*, **22**, 1637–42.

Buskirk, E.R., Thompson, R.H. and Whedon, G.D. (1963) Metabolic response to cold air in men and women in relation to total body fat content. *J. Appl. Physiol.*, **18**, 603–12.

Carnie, J.A. and Smith, D.G. (1978) Release of fatty acids from adipose tissue in genetically obese (ob/ob) mice. *FEBS Lett.*, **90**, 132–4.

Chlouverakis, C. (1970) Induction of obesity in obese-hyperglycemic mice (ob/ob) on normal food intake. *Experientia*, **26**, 1262–3.

Clausen, T. and Hansen, O. (1982) The $Na^+ - K^+$ pump, energy metabolism, and obesity. *Biochem. Biophys. Res. Commun.*, **104**, 357–62.

Coleman, D.L. (1978) Diabetes and Obesity: Thrifty Mutants? *Nutr. Rev.*, **36**, 129–32.

Contaldo, F. (1981) The development of obesity in genetically obese rodents. in *The Body Weight Regulatory System: Normal and Disturbed Mechanisms* (eds L.A. Cioffi, W.P.T. James and T.B. Van Itallie), Raven Press, New York, pp. 237–42.

Contaldo, F., Gerber, H., Coward, W.A. and Trayhurn, P. (1981) Milk intake in pre-weanling genetically obese (ob/ob) mice. in *Obesity: Pathogenesis and Treatment* (eds G. Enzi, G. Crepaldi, G. Pozza and A.E. Renold), Academic Press, London and New York, pp. 319–22.

Cox, J.E. and Powley, T.L. (1977) Development of obesity in diabetic mice pair-fed with lean siblings. *J. Comp. Physiol. Psychol.*, **91**, 347–58.

Dauncey, M.J. (1981) Influence of mild cold on 24 h energy expenditure, resting metabolism and diet-induced thermogenesis. *Br. J. Nutr.*, **45**, 257–67.

Davis, T.R.A. and Mayer, J. (1954) Imperfect homeothermia in the hereditary obese-hyperglycemic syndrome of mice. *Am. J. Physiol.*, **177**, 222–6.

Deb, S., Martin, R.J. and Hershberger, T.V. (1976) Maintenance requirement and energetic efficiency of lean and obese Zucker rats. *J. Nutr.*, **106**, 191–7.

Doi, J., Ohno, T., Kurahashi, M. and Kuroshima, A. (1979) Thermoregulatory non-shivering thermogenesis in men with special reference to lipid metabolism. *Jap. J. Physiol.*, **29**, 359–72.

Dubuc, P. (1976) Effects of limited food intake in the obese-hyperglycemic syndrome. *Am. J. Physiol.*, **230**, 1474–9.

Felber, J.-P., Meyer, H.U., Curchod, B., Iselin, H.U., Rousselle, J., Maeder, E., Pahud, P. and Jéquier, E. (1981) Glucose storage and oxidation in different degrees of human obesity measured by continuous indirect calorimetry. *Diabetologia*, **20**, 39–44.

Festing, M.F.W. (1979) The inheritance of obesity in animal models of obesity, in *Animal Models of Obesity* (ed. M.F.W. Festing), Macmillan Press, London and Basingstoke, pp. 15–37.

Foster, D.O. and Frydman, M.L. (1978) Non-shivering thermogenesis in the rat II. Measurements of blood flow with microspheres point to brown adipose tissue as the dominant site of the calorigenesis induced by noradrenaline. *Can. J. Physiol. Pharmacol.*, **56**, 110–22.

Foster, D.O. and Frydman, M.L. (1979) Tissue distribution of cold-induced thermogenesis in conscious warm- or cold-acclimated rats re-evaluated from changes in tissue blood flow. The dominant role of brown adipose tissue in the replacement of shivering by non-shivering thermogenesis. *Can. J. Physiol. Pharmacol.*, **57**, 257–70.

Fraser, D.R. and Trayhurn, P. (1981) Calcium transport by brown adipose tissue mitochondria from lean and genetically obese (ob/ob) mice. *Biochem. Soc. Trans.*, **9**, 470–1.

Garrow, J.S. (1978) The regulation of energy expenditure in man, in *Recent Advances in Obesity Research:* II (ed. G.A. Bray), Newman Publishing, London, pp. 200–10.

Girardier, L. (1981) Brown adipose tissue as energy dissipator: a physiological approach, in *Obesity: Pathogenesis and Treatment* (eds G. Enzi, G. Crepaldi, G. Pozza and A.E. Renold), Academic Press, London and New York, pp. 55–72.

Godbole, V., York, D.A. and Bloxham, D.P. (1978) Developmental changes in the fatty (fafa) rat: Evidence for defective thermogenesis preceding the hyper-lipogenesis and hyperinsulinaemia. *Diabetologia*, **15**, 41–4.

Goodbody, A.E. and Trayhurn, P. (1981) GDP binding to brown adipose tissue mitochondria of diabetic-obese (db/db) mice: Decreased binding in both the obese and pre-obese states. *Biochem. J.*, **194**, 1019–22.

Goodbody, A.E. and Trayhurn, P. (1982) Studies on the activity of brown adipose tissue in suckling, pre-obese, ob/ob mice. *Biochim. Biophys. Acta*, **680**, 119–26.

Guernsey, D.L. and Morishige, W.K. (1979) Na^+ pump activity and nuclear T_3 receptors in tissues of genetically obese (ob/ob) mice. *Metabolism*, **28**, 629–32.

Heaton, J.M. (1972) The distribution of brown adipose tissue in the human. *J. Anat.*, **112**, 35–9.

Heim, T. (1971) Thermogenesis in the newborn infant. *Clin. Obstet. Gynecol.*, **14**, 790–820.

Hey, E. (1975) Thermal neutrality. *Br. Med. Bull.*, **31**, 69–74.

Himms-Hagen, J. and Desautels, M. (1978) A mitochondrial defect in brown adipose tissue of the obese (ob/ob) mouse: Reduced binding of purine nucleotides and a failure to respond to cold by an increase in binding. *Biochem. Biophys. Res. Commun.*, **83**, 628–34.

Hogan, S. and Himms-Hagen, J. (1980) Abnormal brown adipose tissue in obese mice (ob/ob): Response to acclimation to cold. *Am. J. Physiol.*, **239**, 301–9.

Hollifield, G. and Parson, W. (1958) Body composition of mice with gold thioglucose and hereditary obesity after weight reduction. *Metabolism*, **7**, 179–83.

Huttenen, P., Hirvonen, J. and Kinnula, V. (1981) The occurrence of brown adipose tissue in outdoor workers. *Eur. J. Appl. Physiol.*, **46**, 339–45.

James, W.P.T., Davies, H.L., Bailes, J. and Dauncey, M.J. (1978) Elevated metabolic rates in obesity. *Lancet*, **i**, 1122–5.

James, W.P.T. and Trayhurn, P. (1976) An integrated view of the metabolic and genetic basis for obesity. *Lancet*, **ii**, 770–3.

James, W.P.T. and Trayhurn, P. (1981*a*) Thermogenesis and obesity. *Br. Med. Bull.*, **37**, 43–8.

James, W.P.T. and Trayhurn, P. (1981*b*) Obesity in mice and men. in *Nutritional factors: Modulating effects on metabolic processes* (eds R.F. Beers, Jr. and E.G. Basset), Raven Press, New York, pp. 123–138.

James, W.P.T., Trayhurn, P. and Garlick, P.J. (1981) The metabolic basis of subnormal thermogenesis in obesity. in *Recent Advance in Obesity Research* III (eds. P. Bjorntorp, M. Cairella and A.N. Howard), John Libbey, London, pp. 220–7.

Jessen, K., Rabol, A. and Winkler, K. (1980) Total body and splanchnic thermogenesis in curarized man during a short exposure to cold. *Acta Anaesth. Scand.*, **24**, 339–44.

Joy, R.J.T. (1963) Responses of cold-acclimatized men to infused norepinephrine. *J. Appl. Physiol.*, **18**, 1209–12.

Jung, R.T., Shetty, P.S., James, W.P.T., Barrand, M. and Callingham, B.A. (1979) Reduced thermogenesis in obesity. *Nature*, **279**, 322–3.

Kaplan, M.L. and Leveille, G.A. (1974) Core temperature, O_2 consumption, and early detection of ob/ob genotype in mice. *Am. J. Physiol.*, **227**, 912–5.

Kaplan, M.L. and Leveille, G.A. (1976) Calorigenic response in obese and nonobese women. *Am. J. Clin. Nutr.*, **29**, 1108–13.

Levin, B.E., Triscari, J. and Sullivan, A.C. (1980) Abnormal sympatho-adrenal function and plasma catecholamines in obese Zucker rats. *Pharmacol. Biochem. Behav.*, **13**, 107–13.

Lin, M.H., Romsos, D.R., Akera, T. and Leveille, G.A. (1978) Na$^+$, K$^+$-ATPase enzyme units in skeletal muscle from lean and obese mice. *Biochem. Biophys. Res. Commun.*, **80**, 398–404.

Lin, M.H., Romsos, D.R., Akera, T. and Leveille, G.A. (1979*a*) Na$^+$, K$^+$-ATPase enzyme units in skeletal muscle and liver of 14-day-old lean and obese (ob/ob) mice. *Proc. Soc. Exp. Biol. Med.*, **161**, 235–8.

Lin, P.-Y., Romsos, D.R. and Leveille, G.A. (1977) Food intake, body weight gain, and body composition of the young obese (ob/ob) mouse. *J. Nutr.*, **107**, 1715–23.

Lin, P.-Y., Romsos, D.R., Vander Tuig, J.G. and Leveille, G.A. (1979*b*) Maintenance energy requirements, energy retention and heat production of young obese (ob/ob) and lean mice fed a high-fat or a high-carbohydrate diet. *J. Nutr.*, **109**, 1143–53.

Mayer, J. and Barnett, R.J. (1953) Sensitivity to cold in the hereditary obese-hyperglycemic syndrome of mice. *Yale J. Biol. Med.*, **26**, 38–45.

Miller, B.G., Grimble, R.F. and Taylor, T.G. (1977) Liver protein metabolism response to cold in genetically obese (ob/ob) mice. *Nature,* **266,** 184–6.

Miller, B.G., Otto, W.R., Grimble, R.F., York, D.A. and Taylor, T.G. (1979) The relationship between protein turnover and energy balance in lean and genetically obese (ob/ob) mice. *Br. J. Nutr.,* **42,** 185–99.

Miller, D.S. (1975) Thermogenesis in everyday life. in *Regulation of Energy Balance in Man* (ed E. Jéquier), Editions Medecine et Hygienes, Geneva, pp. 198–208.

Neel, J.V. (1962) Diabetes mellitus: A 'thrifty' genotype rendered detrimental by 'progress'. *Am. J. Human Genet.,* **14,** 353–62.

Newsholme, E.A., Brand, K., Lang, J., Stanley, J.C. and Williams, T. (1979) The maximum activities of enzymes that are involved in substrate cycles in liver and muscle of obese mice. *Biochem. J.,* **182,** 621–4.

Ohtake, M., Bray, G.A. and Azukizawa, M. (1977) Studies on hypothermia and thyroid function in the obese (ob/ob) mouse. *Am. J. Physiol.,* **233,** R110–5.

Perkins, M.N., Rothwell, N.J., Stock, M.J. and Stone, T.W. (1981) Activation of brown adipose tissue thermogenesis by the ventromedial hypothalamus. *Nature,* **289,** 401–2.

Pittet, P., Chappuis, P., Acheson, K., de Techtermann, F. and Jéquier, E. (1976) Thermic effect of glucose in obese subjects studied by direct and indirect calorimetry. *Br. J. Nutr.,* **35,** 281–92.

Pullar, J.D. and Webster, A.J.F. (1974) Heat loss and energy retention during growth in congenitally obese and lean rats. *Br. J. Nutr.,* **31,** 377–92.

Pullar, J.D. and Webster, A.J.F. (1977) The energy cost of fat and protein deposition in the rat. *Br. J. Nutr.,* **37,** 355–63.

Rath, E.A. and Thenen, S.W. (1979) Use of tritiated water for measurement of 24-hour milk intake in suckling lean and genetically obese (ob/ob) mice. *J. Nutr.,* **109,** 840–7.

Rennie, D.W., Corino, B.G., Howell, B.J., Song, S.H., Kang, B.S. and Hong, S.A. (1962) Physical insulation of Korean diving women. *J. Appl. Physiol.,* **17,** 961–6.

Rolls, B.J., Rowe, E.A. and Turner, R.C. (1980) Persistent obesity in rats following a period of consumption of a mixed, high energy diet. *J. Physiol.,* **298,** 415–27.

Rothwell, N.J., Saville, M.E. and Stock, M.J. (1981) Acute effects of food, 2-deoxy-D-glucose and noradrenaline on metabolic rate and brown adipose tissue in normal and atropinised lean and obese (fa/fa) Zucker rats. *Pflügers Arch.,* **392,** 172–7.

Rothwell, N.J. and Stock, M.J. (1979a) Regulation of energy balance in two models of reversible obesity in the rat. *J. Comp. Physiol. Psychol.,* **93,** 1024–34.

Rothwell, N.J. and Stock, M.J. (1979b) A role for brown adipose tissue in diet-induced thermogenesis. *Nature,* **281,** 31–5.

Rothwell, N.J. and Stock, M.J. (1981) Influence of noradrenaline on blood flow to brown adipose tissue in rats exhibiting diet-induced thermogenesis. *Pflügers Arch.,* **389,** 237–42.

Sclafani, A. and Springer, D. (1976) Dietary obesity in adult rats: similarities to hypothalamic and human obesity syndromes. *Physiol. Behav.,* **17,** 461–71.

Seydoux, J., Assimacopoulos-Jeannet, F., Jeanrenaud, B. and Girardier, L. (1982) Alterations of brown adipose tissue in genetically obese (ob/ob) mice. I. Demonstration of loss of metabolic response to nerve stimulation and catecholamines and its partial recovery after fasting or cold adaptation. *Endocrinology*, **110**, 432–8.

Seydoux, J., Rohner-Jeanrenaud, F., Assimacopoulos-Jeannet, F., Jeanrenaud, B., and Girardier, L. (1981) Functional disconnection of brown adipose tissue in hypothalamic obesity in rats. *Pflügers Arch.*, **390**, 1–4.

Shetty, P.S., Jung, R.T., James, W.P.T., Barrand, M.A. and Callingham, B.A. (1981) Postprandial thermogenesis in obesity. *Clin. Sci.*, **60**, 519–25.

Shimazu, T. and Takahashi, A. (1980) Stimulation of hypothalamic nuclei has differential effects on lipid synthesis in brown and white adipose tissue. *Nature*, **284**, 62–3.

Sims, E.A.H. (1976) Experimental obesity, dietary-induced thermogenesis, and their clinical implications. *Clinics Endocrinol. Metab.*, **5**, 377–95.

Sims, E.A.H., Danforth, E.Jr., Horton, E.S., Bray, G.A., Glennon, J.A. and Salans, L.B. (1973) Endocrine and metabolic effects of experimental obesity in man. *Recent Prog. Hormone Res.*, **29**, 457–96.

Tanuma, Y., Ohata, M., Ito, T. and Yokochi, C. (1976) Possible function of human brown adipose tissue as suggested by observation on perirenal brown fat from necropsy cases of variable age groups. *Arch. Histol. Jap.*, **39**, 117–45.

Tanuma, Y., Yamamoto, M., Ito, T. and Yokochi, C. (1975) The occurrence of brown adipose tissue in perirenal fat in Japanese. *Arch. Histol. Jap.*, **38**, 43–70.

Thurlby, P.L. and Trayhurn, P. (1979) The role of thermoregulatory thermogenesis in the elevated energy gain of obese (ob/ob) mice pair-fed with lean siblings. *Br. J. Nutr.*, **42**, 377–85.

Thurlby, P.L. and Trayhurn, P. (1980) Regional blood flow in genetically obese (ob/ob) mice: The importance of brown adipose tissue to the reduced energy expenditure on non-shivering thermogenesis. *Pflügers Arch.*, **385**, 193–201.

Trayhurn, P. (1979) Thermoregulation in the diabetic-obese (db/db) mouse: The role of non-shivering thermogenesis in energy balance. *Pflügers Arch.*, **380**, 227–32.

Trayhurn, P. and Fuller, L. (1980) The development of obesity in genetically diabetic-obese (db/db) mice pair-fed with lean siblings: The importance of thermoregulatory thermogenesis. *Diabetologia*, **19**, 148–53.

Trayhurn, P. and James, W.P.T. (1978) Thermoregulation and non-shivering thermogenesis in the genetically obese (ob/ob) mouse. *Pflügers Arch.*, **373**, 189–93.

Trayhurn, P., Jones, P.M., McGuckin, M.M. and Goodbody, A.E. (1982) Effects of overfeeding on energy balance and brown fat thermogenesis in obese (ob/ob) mice. *Nature*, **295**, 323–5.

Trayhurn, P., Thurlby, P.L. and James, W.P.T. (1976) A defective response to cold in the obese (ob/ob) mouse and the obese Zucker (fa/fa) rat. *Proc. Nutr. Soc.*, **35**, 133A.

Trayhurn, P., Thurlby, P.L. and James, W.P.T. (1977) Thermogenic defect in pre-obese ob/ob mice. *Nature*, **266**, 60–2.

Trayhurn, P., Thurlby, P.L., Woodward, C.J.H. and James, W.P.T. (1979) Thermoregulation in genetically obese rodents: The relationship to metabolic efficiency. in *Animal Models of Obesity* (ed. M.F.W. Festing), Macmillan Press, London and Basingstoke, pp. 191–203.

Turner, M.L. (1948) Hereditary obesity and temperature regulation. *Am. J. Physiol.*, **152**, 197–204.

VanderTuig, J.G., Knehans, A.W. and Romsos, D.R. (1982) Reduced sympathetic nervous system activity in rats with ventromedial hypothalamic lesions. *Life Sci.*, **30**, 913–20.

Wade, A.J. (1980) Glucose metabolism and recycling of radioactively labelled glucose in the Zucker genetically obese rat (fa/fa). *Biochem. J.*, **186**, 161–8.

Welton, R.F., Martin, R.J. and Baumgardt, B.R. (1973) Effects of feeding and exercise regimens on adipose tissue glycerokinase activity and body composition of lean and obese mice. *J. Nutr.*, **103**, 1212–9.

Wise, P.H. (1977*a*) Significance of anomalous thermoregulation in the pre-diabetic spiny mouse *(Acomys cahirinus):* Oxygen consumption and temperature regulation. *Aust. J. Exp. Biol. Med. Sci.*, **55**, 463–73.

Wise, P.H. (1977*b*) Significance of anomalous thermoregulation in the pre-diabetic spiny mouse *(Acomys cahirinus):* Cold tolerance, blood glucose and food consumption responses to environmental heat. *Aust. J. Exp. Biol. Med. Sci.*, **55**, 475–84.

Woodward, C.J.H., Trayhurn, P. and James, W.P.T. (1977) Costs of maintenance and growth in genetically obese (ob/ob) mice. *Proc. Nutr. Soc.*, **36**, 115A.

Yen, T.T., Fuller, R.W. and Pearson, D.V. (1974) The response of 'obese' (ob/ob) and 'diabetic' (db/db) mice to treatments that influence body temperature. *Comp. Biochem. Physiol.*, **49A**, 377–85.

York, D.A., Bray, G.A. and Yukimura, Y. (1978) An enzymatic defect in the obese (ob/ob) mouse: Loss of thyroid-induced sodium – and potassium-dependent adenosinetriphosphatase. *Proc. Natl. Acad. Sci. USA*, **75**, 477–81.

Young, R.A., Tulp, O.L. and Horton, E.S. (1980) Thyroid and growth responses of young Zucker obese and lean rats to a low protein-high carbohydrate diet. *J. Nutr.*, **110**, 1421–31.

Zed, C.A. and James, W.P.T. (1982) Thermic response to fat feeding in lean and obese subjects. *Proc. Nutr. Soc.*, **41**, 32A

Zucker, L.M. (1975) Efficiency of energy utilization by Zucker hereditarily obese fatty rat. *Proc. Soc. Exp. Biol. Med.*, **148**, 498–500.

Chapter Nine

Hypermetabolism in Trauma

L. Howard Aulick and Douglas W. Wilmore

9.1 INTRODUCTION

Body injury, whether it be the result of a carefully planned surgical procedure or due to some accidental means, alters the normal homeostatic
balance of the organism and initiates a well-integrated, total body response.
The metabolic components of this response were first documented in the
early 1900s when it was reported that haemorrhage and operation increased
urinary nitrogen excretion (Hawk and Gies, 1904; Haskins, 1907). Evolution of this field of investigation has recently been described by Sir David
Cuthbertson (1976), a man who over the last fifty years, has been its single
major contributor.

The metabolic response to injury is a universal response; common
features have been identified in plants (Adams and Rowan, 1970) and across
the animal kingdom from crabs (Needham, 1955) and earthworms (Needham, 1958) to man. The immediate response is a generalized depression of
all physiological activity, called the 'shock' or 'ebb' phase of injury. If the
animal can be adequately resuscitated and survives, the depressed state
rapidly gives way to a period of accelerated function. This phase of recovery
has been called the 'flow' or 'catabolic' phase and, once again, appears to
involve all major systems. During this time, cardiopulmonary function
increases, body protein and fat stores are rapidly broken down, hepatic
glucose production is markedly increased and metabolic heat production
and body temperatures are elevated. Cuthbertson suggests that the
accelerated release of endogenous fuels may be part of a primitive survival
reflex designed to provide energy substrates during a time when the injured
animal would be unable to search for and obtain an adequate food supply
(Cuthbertson, 1980). The degree of heightened energy turnover and the
associated circulatory and thermal effects are primarily determined by the
nature and extent of injury. The magnitude of the injury response varies
with the changing character of the wound, reaching its greatest level in the
early stages of repair and then gradually returning to normal with wound
healing and restoration of function. Age, sex, body habitus, associated

injury, infection or disease, level of nutrition, treatment regime and other factors also affect the character of the response.

This chapter will describe the metabolic, circulatory and thermoregulatory alterations which occur during both the ebb and flow phases of injury. It will outline the afferent and efferent pathways involved and discuss some of the more important modifiers of the basic metabolic response, i.e. infection, thermal environment, and level of nutrition. Finally, the clinical implications of these alterations will be considered. Most studies cited in this review were performed on previously healthy young men or animals. Treatment, in many cases, had to be modified in order to satisfy research requirements. So, while the basic character of the injury response can be defined by this investigative approach, the reader must continually keep in mind that the actual clinical manifestation of injury is frequently altered by the interaction of the various modifiers listed above.

9.2 AFFERENT MEDIATORS OF METABOLIC ALTERATIONS

The metabolic alterations following injury result from a series of integrated neurohormonal signals which for the most part originate in the central nervous system. The magnitude of these signals depends primarily on the extent of stress, related to the mass of tissue damage, although these signals and the resultant tissue responses may be modulated by the age and sex of the patient, physiological reserve (i.e. stress capacity), nutritional status and underlying disease processes.

The causative factors and afferent stimuli which evoke the stress response in the injured patient appear specific and are related in time following the initial insult. During the 'ebb' phase of injury, three general types of stimuli signal the central nervous system to initiate homeostatic adjustment:

9.2.1 Fluid loss

Fluid loss from the vascular compartment results in stimulation of volume and pressure receptors, initiating a series of CNS-mediated cardiovascular adjustments. Cardiac output falls, peripheral resistance increases and blood flow is redistributed to vital organs to maintain function. With progressive volume loss into the area of injury, the resulting hypoperfusion reduces tissue oxygenation and alters acid–base equilibrium. Chemoreceptor stimulation thus serves as additional afferent input to both vasomotor and respiratory centres during hypovolaemia. Because loss of fluid volume following injury is closely related to the extent of tissue damage, these specific mechanisms provide afferent signals quantitatively related to the extent of injury.

9.2.2 Afferent nerve fibres

Afferent nerve fibres provide the most direct and quickest route for signals to reach the central nervous system following stress. It has frequently been suggested that pain may serve as the initial afferent signal following injury, and a variety of studies suggest that neurogenic afferents from the injured area are essential for the stimulation of the pituitary–adrenal axis. The adrenocortical response to injury was not observed in animals after section of the peripheral nerves to the area of injury, transection of the spinal cord above the injury or section through the medulla oblongata (Hume and Egdahl, 1959). A similar pattern of response to denervation before injury has been described in man. Both growth hormone (GH) and ACTH levels in the serum rise within one hour following incision in patients receiving general anaesthesia and undergoing cholecystectomy or inguinal herniorrhaphy. However, this hormonal response did not occur in patients undergoing herniorrhaphy who received spinal anaesthesia (Newsome and Rose, 1971), nor did the usual rise in serum cortisol occur in patients undergoing abdominal procedures when epidural blockade was utilized in conjunction with the general anaesthetic (Brandt *et al.*, 1976; Bromage, Shibata and Willoughby, 1971). Studies of the pituitary–adrenal axis following operation in paraplegic patients demonstrate a markedly diminished cortisol response when the operation is performed in a denervated area (Hume, 1969). Nervous afferents also appear to stimulate the elaboration of antidiuretic hormone following trauma (Ukai, Moran and Zimmerman, 1968). In addition, a number of factors which accompany the 'stress' of critical illness – restraint, immobilization, environmental disturbances – most likely alter nervous afferent impulses and affect the response to injury.

9.2.3 Circulating substances

Circulating substances may directly or indirectly stimulate the central nervous system and set into motion various components of the injury response. Alterations in serum electrolytes, release of cell breakdown products (Haist and Hamilton, 1944), changes in the amino acid pattern and formation of endogenous pyrogens, all originating from a direct result of the wound, may initiate homeostatic adjustments during the early phase of injury.

With non-fatal injuries, the 'ebb' phase evolves into the 'flow' phase response, which is characterized by hypermetabolism and increased loss of nitrogen and other intracellular constituents from the body. During the 'flow' phase, the increased vascular permeability resolves and blood volume and composition of other fluid compartments stabilize. Hypovolaemia or abnormalities in acid–base composition of the blood disappear and are not signals which explain the physiological alterations which occur during this hypermetabolic phase of injury.

To determine the role of afferent nervous signals from the area of injury during the 'flow' phase, a variety of clinical studies have been conducted in injured patients. First, a patient with traumatic spinal cord transection and burns of the lower extremities has been studied: hypermetabolism and the associated metabolic responses occurred despite denervation of a major portion of the wound (Taylor *et al.*, 1976). A topical anaesthetic was applied to the wounds of other burn patients to achieve anaesthesia and insure that stimulation from pain receptors was blocked in the injured area. No alteration in metabolic rate or body temperature occurred for up to six hours following the application of the topical anaesthetic, although most patients were rendered pain free and slept throughout the study. Finally, a spinal anaesthetic was placed and maintained in a patient with multiple long-bone fractures and 33% total body surface burns over the lower extremities. No significant effect on metabolic rate or core temperature was detected following denervation of the injured area. Therefore, there is little evidence from these studies that sensory nerves play a major role in the afferent limb of the stress response during the 'flow' phase of injury.

At one time, it was thought that the increased evaporative water loss which occurs from the damaged surface of burn patients stimulated cold receptors causing a rise in metabolic heat production. Subsequent studies have demonstrated that the hypermetabolism following thermal injury is temperature sensitive but not temperature dependent (Wilmore *et al.*, 1975). Although heat production can be minimized by treating burn patients in a warm environment, the marked elevation in metabolic rate does not return to normal with external heating (Aulick *et al.*, 1979). The thermoregulatory impact of injury will be described in greater detail later in this chapter.

Although nervous afferent stimulation may not be responsible for the 'flow' phase response, pain following treatment and patient manipulation will increase metabolism above the already elevated level which occurs following injury. However, when burn patients were studied in ambient conditions of comfort and allowed to sleep with or without analgesics, the hypermetabolic response was not abated and metabolic rates were maintained at levels 50% to 80% above normal. Patient care should be oriented to minimize all painful stimuli: judicious use of analgesics and tranquillizers may be necessary in the treatment of critically ill patients.

Because of the inability to identify specific nervous afferent stimuli as the initiators and propagators of the 'flow' phase injury response, a search for circulating afferent signals has begun. In one study, heparinized blood was collected from burn patients and normals in pyrogen-free syringes (Wilmore, 1976). A micro-aliquot of each sample was injected through indwelling chronic cannulae in rabbits, placed by standard stereotactic technique so that the distal tip lay in the preoptic area of the hypothalamus. Injection of normal serum from six control subjects resulted in no more than a 0.1°C rise in rectal temperature. Hypothalamic injection of serum from 9 of the 13 patients elicited a febrile response (0.63–0.93°C over two hours). Limulus lysate assay for endotoxin was negative in all these samples. After heat treatment of the serum, the febrile response was attenuated, suggesting that endogenous pyrogens mediated this response.

Prostaglandins (PG) are known to affect hypothalamic function, and increased concentrations of these substances are found in lymph from areas of injury and in exudate from burn wounds (Arturson, 1978). To evaluate these substances as possible 'wound hormones', arterial and venous concentrations of PGA, E and F were determined, using specific antibody assay techniques (Wilmore, 1976). Twenty-one patients were studied and blood was drawn specifically from the femoral vein in 15 of these subjects with burned lower extremities to determine the contribution of injured tissue to the prostaglandin level. Arterial and venous concentrations of prostaglandins A, E and F were similar to those observed in normal subjects: patients with and without leg burns had similar concentrations of these substances in femoral vein blood when studied between the third and thirty-first day post-injury. So, while these products of tissue injury may have profound local metabolic and circulatory effects in the wound, it does not appear that they exert a significant systemic effect.

9.3 NEUROENDOCRINE RESPONSES

9.3.1 Central nervous system adjustments

The central and autonomic nervous systems are essential to the hypermetabolic response to injury: patients with 'brain death' and associated soft tissue injury failed to mount a 'flow' phase response (Taylor *et al.*, 1976). Similarly, morphine anaesthesia, which markedly reduced hypothalamic function, resulted in a prompt decrease in hypermetabolism, rectal temperature, and cardiac output in severely burned patients. In contrast, however, quadriplegic patients with high spinal cord transections which totally interrupted sympathetic efferent activity failed to generate a febrile response to infection, but were able to increase the leukocyte count and blood glucose concentration during sepsis. Moreover, when low-molecular-weight extracts from granulation tissue have been injected (both intravenously and subcutaneously) into normal animals, fibroblast proliferation and collagen biosynthesis were observed (Lerman *et al.*, 1977). These findings suggest that several circulating mediators with direct and specific cellular effects may exist and contribute to the metabolic response to injury. However, in all patients with intact central nervous systems, a variety of adjustments are observed within the hypothalamus and pituitary gland.

These alterations in neurohumoral control appear to be specific compensatory adjustments to stress and impact on thermoregulation, substrate mobilization and intraorgan energy transfer.

9.3.2 Endocrine elaboration

(a) Human growth hormone

Several alterations in hypothalamic and pituitary activity have been noted during the flow phase of injury. Pituitary function has been assessed by examining specific hormonal responses to various provocative stimuli. Human growth hormone (GH) response to insulin hypoglycaemia and arginine infusion was measured in nine burn patients and five normal controls (Wilmore *et al.*, 1975). Initial studies were carried out between day 3 and day 24 post-burn and were repeated in surviving patients after wound closure. Provocative tests for human growth hormone release were performed on consecutive days in the early morning. On the first day of the study, 0.2 units insulin (kg body weight)$^{-1}$ were administered to burn patients, and 0.5 units (kg body weight)$^{-1}$ to recovered patients and normal individuals. Serial blood samples were obtained for estimation of blood glucose and GH concentrations. On the subsequent morning, 30 g of a 10% solution of arginine hydrochloride were infused intravenously over 30 min, and blood was serially assayed for glucose, urea nitrogen, GH and insulin.

Fasting GH was significantly elevated above normal in burn patients during the hypermetabolic phase of injury and during recovery, and elevated GH levels occurred during the period of acute injury despite the associated fasting hyperglycaemia. The GH response to hypoglycaemia was more rapid, but the peak response was diminished in burn patients when compared with recovered patients and controls. An attenuated response also was observed after arginine infusion (Table 9.1). This work then demonstrates that there is a chronic elevation in GH levels, but responses to provocative signals appear to be blunted. This does not appear to be due to a lack of pituitary stores, for examination of the pituitary gland on autopsy demonstrates anatomical confirmation of available growth hormone stores.

(b) Thyroid hormones

Alterations in thyroid kinetics are known to occur in acute medical and surgical illnesses, but the precise role of these changes remains unknown. Serum concentrations of thyroid hormone have been determined sequentially for the first two weeks after injury (Becker *et al.*, 1976). Thyroxin (T_4) and tri-iodothyronine (T_3) concentrations were significantly reduced, and reverse T_3 (rT_3) concentrations became elevated, findings consistent with the observations in other disease processes. These alterations in thyroid

Table 9.1 GH response after provocative stimulation in thermally injured patients (mean±SEM, ng ml^{-1})

	Normal Subjects ($n = 5$)		Acute Burn Patients ($n = 9$)		Recovered Burn Patients ($n = 7$)	
	Basal*	Peak response	Basal	Peak response†	Basal	Peak response
Insulin hypoglycaemia	0.8±0.1	32.6±7.6	1.7±0.2	12.6±2.8	1.7±0.4	27.8±12.0
Arginine infusion	1.0±0.1	10.1±1.4	1.8±0.3	3.9±0.9	1.8±0.5	9.3± 5.1

*Basal level of normal individuals different from acute and recovered patients, $P < 0.001$.
†Mean peak response of acute burn patients different from normal individuals and recovered patients, $P < 0.01$.

hormone concentrations occurred while thyroid stimulating hormone (TSH) remained normal. To assess the pituitary TSH responsiveness, 12 additional, injured patients were studied (Wilmore, Aulick and Pruitt, 1978). Thyrotropin releasing hormone (TRH), the hypothalamic hormone which stimulates TSH release, was given as a 400 μg intravenous bolus, and serial TSH levels were monitored. In four critically ill patients requiring dopamine infusion for cardiocirculatory support, the TSH response to TRH was attenuated, a finding consistent with the pituitary suppression known to occur with dopaminergic stimulation of the pituitary gland (Table 9.2). Five patients without complications demonstrated a normal rise in TSH after administration of TRH. In contrast, three patients with bacteraemia and hypothermia demonstrated an exaggerated TSH response. These persons had significantly higher cortisols and rT_3 concentrations when compared with the patients without complications. However, the bacteraemic patients were relatively hypothermic, a stress that is known to increase pituitary release of ACTH and TSH. Because the hyperresponsive patients had lower body temperatures at the time of testing, this study suggests that the inability to maintain core temperature provides physiological stimulation of the pituitary gland, resulting in concomitant release of ACTH (reflected by the elevated cortisol response) and an augmented TSH response. Thus, the pituitary TSH response to TRH was appropriate and normal in the patients without complications; TSH response was suppressed with dopamine and accelerated by cold exposure. Alterations in peripheral thyroid hormone concentrations, low T_4 and T_3, high rT_3 could not be explained by alterations in TSH or changes in the pituitary control of the thyroid gland. The appearance of rT_3 with complicated illnesses suggests that it is associated with catabolic processes and, like its precursor amino acids (thyrosine and phenylalanine), serves as another metabolic marker of body catabolism.

(c) ACTH–cortisol

With stress and hypothalamic stimulation, increased pituitary secretion of adrenal corticotrophin hormone (ACTH) occurs, causing liberation of glucocorticoids from the adrenal cortex. The hypothalamic centres which stimulate ACTH are under a variety of controls. During normal conditions, ACTH fluctuates inversely with concentration of corticosteroids in the plasma, and plasma concentrations are thought to provide feedback signals to maintain adrenal control. Similarly, administration of exogenous corticosteroids suppress ACTH release and result in adrenocortical atrophy. It is clear, however, that the major physiological cause for liberation of ACTH is stress, either psychological or physiological. In 1950, it was demonstrated that the discharge of adrenal ascorbic acid (a marker to assay corticosteroid secretion) after fracture or mild scald burn was minimal in rats, if the traumatized limb was denervated (Gordon, 1950). However,

Table 9.2 TSH (μIU ml^{-1}) response of 12 burn patients to TRH (400 μU i.v., mean$\pm$SEM)

Patient group	No.	Time Following TRH (min)						Peak response	Response	Integrated response above basal (μIU min^{-1} ml^{-1})
		0	15	30	45	60	90			
Without complications	5	3.6±1.0	10.3±1.8	12.1±1.9	12.3±1.9	12.9±2.1	9.3±1.6	13.8±2.0	10.2±1.5	593±90
Bacteraemia	3	6.2±1.4	28.7±13.5	21.3±5.2	30.5±5.8	25.0±6.1	23.8±3.2	42.2±6.9*	36.0±8.2*	1618±102†
Bacteraemia (dopamine infusion)	4	2.0±0.2	2.4±0.4	3.0±0.5	3.1±0.4	2.5±0.3	3.1±0.6	1.6±0.4*	1.6±0.4*	66±25†

* $P < 0.01$ when compared to patients without complications.
† $P < 0.001$ when compared to patients without complications.

denervation had no effect on discharge of adrenal ascorbic acid if a more severe scald burn occurred. In a series of classic studies, Hume and Egdahl (1959) measured adrenal venous 17-hydroxycorticosteroids in experimental animals and demonstrated that adrenal cortical secretion was not elicited after trauma to the denervated hindlimb. Adrenocortical response to injury was not observed following operative transection of the afferent pathway between the injury and the hypothalamus. Results of other studies demonstrated that electrical lesions in the anterior medial eminence of the hypothalamus abolished post-traumatic elevations of ACTH and adreno-corticoids in animals subjected to standard operative trauma (Hume, 1969). Removal of the cortex of the brain did not ablate the ACTH response to injury.

That afferent nervous stimuli provide signals for the outpouring of ACTH and hence the liberation of cortisol has been demonstrated in humans. Kehlet and associates in Copenhagen studied two groups of patients under-going lower abdominal procedures requiring general anaesthesia (Kehlet, Brandt and Rem, 1980). In one group, epidural anaesthesia was also given, thereby blocking transmission of impulses via the spinal cord, either to hypothalamic centres or more regional reflex mechanisms. In the patients with general anaesthesia alone, there was a brisk elevation in blood glucose which paralleled the increase in serum cortisol. In contrast, the patients with the neurogenic blockade maintained euglycaemia and did not show signifi-cant elevations in serum cortisol concentrations. Using the same experi-mental protocol, other studies demonstrated that the addition of a neurogenic blockade reduced the negative nitrogen balance of these patients below that observed in the other postoperative group. Therefore, it appears that nervous afferent stimuli are dominant factors for the notifi-cation and stimulation of ACTH output immediately after injury. This, in turn, signals the release of cortisol, which itself exerts a host of other metabolic effects.

With marked hypothalamic stimulation, primarily from afferent nerves, the initial blood concentrations of corticoids are elevated and return slowly to normal levels after the acute phase of injury. Measurements of cortisol turnover during severe infection indicate that adrenal cortical steroid secretion may increase up to two to five times normal and return to baseline levels during convalescence (Beisel, 1975). Elevated cortisol concentrations are associated with a period of cardiovascular instability and hypotension, and increased serum concentrations may also occur during hepatic dysfunc-tion when deconjugation processes are impaired. During stable and prolonged infectious and traumatic illnesses, the serum cortisol remains elevated or in the high normal range, depending upon the severity of injury. In this stable state, circadian variation of cortisol has been described, although serum concentrations are re-set at a higher level.

The effects of steroids can be classified into two general categories: (1)

those concerned with organic metabolism, i.e. metabolism of fat, carbohydrate, and protein, inflammation, wound healing and myocardial metabolism; and (2) those affecting mineral metabolism. The primary metabolic effects of steroids are to initiate substrate flux and process the metabolism of body fuels. Glucocorticoids provide specific signals which augment hepatic gluconeogenesis by stimulating enzymes which direct conversion of 3-carbon fragments into synthesis of new glucose (Exton, 1972). Glucocorticoids augment other hormonal signals which call for the production of new glucose, and increased gluconeogenesis occurs when glucocorticoids are administered in conjunction with glucagon and catecholamines. In addition, glucocorticoids promote storage of carbohydrate as glycogen, but this could occur because of increased glucose synthesis. In adrenalectomized animals, for example, there is a reduction of urinary nitrogen, blood glucose and liver glycogen, and these concentrations are restored to normal with steroid administration.

Another major function of glucocorticoids is the stimulation of body protein breakdown and the mobilization of amino acid fragments to the liver for the synthesis of acute phase protein and glucose. In the stable, adrenalectomized animal, amino acids cannot be mobilized from skeletal muscle. However, there is no impairment in utilization of free amino acids provided in the diet or by intravenous administration. Accelerated myofibrillar protein breakdown in skeletal muscle only occurs in experimental animals when plasma levels of glucocorticoids exceed normal concentrations. If these data are applicable to critically ill patients, then corticosteroid mediated breakdown of muscle protein would only be associated with more severe injury and infection which elevate serum cortisol concentrations. Glucocorticoids facilitate amino acid mobilization from the periphery, favour protein biosynthesis in the liver and result in overall transfer of protein from the carcass to the visceral organs. This is an important process for, as will be discussed later, the mobilization, the increased breakdown and the synthesis of protein produce additional heat, and glucocorticoids may participate in a variety of signals that initiate this process.

(d) Catecholamines

Post-traumatic hypermetabolism appears to be the result of increased sympathetic outflow, which is associated with the elaboration of high quantities of catecholamines. Catecholamines are elevated after injury and adrenergic activity has been related to the extent of the injury and to the oxygen consumption of the patient (Harrison, Saton and Feller, 1967; Wilmore *et al.*, 1974). Carefully controlled adrenergic blockade in patients with large surface area burns has demonstrated a consistent decrease in metabolic rate with combined α- and β- or β-adrenergic blockade alone (Wilmore *et al.*, 1974a). This evidence suggests that catecholamines (increased adrenergic activity) are the major calorigenic mediators responsible for the hypermetabolic response following injury. Limited increases in calorigenesis have also been noted, however, with growth hormone administration and infusion of glucagon. The physiological significance of these effects has yet to be determined, but these hormones may act to augment or potentiate the catecholamine-directed heat production in injured patients.

Catecholamine calorigenesis depends upon the availability of catecholamine reserves and the ability of tissues to respond. Catecholamine stores in patients who die after injury and stress are depleted in the adrenal medulla, sub-hepatic nerve endings, sympathetic ganglia and heart (Goodall and Moncrief, 1965). Dopamine turnover in burn patients is increased markedly, possibly reflecting substrate limitation (Goodall and Alton, 1969). Patients with burn injuries of more than 40% of their body surface appear to maintain near maximal rates of catecholamine synthesis and utilization (Wilmore *et al.*, 1974*a*). Exposing these patients to a cool environment (21°C) results in a mild cold stress and stimulates the sympathetic nervous system. Patients who eventually survive respond by increasing heat production as a result of greater elaboration of catecholamines. In contrast, patients who lack catecholamine reserves or tissue responsiveness fail to generate additional heat in a 21°C environment and become hypothermic. All of the non-responders in the study described subsequently died from complications of their injuries. This suggests that the capacity to withstand injury and many of its associated stresses (thermal, infection and haemorrhage) depend on the availability of and sensitivity to catecholamines. The complete expression of this calorigenic effect also depends on tissue oxygenation which, in turn, depends on cardiopulmonary function: when oxygenization is inadequate (as in ebb phase), calorigenesis is limited.

The metabolic effects of increased sympathetic outflow are well known. During the ebb and flow phases of injury, the marked outpouring of catecholamines results in the increased mobilization of glucose, the stimulation of lipolysis and, either through direct or indirect mechanisms, accelerated protein degradation. Specific mechanisms of the latter are not known, for *in vitro* studies demonstrate that beta stimulation of skeletal muscle facilitates protein synthesis, not degradation (Garber, Karl and Kipnis, 1976). In addition, sympathetic outflow regulates the endocrine pancreas and most probably has a variety of effects upon hepatic function, since hepatocytes possess a dense autonomic innervation.

The sympathetic nervous system is stimulated by a variety of signals, including pain, anxiety, anaesthesia, dehydration, blood loss, operation, infection, hypoglycaemia, increased intracranial pressure and alterations in tissue perfusion and metabolism which affect the chemical environment of the body. All of these factors and possibly other unknown stimuli may cause the sympathetic stimulation following injury. The hypothalamus receives most of these afferent signals and must integrate this input in a meaningful response. This is accomplished by a variety of interactions which occur in various hypothalamic nuclei which are all located in close proximity to each other and all interconnected by numerous nervous pathways. Discrete areas of the hypothalamus regulate temperature, blood glucose and other fuels, appetite, thirst, blood pressure and respiration. All these nuclei are interconnected and also send fibres to an area in the posterior hypothalamus through which all sympathetic nerve traffic flows (often referred to as the 'sympathetic centre' of the brain). But, rather than a discrete control centre, this area in the posterior hypothalamus serves only as a relay station. The hypothalamic nuclei are also intimately related to pituitary gland function via neurons in the pituitary portal system. Thus, sympathetic outflow results from the stimulation and integration of numerous, poorly understood afferent signals in the hypothalamus and is responsible for many of the thermoregulatory and metabolic responses that occur following major injury.

(e) Insulin and glucagon

The peptide hormone insulin is synthesized by β-pancreatic cells and is essential to glucose homeostasis. Pancreatectomized man requires insulin for survival. Insulin facilitates glucose entry into many tissues, increasing the flow of glucose along all pathways concerned with intracellular glucose metabolism. In addition to facilitating glucose entry into skeletal muscle, insulin induces key enzymes in the liver which favour storage of glycogen, suppress hepatic enzymes concerned with glycolysis and gluconeogenesis and reduce net hepatic glucose production. Insulin also participates with catecholamines in controlling lipolysis. Catecholamines act as the major stimulus for lipolysis, and this is counteracted primarily by the presence of insulin. Finally, when insulin concentrations fall or insulin is absent, the efflux of amino acids from skeletal muscle increases. Insulinization of skeletal muscle reverses this effect by augmenting the transport of amino acids into the muscle cells and stimulating protein synthesis.

The output of insulin from the pancreas is generally proportional to the hyperglycaemia which occurs and to the duration of the elevated blood glucose. This response, however, can be modulated by other factors, including output from the sympathetic nervous system; the infusion of epinephrine will suppress the insulin elaboration which follows a pro-vocative glucose load. Catecholamines thus serve as set point controllers for insulin release, and the increased sympathetic nervous system activity appears to be responsible for the insulin suppression and glucose intolerance observed during operation, volume depletion, mild or severe infection and severe shock (Porte and Robertson, 1973). The insulin inhibitory effect of the sympathetic nervous system is mediated by α-receptors, while adrenergic β-receptor stimulation augments insulin elaboration. During periods of marked sympathetic nervous system discharge, such as shock, myocardial infarction or systemic infection, the α-receptor effect appears to dominate, resulting in glucose intolerance. However, hypermetabolism secondary to adrenergic activity is characterized by increased mass flow of glucose from the liver to peripheral tissues, which is related to increased insulin secretion. β-Adrenergic receptors may augment insulin elaboration during the flow phase of injury (Iversen, 1973).

Glucagon is an insulin counter-regulatory hormone which is produced in the α-2 cells of the islets of Langerhans in the pancreas. In general, the biological role of glucagon is to act in concert with insulin to insure a steady supply of substrate from the liver under a wide variety of physiological conditions. While insulin has been referred to as the hormone of energy storage, glucagon is viewed as a hormone which regulates energy release. Glucagon is suppressed following ingestion of a meal containing carbohydrate and rises during starvation. Unger (1971) has proposed that the insulin:glucagon ratio (I/G ratio) be utilized in a quantitative and qualitative sense to describe hepatic glucose balance in fed and fasting patients and in diabetic individuals. Anabolism and protein conservation occur when insulin is increased relative to glucagon (I/G ratio > 5), a hormonal environment which favours energy storage, limits gluconeogenesis, and increases protein biosynthesis and decreases urea nitrogen excretion. The infusion of glucagon into fasting man increases glucagon relative to insulin (I/G ratio > 3), and this hormonal milieu is associated with increased glycogenolysis, gluconeogenesis and urogenesis at the expense of protein biosynthesis.

Glucagon levels are markedly elevated in critically ill patients, even in the face of glucose administration and hyperglycemia (Wilmore *et al.*, 1974*b*). The close relationship between glucagon and catecholamines in severely injured patients suggests increased adrenergic activity which may contribute to this combination of hyperglycaemia and hyperglucagonaemia during critical illness. Glucagon and catecholamine elaboration gradually return to normal with wound healing. In traumatized patients, the autonomic nervous system may direct pancreatic islet cells to elaborate glucagon and insulin, that is, adjust set points for the hormonal responses of the endocrine pancreas. Glucagon in turn affects the deposition of key body fuel substrates. It acts on the liver primarily to augment and amplify catecholamine-directed, cortisol-mediated hepatic gluconeogenesis. Glucagon does not contribute directly to the efflux of amino acids from skeletal muscle. Glucagon does not appear to be a 'primary' stress hormone, but is augmentive in its actions on the liver.

9.4 METABOLIC RESPONSES

9.4.1 Aerobic metabolism

Increased oxygen consumption is a characteristic, systemic metabolic response to injury, and the rate of oxygen utilization is related to the extent of trauma. In normal man, approximately two-thirds of resting metabolic heat production takes place in the head and trunk. Splanchnic oxygen consumption in burned patients is increased 50 to 60% above normal resting levels (Aulick *et al.*, 1981; Gump, Price and Kinney, 1970). Peripheral oxygen consumption has been determined in burned and unburned legs of patients by measuring limb blood flow and femoral arteriovenous oxygen differences (Wilmore *et al.*, 1977). Leg oxygen consumption was unaffected by the local presence of a burn wound, but remained a relatively constant proportion of the total body oxygen consumption (5–6%) in both hypermetabolic burned patients and normal controls. This relationship between limb and total body aerobic metabolism is in good agreement with the estimates of 5.9% in normal subjects (Stolwijk, 1970). Since splanchnic oxygen consumption increases following thermal injury and peripheral oxygen uptake remains a fixed portion of total aerobic metabolism, burn hypermetabolism appears to be a generalized or systemic response involving the entire body. Consequently the general increase in body heat production appears to be distributed in a relatively normal fashion – two-thirds of the heat being produced in the visceral tissues and the other one-third in the extremities (Fig. 9.1).

Following injury, the heat produced in the fasted patient is primarily a result of fat oxidation. Because the patients are febrile, some of the increase in oxygen consumption is related to the elevated body temperature (Q_{10} effect). However, this effect of temperature on metabolism is small and only accounts for 10 to 20% of the extra oxygen consumed. The major portion of the heat produced is the result of biochemical inefficiency. Once high energy phosphate bonds are synthesized, they are utilized for mechanical, transport or synthetic work. Some increase in mechanical work occurs in injured patients because of increased respiratory and cardiac activity, but this accounts for a minor portion of the total heat produced. Similarly, transport work may be increased slightly (probably a result of the initial sodium load administered during resuscitation and/or the large solute load handled by the kidneys during catabolism). However, synthetic work is the most clearly identifiable requirement for extra energy; the thermally injured patient produces new glucose, acute phase proteins, albumin and leukocytes and heals a large surface wound. The increased energy requirement for protein synthesis may account for the major quantity of energy utilized for synthetic purposes. The energy yield from the conversion of glucose into lactate in the wound is inefficient when compared with the complete glucose

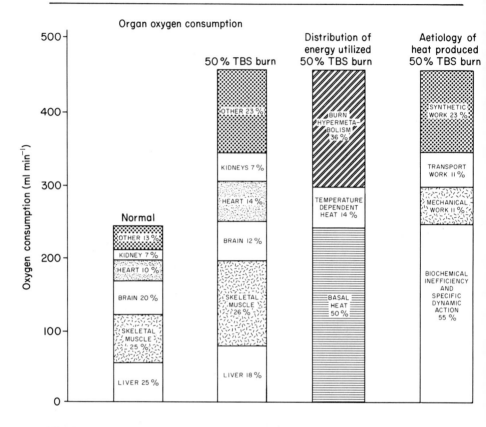

Fig. 9.1 Partition of the hypermetabolism following thermal injury.

oxidation. The relative contribution of all these synthetic processes to the production of body heat is unknown at this time.

The extra heat given off by these relatively inefficient metabolic processes in the injured patient is sufficient to raise body temperatures. Increased rates of wound blood flow channel body heat to the injured area where it raises local temperatures. Not only will this increase in wound temperature accelerate healing, but the generalized febrile response to injury is thought to be beneficial to the host, especially following exposure to infectious organisms (Kluger, 1978). Resistance to infection in experimental animals is favourably influenced by raising body temperature, either by elevating ambient temperature or by supporting the organisms' fever mechanisms. However, the increased heat production imposes a metabolic cost to the patient which results in accelerated tissue catabolism and disruption of body mass. The increased energy demands must be met by a vigorous feeding programme if the complication of catabolism and weight loss are to be averted and the accelerated thermogenic mechanisms supported.

9.4.2 Protein metabolism

Fifty years ago, Cuthbertson (1930) described the extensive urinary nitrogen loss which occurred following long-bone fracture. Because of the magnitude of these losses, the progressive wasting of the skeletal muscle mass and the associated muscle weakness, he attributed the nitrogen loss to a generalized and accelerated breakdown of muscle protein. The loss of nitrogen following injury and infection has been studied in detail and is related to the extent of the trauma, the previous nutritional state of the individual and the age and sex of the patient (Moore, 1959).

Nitrogen equilibrium is maintained by a careful balance between rates of protein synthesis and degradation (Waterlow, Garlick and Millward, 1978). Nitrogen balance studies, however, reflect only the *net* gain or loss of protein from the body and do not reflect actual changes in rates of protein synthesis or breakdown. Thus, negative nitrogen balance can occur if the breakdown rate increases and protein synthesis remains the same or if the protein breakdown rate remains the same and the rate of synthesis decreases. The use of isotopically labelled, non-radioactive amino acids allows quantification of the alterations in synthesis and breakdown rates associated with a wide variety of disease processes. These studies have increased our understanding of the mechanisms underlying the altered nitrogen balance following trauma and sepsis.

A variety of factors which influence the nitrogen balance in critically ill patients have been evaluated carefully in controlled studies and then translated to patients following injury or severe infection. Herrmann *et al.* (1980) administered [^{15}N]glycine to achieve a steady state and measured [^{15}N]urea nitrogen enrichment using the two-pool model of Picou and Taylor-Roberts.(1969; discussed in Waterlow, Garlick and Millward, 1978). Turnover rates of protein were measured and rates of synthesis and catabolism calculated during fed and fasted states in normal subjects. During feeding, synthesis and catabolism were equal and the subjects maintained nitrogen equilibrium. Restriction of food intake caused a marked reduction in synthesis with minimal impact on rates of protein catabolism. Schønheyder, Heilskov and Olesen (1954) found that protein synthesis and catabolism remained unchanged during immobilization in normal individuals. Crane *et al.* (1977) and O'Keefe, Sender and James (1974) reported that changes in the nitrogen equilibration of patients undergoing elective orthopaedic operations were similar to those described for bedrest and starvation. Birkhahn *et al.* (1981) described protein kinetics in four individuals following multi-system injury, including long-bone fractures. In contrast to patients undergoing elective orthopaedic operations, these patients demonstrated a marked increase in catabolic rate and a slight increase in synthesis. Because catabolism outstripped synthesis, the patients were in marked negative nitrogen balance while on standard therapy of intravenous 5% dextrose in water. Similar studies in septic patients demonstrated comparable results, although the magnitude of change was not as great (Long *et al.*, 1977). Herrmann *et al.* (1980) studied both trauma and septic patients while receiving other intravenous feedings. Although catabolic rates were markedly increased following injury or infection, the synthesis rate could be greatly enhanced by the provision of foodstuffs, which provided maintenance energy and nitrogen requirements. Thus, trauma and sepsis accelerate nitrogen flux; both protein synthesis and breakdown are increased. In the unfed patients, breakdown rate exceeds synthesis and negative nitrogen balance results. Providing exogenous calories and nitrogen increases synthesis and when adequate nutrients are provided, the two rates are matched and nitrogen balance is maintained (Table 9.3).

Table 9.3 Alterations in rates of protein synthesis and catabolism which may affect hospitalized patients

	Synthesis*	Catabolism*	Reference
Normal – starvation	↓	0	Waterlow *et al.* (1978); Herrmann *et al.* (1980)
Normal – fed – bedrest	↓	0	Schǿnheyder *et al.* (1954)
Elective Surgical Procedure	↓	0	Crane *et al.* (1977); O'Keefe *et al.* (1974)
Injury/Sepsis – i.v. dextrose	↑ ↑	↑ ↑ ↑	Birkhahn *et al.* (1981); Long *et al.* (1977)
Injury/Sepsis – fed	↑ ↑ ↑	↑ ↑ ↑	Herrmann *et al.* (1980)

*0 = No change; ↓ = Decrease; ↑ = Increase.

That muscle is the origin of the nitrogen which is lost in the urine following extensive injury was initially suggested by Cuthbertson (1930). Moreover, in his patients with long-bone fractures, he suggested that this response was a generalized or systemic response to injury and the nitrogen was not lost solely from damaged muscle at the site of injury. This hypothesis was proposed because of the magnitude of the nitrogen loss and the generalized muscle wasting that occurred following severe injury. This concept has been supported both in injury and extensive infection by a variety of studies which measured important markers of muscle catabolism such as creatinine, creatine, zinc and, more recently, 3-methylhistidine (Long *et al.*, 1975; Threlfall, Stoner and Galasko, 1981).

Further evidence of *net* skeletal muscle breakdown has been demonstrated by quantifying the loss of amino acids from extremities of severely injured patients. Using plethysmographic techniques to measure leg blood flow and determining arterial and femoral venous amino acid concentrations, Aulick and Wilmore (1979) found a three- to fourfold increase in amino acid flux from the extremities of injured patients when compared with normal subjects. Alanine efflux was the most significantly elevated of the amino acids measured (glutamine was not measured). The increase in alanine release from the legs of the severely traumatized patients was generally related to the extent of injury and the oxygen consumption of the patient but was not related to the size of the limb injury or to the leg blood flow. Accelerated rates of alanine release from the limbs of these patients appeared to be a generalized catabolic effect of injury rather than a response to local inflammatory or metabolic events in the injured extremities.

In subsequent studies, this same group of investigators measured amino acid uptake across the splanchnic bed and kidney (Wilmore *et al.*, 1980). An accelerated splanchnic uptake of amino acids occurred in the non-infected burn patients when compared to hepatic amino acid uptake in post-absorptive normals. Alanine, which quantitatively was the major nitrogen transport compound from skeletal muscle to liver and which provides 3-carbon skeletons as glucose precursors, was taken up at an average of 124 μmol min^{-1} m^{-2} in the non-infected burn patients, rates three to four times those reported for post-absorptive normal subjects. Since arterial concentrations of alanine in this group of patients were within normal ranges and the percentage of amino acid extracted was comparable to the levels reported in normal subjects (approximately 36%), the mechanism for the augmented alanine uptake was dependent upon the increase in delivery of amino acids to the liver via the elevated splanchnic blood flow. While amino acid delivery to the liver could account for a large portion of glucose being produced in the non-infected and infected trauma patients, alterations in plasma concentrations of acute phase reactant proteins, also synthesized in the liver, could account for the utilization of some of these amino acid precursors in the synthesis of plasma proteins.

In addition to these regional changes in protein turnover following extensive injury and injury complicated by infection, alterations also occur in the concentrations of amino acid in the free amino acid pools within cells. Skeletal muscle, the largest tissue mass in the body, is rich in myofibrillar protein and contains large intracellular stores of free amino acids, with glutamine representing the major intracellular amino acid present in the body (Giacometti, 1979). With brief starvation, bedrest, severe injury or injury complicated by infection, muscle biopsy data reveal a marked fall in intracellular glutamine concentration; the extent of the decrease in intracellular concentration is generally related to the extent of illness. The cause of the fall in intracellular glutamine concentration following illness is unknown but may occur because of failure of glutamine synthesis, increase in intracellular glutamine degradation or net efflux of glutamine from the intracellular store into the bloodstream with subsequent utilization at another site.

These mechanisms of increased protein synthesis and breakdown are important, for peptide bond formation requires a large quantity of ATP, energy which is not recovered during protein degradation. The rapid flux of amino acids from skeletal muscle to visceral organs and the concomitant breakdown and synthesis of protein is a mechanism which could convert energy into heat and may account for a significant portion of the hypermetabolism in the traumatized patient.

9.4.3 Carbohydrate metabolism

Alterations in carbohydrate metabolism occur following severe injury. These adjustments appear central to the metabolic response to trauma and trauma complicated by infection. With the stress of critical illness, blood glucose usually rises. In early studies oral and intravenous glucose tolerance tests were performed and glucose intolerance was demonstrated, thus prompting such terms as 'diabetes of injury' or 'stress diabetes' to describe the glucose dynamics following severe injury. This concept was appealing, for the hyperglycaemia was associated with increased urinary nitrogen loss, alterations which were also observed in the insulin-deficient state and could be corrected in diabetic patients with insulin administration. Work over the past ten years, however, has not confirmed insulin lack associated with injury or infection, although the elevations in other glucoregulatory hormones such as glucocorticoids, glucagon and catecholamines may provide for this 'relative' insulin-deficient state.

The hyperglycaemia which accompanies the hypermetabolic phase of extensive injury is the result of increased hepatic glucose production, not impaired glucose clearance. Studies over the past several years have confirmed the increase in hepatic glucose production and gluconeogenesis in severely injured patients (Wilmore *et al.*, 1980). Following a six-hour, overnight fast, simultaneous arterial and hepatic venous blood samples were drawn and estimates of splanchnic blood flow were determined by the clearance of indocyanine green dye. Glucose production in the non-infected, injured patients was markedly elevated. While normal individuals produced approximately 200 g of glucose per day, the thermally injured, non-infected patient produces approximately 320 g of glucose per day. This direct measure of increased net splanchnic glucose production is consistent with data derived from tracer studies which suggest increased glucogenesis following injury (Wolfe *et al.*, 1979). Moreover, the hepatic uptake of 3-carbon precursors (lactate, pyruvate and amino acids) was significantly greater than in normal subjects following a brief fast, suggesting that increased gluconeogenesis accounts for the major portion of hepatic glucose produced.

To determine the peripheral tissue which utilized the large quantity of glucose produced by the liver, substrate flux was measured across injured and uninjured extremities of severely burned patients, matched for age, weight and extent of total body surface burn (Wilmore *et al.*, 1977). Net glucose flux across uninjured extremities was low, suggesting that fat and not glucose was the primary fuel for skeletal muscle, a finding similar to that observed in normal subjects. However, increased glucose uptake and lactate production occurred in the extensively injured extremity, suggesting that little or no oxygen was utilized for glucose metabolism in extensively injured limbs. The increased lactate release from the injured extremities accounted for as much as 80% of the glucose consumed. This is also in accordance with our knowledge of the biochemistry of the highly specialized cells of the wound and inflammatory tissue (fibroblasts, macrophages, leukocytes) which are all glycolytic and demonstrate a major capacity for anaerobic metabolism. Additional measurements of blood flow and substrate concentration differences across the kidney (Wilmore *et al.*, 1980) and brain (Goodwin *et al.*, 1980) allow further characterization of regional glucose metabolism in severely injured patients. Approximately 30% to 50% of hepatic glucose produced is synthesized from lactate and pyruvate as the wound metabolizes glucose via the Cori cycle. The central nervous system in the injured patient appears to take up approximately 120 g (roughly 35–40% of hepatic production), while the kidney's consumption of glucose is twice normal, consuming about 75 g per day; little, if any, glucose appears to be taken up by the resting skeletal muscle (Fig. 9.2).

Studies over the past five years have extended our knowledge of endocrine control of glucose metabolism following severe injury and infection. Glucocorticoid, glucagon and catecholamine excess appear to be an

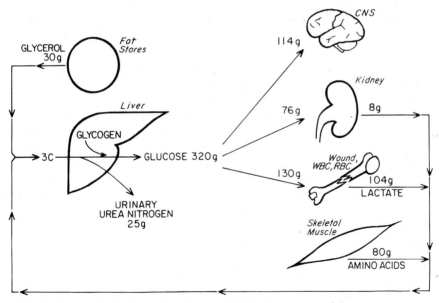

Fig. 9.2 24-hour flux of 6- and 3-carbon units following severe injury.

integrated signal that promotes gluconeogenesis (Wilmore, 1976; Shamoon, Hendler and Sherwin, 1980). In the early ebb phase of injury, when insulin concentration is low, these counter-regulatory hormones are markedly elevated. Following resuscitation and restoration of blood volume, insulin concentration returns to normal or is even elevated, while the counter-regulatory hormones fall slowly with convalescence and reach normal levels after wound healing has been achieved (Wilmore, 1976).

To characterize further the alterations in glucose and insulin kinetics during the flow phase of injury, six non-septic, traumatized patients were studied five to ten days post-injury, using the hyperglycaemic glucose 'clamp' technique, and the results compared with 11 age-matched controls (Black *et al.*, 1982). After an overnight fast, 20% glucose was infused intravenously to elevate acutely plasma glucose concentrations 125 mg d^{-1} above basal. This elevation was maintained for two hours using bedside glucose monitoring and negative feedback servo-control of the infusion rate. The quantity of glucose infused per unit time is equal to the rate of transfer of glucose from the extracellular to the intracellular compartment and reflects the metabolic clearance rate of glucose (M, mg kg^{-1} min^{-1}), and the glucose clearance rate per unit serum insulin (M/I ratio) indicates insulin sensitivity. The results demonstrated that there was a progressive increase in glucose disposal with time in normal controls, while patients maintained a constant glucose clearance throughout the study (Table 9.4).

Table 9.4 Effects of fixed hyperglycaemia on insulin response in normals and trauma patients (means ± SEM)

	Controls	Patients
N	11	6
Basal values		
Glucose (mg 100 ml^{-1})	95±3	102±5
Insulin (μU ml^{-1})	9.5±1.2	22.3±4.6*
Glucagon (pg ml^{-1})	108±22	200±50
'Clamp' values		
M (mg kg^{-1} min^{-1})	9.66±0.83	6.23±0.87*
Insulin (μU ml^{-1})	42±8	145±45
M/I	25.9±2.2	9.3±3.4*

* $P < 0.05$.

Moreover, the quantity of insulin elaborated by the patients was significantly greater than controls, yet these rising insulin concentrations failed to exert as great an effect on clearance of glucose in the patients as it did in the uninjured controls. The index of tissue sensitivity (M/I ratio) in the trauma patients was significantly below normal. This exaggerated pancreatic β-cell response to fixed hyperglycaemia combined with tissue insensitivity to insulin reflects the marked insulin resistance that occurs following injury, and this effect may be central to the post-traumatic metabolic response.

Thus, profound insulin insensitivity has been noted in injured patients, and preliminary data suggest this abnormality probably occurs at the post-receptor level. This defect would contribute to the marked catabolism of skeletal muscle protein and peripheral efflux of amino acids. The cause of this marked insulin insensitivity is presently unknown. The mediating mechanism may be hormonal or effected through substrate interaction such as has been suggested with the increased utilization of fatty acids by skeletal muscle tissue (Barrett *et al.*, 1981).

9.4.4 Fat metabolism

In spite of the marked increase of hepatic gluconeogenesis, injured patients and trauma patients with infection demonstrate respiratory exchange ratios (RQ) which approach 0.7, indicating that the major oxidized fuel is fat, not carbohydrate. Mobilization and utilization of free fatty acids from triglyceride stores are controlled by an interaction between adrenergic activity which promotes breakdown, and insulin which promotes fat storage. In the post-traumatic period, free fatty acid concentrations in the bloodstream are highly variable, but a variety of studies suggest that free fatty acid release and disappearance from the plasma compartment is markedly increased (Shamoon *et al.*, 1980). Moreover, suppression of lipolysis with glucose infusion can be demonstrated in normal subjects but does not occur in trauma and septic patients.

Mobilization of free fatty acids is coupled with enhanced hepatic uptake which, during simple starvation, promotes hepatic ketogenesis and triglyceride formation. Evidence from animal experiments suggests that triglyceride formation may continue following injury and infection, but a variety of studies demonstrates that ketone production is attenuated (Beisel and Wannemacher, 1981). Failure of appearance of these unique compounds (β-hydroxybutyrate and acetoacetate) is one hypothesis used to explain why the body fails to compensate to starvation associated with injury and infection. The specific causes for the failure of ketogenesis are unknown, but it has been speculated that the increased secretion of insulin inhibits ketogenesis during infection and injury and favours lipogenesis by the hepatocytes rather than ketogenesis (Beisel and Wannemacher, 1981). Further work in this area is needed.

The effects of ketone production on the regulation of gluconeogenesis and amino acid mobilization have been proposed, and the role of exogenous β-hydroxybutyrate administration was studied in an infected animal model

(Radcliffe *et al.*, 1981). Because of the marked interrelationship between carbohydrate and fat metabolism, it has been speculated that the provision of exogenous ketones would further suppress gluconeogenesis and proteolysis as has been demonstrated in fasting humans (Sherwin, Hendler and Felig, 1975). In an infected sheep model, ketogenesis did not occur and gluconeogenesis was markedly accelerated when compared to fasting, non-infected animals. While the infusion of β-hydroxybutyrate attenuated glucose production in normal or starved animals, the effect did not occur in the infected animals. Utilization of β-hydroxybutyrate was evident with the appearance of acetoacetate and clearance of both of these substances from the bloodstream was documented. Therefore, it appears that the hyper-catabolic state of injury associated with infection is not regulated by simple substrate interaction.

9.5 CARDIOVASCULAR RESPONSES

Since energy turnover in the body is ultimately dependent on the transport capacity of the cardiovascular system, any discussion of the metabolic response to injury must include a review of the associated circulatory adjustments. Clinically, the acute, systemic response to injury is termed 'traumatic shock' and is characterized by a reduction in blood volume, a fall in arterial blood pressure and cardiac output and an increase in total peripheral resistance. With restoration of blood volume, cardiac output rises above normal during the subsequent 'flow' or hypermetabolic phase of injury. To appreciate these systemic cardiovascular adjustments to injury better, we will examine first the changes in wound perfusion and describe how these local microcirculatory events affect total body circulation and metabolism.

9.5.1 Ebb phase circulation

The immediate microcirculatory disturbances following trauma depend on the nature and extent of the insult. With mechanical soft tissue injury, physical disruption of the vascular architecture and bleeding are the initial microcirculatory events. While vascular destruction can be just as severe in thermal burns, rapid coagulation necrosis limits the initial blood loss. Immediately adjacent to the region of vascular destruction, arterioles dilate. The intensity and duration of this dilatation is a function of the severity of the injury (Rybeck, 1974). Originally described as 'shunt' flow, this acute increase in wound perfusion has subsequently been identified as increased flow through a wide-open capillary bed (Lewis and Lim, 1970), which develops from active dilatation of resistance vessels rather than a simple reduction of constrictor tone. Cross-perfusion studies have demonstrated the presence of potent vasodilators in the venous effluent from the wound

(Rybeck *et al.*, 1974) and numerous locally released substances have been implicated (i.e. kinins, extravascular proteins modified by injury, prostaglandins, thromboxanes, etc.) (Arturson, 1978; Jonsson, Granström and Hanberg, 1979; Anggård and Jonsson, 1971; Edery and Lewis, 1963; Lewis, 1981). The biological significance of this early vasodilation is not well understood, since it would promote haemorrhage, occurs well in advance of any recovery or healing and does not increase substrate utilization by the damaged tissue. One suggested benefit is that it may ensure the initial delivery of white blood cells for host defence (Lewis, 1981).

Much of the early vasodilation rapidly disappears and is replaced by a period of reduced vasomotor reactivity and capillary perfusion. Wound blood flow is reduced by sustained arteriolar constriction, swelling of capillary endothelium, microthrombi formation and leukocytes sticking to vessel walls. During the same period, capillaries and small venules become increasingly permeable, and plasma begins to leak into the extravascular space. As blood flow becomes more sluggish, ischaemic necrosis may occur and increase the size of the original wound.

The acute, systemic, cardiovascular response to injury is principally the result of fluid loss across the wound, but other factors, such as anxiety, pain and changes in blood composition also play a role (Stoner, 1976). When fluid loss is sufficient to affect venous return, cardiac output drops and the subsequent hypotension reduces baroreceptive inhibition of the sympathetic efferent outflow from vasomotor centres in the medulla. The resultant increase in sympathoadrenal activity increases heart rate and contractility, total peripheral vascular resistance and blood pressure. The magnitude of this sympathetic response is proportional to the extent of injury and related to both the amount and rate of fluid loss.

The dominant peripheral vascular response during traumatic shock is a generalized vasoconstriction, but changes in vasomotor tone are by no means uniform throughout. Slater *et al.* have utilized radiomicrospheres to describe the changes in the distribution of the cardiac output of unanaesthetized dogs during progressive haemorrhagic shock. Soon after reducing arterial pressure to about 50 mm Hg and cardiac output by 50%, vasodilation in the brain and heart was sufficient to maintain flow at control levels. Liver blood flow also remained unchanged over this same period. This was achieved by hepatic artery vasodilation, offsetting a simultaneous drop in portal vein flow. During haemorrhage, blood flow decreased moderately in the gut and skeletal muscle and to a greater extent in the skin, kidneys and pancreas. Using comparable techniques, Asch *et al.* found that unresuscitated, awake dogs experienced a 35% reduction in cardiac output during the first hour after a 40% total body burn. Distribution of the circulating blood volume in these burned animals was essentially as described following haemorrhage with one notable exception, i.e. liver blood flow was reduced in burn shock despite the maintenance of gastrointestinal circulation. This difference may reflect the subtle effects of additional neurogenic and circulating afferents in the injured animal, which were superimposed on the common baroreceptor drive. Other factors which may contribute to these regional differences in vasomotor tone during shock are (1) variations in the threshold of central autonomic neurons to incoming baroreceptor and chemoreceptor afferents (Folkow, 1962), (2) differences in the degree of α- or β-adrenergic innervation in the various tissues, and (3) competing effects of locally produced and circulating vasodilators (Tysse-botn and Kirdebø, 1977; Hillman and Lundvall, 1980). Irrespective of these regional differences, the net effect is to promote fluid reabsorption in vasoconstricted beds while at the same time redirecting the reduced circulatory blood volume to the more critical tissues. These compensatory adjustments in regional blood flow are usually adequate provided the overall fluid loss does not exceed 3–5% of the body weight. During shock, treatment goals include the elimination of fluid loss and restoration of the circulating blood volume.

Hypoperfusion of metabolically active tissues is primarily responsible for the fall in energy turnover immediately after injury. This is particularly true for the relatively large skeletal muscle mass which is consistently vasoconstricted during traumatic shock (Asch *et al.,* 1971; Miller and Fegerson, 1977). The reduced liver blood flow after burn injury may further amplify ebb phase hypometabolism, but since hepatic circulation is main-tained during simple haemorrhage, any immediate change in liver metab-olism during haemorrhagic shock may not be flow dependent. Liver hypo-perfusion is not apparently responsible for the initial metabolic depression following bilateral hindlimb ischaemia. Stoner (1969*b*, 1970, 1971, 1972) has provided evidence that ongoing thermoregulatory behaviour is inhibited during this type of ischaemic shock and attributes some of the hypo-metabolism observed to the inhibition of non-shivering thermogenesis and the Q_{10} effects of decreasing body temperatures. The role of these thermoregulatory adjustments in ebb phase hypometabolism following injury is unknown, but the reduction of tissue metabolic demands during hypoperfusion may be of significant value.

There is another important change in regional circulation during ebb phase which deserves mention. In a recent review, Kovách and Sándor (1976) describe changes in cerebral blood flow during hypotension and shock. They point out that, while blood flow to the whole brain is usually maintained, there are many areas which appear to be underperfused. This is particularly true in the cerebral cortex, but there are patchy areas of affected tissue throughout. This imbalance between substrate delivery and demand is particularly significant, since unlike most of the rest of the body, brain metabolism increases in many areas during shock (Kovách and Fonyó, 1960). Whether these are initial consequences of shock and therefore contribute to the early changes in metabolism or developed only after prolonged hypotension remains to be evaluated.

9.5.2 Flow phase circulation

Once bleeding stops and vascular permeability approaches normal, restoration of the circulating volume begins, and the patient passes from the 'ebb' to the 'flow' phase of injury. Volume expansion raises venous return and blood pressure, reduces peripheral resistance and increases cardiac output. Total body blood flow continues to rise well above normal and eventually plateaus at a point roughly proportional to the extent of injury. The greatest increase in cardiac output has been recorded in severely burned patients where it may reach levels two to three times normal (Wilmore *et al.*, 1977). In these patients, the hyperdynamic circulation becomes fully developed during the first two weeks after injury and then gradually subsides as the wound is covered and heals. As before, we will consider first the changes in wound vasculature and then describe how these changes affect local vasomotor control and ultimately the systemic circulation. Once these basic adjustments to injury have been outlined, the added effects of infection will be described.

The wound may affect systemic circulation in two ways. First, it can be viewed as an added tissue mass which has its own morphology, metabolic needs and circulatory requirements, i.e. a new 'organ'. Second, injury and wound healing place demands on the entire body, and the associated increase in total energy turnover creates extra demands on the cardiovascular system.

The wound defect is gradually transformed into a highly vascularized bed of granulation tissue. Revascularization begins as early as two to three days post-injury when endothelial sprouts form on pre-existing arterioles, venules and capillaries. These sprouts lengthen to form solid endothelial cords which grow toward the centre of the wound at rates of $0.2-0.4$ mm d^{-1}. They anastomose with each other or with other neighbouring vessels and then canalize to form a fragile, highly permeable network of new capillaries. As soon as continuity is established with patent vessels, red cells and plasma rush in and the new capillaries dilate and begin to pulsate. Some new channels rapidly begin to develop smooth muscle coats and become small arterioles and venules, while others may clot off and eventually disappear. External support for the neovasculature is provided by a mesh network of fibrin imbedded in a collagenous gel. For a more detailed description of this vascularization process, the reader is referred to several excellent reviews (Clark and Clark, 1939; Jennings and Florey, 1970; Eriksson and Zarem, 1977).

Blood flow increases in granulation tissue as endothelial tubes form and make connections with pre-existing, undamaged vessels. Vasomotor control of this new circulatory bed does not begin, however, until the undifferentiated capillaries mature and develop vascular smooth muscle. This can begin as early as 48 hours after the appearance of the new capillary and usually takes another three or four days before definite arterioles are recognizable. The rapid proliferation and differentiation of the neovasculature occurs well in advance of vasomotor nerve regeneration, so neurogenic vasomotor control is rarely evident in the granulating wound in the first three to four weeks post-injury (Clark and Clark, 1934). In some cases, where there is extensive damage and subsequent scarring, vasculature reinnervation may take 3.5 to 7.5 months and is frequently incomplete.

Prior to reinnervation, vascular tone is affected by changes in local and circulating vasoactive substances. Numerous local inflammatory and metabolic vasodilators have been identified in injured tissue (i.e. prostaglandins, kinins, lactic acid, potassium ions, etc.), but the relative significance of each has not been established (Arturson, 1978; Wilmore *et al.*, 1977; Jonsson *et al.*, 1979; Anggård and Jonsson, 1971; Edery and Lewis, 1963). Vascular smooth muscle in granulation tissue displays normal sensitivity to changes in local temperature, but due to the lack of an extrinsic nerve supply, it does not respond reflexly to temperature changes in other parts of the body (Aulick *et al.*, 1977a, b).

The increase in wound circulation is not in response to local tissue hypoxia, as oxygen consumption in surface granulation tissue was found to be like that of normal skin despite a marked increase in blood flow (Wilmore *et al.*, 1977). Increasing the rate of oxygen delivery to the wound had no affect on blood flow (Aulick *et al.*, 1980). The absence of a potent oxygen drive on wound circulation is indeed unique but consistent with the fact that, while granulation tissue is metabolically very active, the fibroblasts, leukocytes and new epithelial cells supported by this circulation demonstrate a major capacity for aerobic glycolysis (Chen and Postlethwait, 1964; Grant and Prockop, 1972). Glucose uptake and lactate production are accelerated in the hyperaemic wound, but there is little evidence to suggest that these particular local metabolic events determine blood flow (Aulick *et al.*, 1980).

As described earlier, a common response to all forms of injury is the increased elaboration of catecholamines. To evaluate the effects of these circulating vasoactive substances, norepinephrine and epinephrine were both infused intravenously and applied directly to a large surface wound (Aulick *et al.*, 1980). The results of this work indicated that wound vasculature possessed α- but not β-adrenergic receptors. The absence of β-receptors suggests that surface granulation tissue, like normal skin, will only vasoconstrict in response to rising levels of circulating catecholamines. The functional significance of these and other humoral vasoconstrictors in granulation tissue is unknown.

While much more work is needed before the control of wound blood flow is fully understood, the available data suggest that much of the neovasculature is denervated and therefore unable to offset the effect of local and/or circulating vasodilators. This form of functional sympathectomy shifts vasomotor control away from higher centres and makes the wound more like other critical tissues (i.e. heart, brain and active skeletal muscle) where blood flow is determined primarily by local conditions rather than as part of some integrated total body circulatory reflex.

Wound perfusion appears to vary with changes in cardiac output and systemic blood pressure. In the hyperdynamic burn patient, measurements of peripheral blood flow and cardiac output have shown that as much as 60% of the extra circulation is directed to the surface wound (Wilmore *et al.*, 1977: Aulick *et al.*, 1978, 1980). Wound perfusion accounted for most, if not all, of the increase in peripheral circulation, since blood flow in the uninjured skin and resting skeletal muscle remained at control levels.

The selective increase in wound blood flow, when other peripheral tissues exhibit heightened vasoconstrictor tone, is most likely the result of the localized sympathectomy just described. Normal levels of cutaneous circulation in the undamaged skin of these hyperthermic, hypermetabolic patients reflect appropriate vasomotor reflexes to limit superficial heat loss and maintain the elevated body temperatures. Increased vasoconstrictor tone is also evident in inactive skeletal muscles, where blood flow remains at resting levels despite an acceleration in oxygen uptake after thermal injury (Wilmore *et al.*, 1977: Aulick *et al.*, 1978). This rather unique separation between skeletal muscle metabolism and blood flow demonstrates the relative dominance of reflex vasoconstrictor influences over those of local dilators.

Renal and splanchnic blood flow studies were recently conducted to determine whether an increase in blood flow to either of these visceral compartments contributed to the rise in cardiac output following burn injury (Aulick *et al.*, 1981). Renal blood flow ranged from normal to twice normal, but on the average, it was not significantly above control values. Individual variations in renal circulation were unrelated to burn size but appeared to parallel changes in 24-hour sodium excretion. These results clearly demonstrated that an increase in renal blood flow was not an obligatory response to injury. The relationship between renal blood flow and sodium excretion is normal and suggested that increased renal perfusion may occur as a basic response to injury, but this response is often overridden by the vasoconstriction which occurs in the face of reduced blood volume. The kidneys of these patients were also hypermetabolic, so like resting skeletal muscle, the increase in local oxygen consumption developed through an increase in renal O_2 extraction rather than any change in regional blood flow.

Splanchnic blood flow, on the other hand, was consistently elevated in a similar group of burn patients (Aulick *et al.*, 1981). It averaged about twice normal and, as such, accounted for about 19% of the cardiac output. Like renal perfusion, splanchnic blood flow did not vary with burn size, but it *did* appear to match changes in local oxygen consumption (arterial-hepatic venous oxygen difference remained in the normal range). The addition of this extra splanchnic blood flow to that going to the wound accounted for about 75% of the increase in cardiac output after burn injury. Other probable recipients of the remaining 25% would most likely include the heart, respiratory muscles, fat depots and possibly portions of the central nervous system.

The systemic cardiovascular response to environmental heat stress most closely mimics that following burn injury, but while superficial blood flow is frequently much greater in the resting hyperthermic subject, cardiac output rarely reaches levels observed in severely burned patients. The basis for this difference resides in the viscera, where there is a compensatory vasoconstriction of the heat-stressed individual. This decrease in visceral blood flow (renal and splanchnic) is the result of graded sympathetic activity and varies with the severity of the heart stress (Rowell, 1974). Since burn patients also exhibit increased sympathetic activity, the increase in splanchnic blood flow and maintenance of renal circulation most likely reflect competing vasodilator influences, which offset or override systemic vasoconstrictor drives. Whether such vasodilators are locally produced metabolites, associated with observed increases in visceral oxygen consumption or other circulating factor(s) remains to be determined. Irrespective of its aetiology, however, the combined increases in visceral and peripheral flows following thermal injury present a unique and very demanding cardiovascular challenge to the severely injured patient.

Hermreck and Thal (1969) have shown that the venous effluent from a 'septic hindlimb' caused vasodilatation in other regional beds. Considering the changes in wound chemistry previously described, the question is whether there are locally produced vasodilators in a noninfected wound that have comparable systemic effects. The results of numerous cross perfusion studies in animals have demonstrated that blood from injured, non-infected tissue has no apparent effect on the vascular resistance of other peripheral or visceral beds (Aulick et al., 1976b). This may be one major distinction between injury and infection, for while blood flow to the affected area increases in both, non-infected wounds do not release significant quantities of circulating vasodilators. The uninfected wound may release substances which affect metabolism in other organs and thus alter blood flow in these tissues indirectly.

Cerra et al. (1979) recently compared the cardiopulmonary patterns of a group of non-infected trauma/general surgery patients with those of a group of surviving septic patients and found that both groups developed the same increase in cardiac output, same drop in total peripheral resistance and the same arterial-central venous oxygen difference. Wilmore et al. (1980) also reported that total body and splanchnic blood flows were comparable in matched groups of non-infected and bacteraemic burn patients. During the hyperdynamic, septic state ('high flow' sepsis), animal and patient studies have shown that blood flow is increased in the infected tissue (Hermreck and Thal, 1969; Cronenwett and Lindenauer, 1979) and splanchnic and renal beds (Wilmore *et al.*, 1980; Gump, Price and Kinney, 1970; Rector *et al.*, 1973), but decreased in the skin (Hermreck and Thal, 1969). The vasomotor response in non-infected skeletal muscle is unclear, as some have reported an increase (Wright *et al.*, 1971; Finley *et al.*, 1975; O'Donnell *et al.*, 1974), while we have found either no change or even a decrease in skeletal muscle perfusion (D.W. Wilmore and L.H. Aulick, unpublished data). Some of this confusion may have a methodological basis or be related to the capacity to achieve a fully relaxed, resting state in conscious, seriously ill subjects.

9.6 THERMOREGULATORY RESPONSE

9.6.1 Ebb phase thermoregulation

The ebb phase of injury is considered that period which begins at the instant the injury is sustained and lasts until there is biochemical and/or physiological evidence of either recovery or necrobiosis. The fall in body temperature during this period is principally the result of a decrease in heat production rather than an increase in heat loss.* Henderson, Prince and Haggard, as early as 1917, reported that oxygen consumption of acutely injured burn patients may drop by as much as 50%. Since that time, a great deal of work has been performed in an effort to establish the basis of ebb phase hypometabolism.

The two standard injury models employed in these studies involved the rat after it had received a 20 to 30%, full-thickness, scald burn or been subjected to a two- to four-hour period of bilateral hindlimb ischaemia. These animals tolerate the burn injury very well, but the mortality rate secondary to the four-hour ischaemic injury is of the order of 70–80%.

Initially, ebb phase hypometabolism and its attendant hypothermia were considered the result of a downward shift in the central reference or 'set point' temperature (Stoner, 1969, 1970, 1971). Injured rats, for example, appeared to be comfortably warm at room temperatures below the thermo-neutral zone for uninjured animals and became markedly hypothermic before making any effort to maintain body temperature. The colonic temperatures of some rats, for instance, did not stabilize until they reached 32°C. These animals appeared to have lower ambient and core temperature thresholds for both shivering and non-shivering thermogenesis, but once these thresholds were reached, the increase in metabolic heat production for a given change in ambient temperature was comparable to that of control animals. The capacity to respond in an appropriate manner but at a reduced

---

*The hypodynamic circulation and a disproportionate decrease in skin blood flow during shock severely restricts the transfer of heat from the body core to the surface.

threshold not only was in keeping with the set point shift hypothesis but indicated that peripheral effector mechanisms for heat production remained intact during shock. This important point was confirmed by other *in vitro* studies which demonstrated that muscle cells harvested from burned animals respired at control rates when incubated in Krebs–Ringer solution (Caldwell *et al.*, 1971).

More extensive work in this area, however, began to suggest that changes in thermoregulation after injury were not the simple result of a downward shift in hypothalamic set point. Stoner (1969) observed, for example, that when injured rats were housed in a 30–34°C environment, they maintained normal basal metabolic rates and core temperatures. If truly thermo-regulating around a lower central reference temperature as proposed, these injured animals should be actively trying to unload heat, but subsequent work demonstrated that the hypothalamic threshold temperature for vaso-dilation of the tail was elevated rather than depressed as predicted (Stoner, 1972). So, rather than normal thermoregulation around a downward shift in the central thermostat, injury appeared to interfere with normal control of both heat gain and heat loss mechanisms.

There is now considerable evidence to suggest that some of the alterations in ebb phase thermoregulation are the result of changes in afferent nerve traffic into the central controller. Through an extensive and elaborate series of studies, Stoner and coworkers (Marshall and Stoner, 1979; Stoner, 1977, 1978; Stoner and Marshall, 1977) have shown that depression of the ambient temperature threshold for shivering following limb ischaemia involves the activity of ascending catecholaminergic nerve fibres which synapse in the posterior hypothalamus of the rat. These inhibitory nerves are not involved in normal thermoregulation of uninjured rats or in the central control of non-shivering thermogenesis. Stitt (1976) has shown that arousal, produced by either electrical stimulation of the brain stem or intraventricular injections of norepinephrine, inhibits normal ongoing thermoregulation in conscious rabbits. He indicates that this inhibition appears to involve areas of the brain which participate in the generalized 'fight or flight' response to stress and suggests that activation of such centres may be a way the autonomic nervous system subjugates less critical bodily functions to those more vital in defence of stress. These centres reside in the same region of the central nervous system as the ascending fibre tracts described by Marshall and Stoner (1979), Stoner (1977, 1978) and Stoner and Marshall (1977). Just how ischaemic trauma triggers this particular afferent pathway and whether it is responsive to other forms of injury remains to be determined.

Local changes in hypothalamic metabolism, circulation and temperature have also been described during the ebb phase of injury and will undoubtedly contribute to alterations in body temperature control. Kovách and Fonyó (1960), for example, found that unlike the rest of the body, brain oxygen consumption increased dramatically during haemorrhagic and ischaemic shock. Autoregulation of hypothalamic blood flow, however, is not as well developed as in other portions of the brain, making this area more vulnerable to any hypotension which may develop during the ebb phase. The effects of hypoxia are evident in various hypothalamically controlled cardiovascular mechanisms where thresholds appear to rise (Kovách and Sándor, 1976). So, provided a sufficient imbalance between oxygen supply and demand develops during ebb phase, similar impairment of other hypothalamic neurons may affect body temperature regulation. Brain temperature of rats and rabbits falls rapidly during the ebb phase, indicating that the observed drop in body temperature of these animals cannot be explained by an increase in the temperature of the central controller (Stoner, 1954, 1958). Taken together, these results suggest that the hypothermia that develops during the acute phase of injury is not the result of normal thermoregulation around a reduced set point temperature but rather is a manifestation of some impairment in central temperature control. Such impairment may be due to the activation of inhibitory neurogenic afferents to the hypothalamus and/or develop subsequent to local metabolic disturbances within the temperature centre itself.

Small mammals survive the ebb phase of injury more readily when housed at temperatures below the thermoneutral zone and their body temperatures permitted to drop several degrees (Haist, 1960). Stoner (1969a) suggests that the beneficial effects of such hypothermia may be to limit the extent of injury and associated fluid loss, as well as to reduce the metabolic requirements during shock. While the local application of cold is a standard first aid procedure, physicians will normally take measures to prevent hypothermia during traumatic shock. Injured humans do not lose body heat or become hypothermic as quickly as the smaller animal. This difference most likely reflects the larger surface-to-mass ratio and lower heat capacity of the small animal. Consequently, any actual reduction in average body temperature represents a more significant heat loss and possibly a greater clinical threat to man than the smaller animal. Therefore, the difference in clinical management of shock between man and smaller animals may be based on physical as well as physiological differences.

9.6.2 Flow phase thermoregulation

During the flow phase of injury, body temperatures rise above normal. Historically, this increase in body heat content has been called 'traumatic or aseptic fever', since it occurred following most forms of injury and was frequently unrelated to systemic infection. From a thermoregulatory point of view, it is important to ask if this is truly fever, the result of changes in central temperature control or simply hyperthermia due to a non-temperature dependent imbalance between metabolic heat production and loss. From a clinical point of view, it is equally important to determine whether the increased energy demands on the injured patient are the cause for or the result of the elevation in body temperature. If metabolic heat production exceeds thermoregulatory requirements, then physical and/or pharmacological measures may be appropriate to reduce body temperature. But, if the patient is truly febrile and thermoregulating around an elevated set point temperature, then efforts must be made to support the febrile state by keeping the individual warm and thereby minimizing the metabolic cost of fever.

The basis of traumatic fever has been extensively studied in burn patients, where thermoregulatory adjustments have more obvious clinical implications than any other form of injury. In these patients, there appears to be good evidence for a change in central temperature control. First, burn patients shiver in what is a comfortable environment for uninjured normal subjects and, despite elevated core and surface temperatures, prefer warmer ambient temperatures to achieve thermal comfort (Wilmore *et al.,* 1975*b*). Second, this increase in body temperature is maintained over a wide range of environments and is defended in a predictable manner (Wilmore *et al.,* 1975*a,b,* 1974*a*). In what is a comfortable environment for the burn patient (30–33°C), leg blood flow in the uninjured limb is normal, in spite of elevated body temperatures (Aulick *et al.,* 1977*b*). Since a comparable degree of hyperthermia in resting normal subjects would cause a four- to fivefold increase in leg blood flow, this suggests that burn patients vasoconstrict normal skin in an effort to conserve body heat and maintain the fever. Only when central body temperatures are elevated above the febrile level by external heating do these patients begin to complain about being too warm and start to vasodilate and sweat from uninjured skin (Aulick *et al.,* 1982).

As mentioned earlier, a heat labile, endogenous pyrogen has been identified in the serum of non-septic burn patients, which when injected into the hypothalamus of rabbits, produced a marked febrile response (Wilmore, 1975–1976). Wilmore (1976), however, reported that aspirin administration had no effect on the core temperature or rate of metabolic heat production of three burn patients. The failure of antipyretics to lower body temperatures of burn patients adds this form of traumatic fever to a growing body of evidence which suggests that prostaglandins are not a basic component of all pyrogen-induced fevers (Kluger, 1979). Herndon *et al.* (1977) found that intraperitoneal injections of indomethacin, R02-5720 (Hoffman–LaRoche), and meclofenamate reduced the hypermetabolic response of a small animal burn model, suggesting either than there are considerable species variations in the role of prostaglandins in traumatic fever or that these drugs acted at a level of prostaglandin metabolism not effected by salicylates.

The thermoregulatory response to traumatically induced changes in the central controller depend on the nature and extent of injury. Consequently, traumatic fever may appear radically different from the febrile response of uninjured man who has normal functioning afferent input into the temperature control centres, as well as effective efferent mechanisms by which to increase body heat content. With traumatic disruption of either the sensory or motor thermoregulatory limbs, the injured patient may have difficulty interpreting environmental conditions and body heat content, as well as taking appropriate actions (heat production and/or conservation) to raise body temperature. In addition, the metabolic and circulatory demands of the wound may compete with the thermoregulatory requirements of the patient. These problems of fever generation and maintenance are especially evident in the burn patients. For, while uninjured man may increase heat storage by a combination of increased heat production and reduced heat loss, the presence of a large surface wound reduces this patient's capacity for heat conservation and places increased demands on heat production to achieve and maintain the febrile state.

When the surface burn is severe enough to damage the upper two-thirds of the epidermis, the normal water-diffusion barrier of the skin is eliminated and the rate of evaporative water loss from the wound accelerated. Lamke, Nilsson and Reitner (1977) recently measured the vapour pressure gradient of air overlying burn wounds of various depths and at different stages of recovery and calculated the rate of evaporative water loss in a 33°C environment. They reported this to be anywhere from 17 to 25 times that of normal skin. At a latent heat of vapourization of 2.43 MJ litre^{-1}, this translates into almost 12.5 MJ or 3000 kcal per day for an average size man with a granulating burn wound covering 50% of the total body surface. This amount of extra heat would either have to be supplied by the patient or the external environment if he or she is to maintain thermal balance.

The high obligatory level of wound perfusion, described earlier, acts to promote healing but further compromises the heat conservation efforts of the febrile burn patients. Wilmore *et al.* (1975*a*), for example, reported that the rate of deep-to-superficial heat flux was twice normal in these patients over a wide range of thermal environments. This increased rate of heat delivery raises surface temperatures and accelerates dry heat loss whenever skin temperature exceeds ambient. It also supplies additional energy to the granulating wound to increase heat loss by evaporation. The inability effectively to constrict wound vasculature creates an insulative deficit for the burn patient, which (1) increases in proportion to the size of surface injury, (2) raises the requirements for metabolic heat production and (3) increases the susceptibility to hypothermia in cool environments.

As described earlier in this chapter, the increased calorigenesis of injury is the result of heightened sympathoadrenal activity and is mediated through the elaboration of increased quantities of catecholamines. Regional blood flow and oxygen consumption measurements indicate that the resultant increase in energy turnover is a generalized phenomenon affecting the entire body and not confined strictly to the healing wound (Wilmore *et al.*, 1977; Aulick *et al.*, 1981). In addition, the increase in metabolic heat production appears to be distributed in a relatively normal fashion with roughly two-thirds of the heat being produced in visceral tissues and the remainder in the extremities. In the burn patient, high obligatory rates of wound blood flow accelerate both dry and wet heat losses from the body surface, but initially, the systemic increase in heat production exceeds heat loss through this insulative deficit, and body temperatures rise. The accelerated rates of peripheral vascular heat loss, combined with increased visceral heat storage exaggerates the differences between core and venous blood temperature returning from injured limbs. In one group of burn patients (Wilmore *et al.*, 1978), for example, blood entered the lower inferior vena cava on the average of 0.8°C below rectal temperature, as compared to only a 0.3°C difference in a group of afebrile controls studied earlier by Eichna *et al.* (1951). The increased visceral heat content of these burn patients, however, rapidly reduced the difference between blood and rectal temperatures until the normal relationship observed in afebrile man was re-established by the time blood reached the pulmonary circulation. The elevated but normal relationship between intravascular and core temperatures indicate that the thermoregulatory adjustments of the burn patient are more than sufficient to offset increased peripheral heat loss. *Can the increased rate of heat production and elevated body temperatures be explained solely as a thermoregulatory response to an upward shift in the central set point temperature?*

While still controversial (Aulick *et al.*, 1979; Danielsson, Artusson and Wennberg, 1976) there is a great deal of experimental evidence to suggest that the increased metabolic heat production in the burn patient is *not* solely to satisfy a thermoregulatory reset. Based on earlier work which demonstrated a reduction in the hypermetabolic response of small animal models upon either blocking their superficial heat loss (Lieberman and Lansche, 1956) or increasing ambient temperature (Caldwell *et al.*, 1959) and another study (Barr *et al.*, 1968) that demonstrated similar effects in burn patients when ambient temperature was increased from 22 to 32°C, it was originally concluded that the increased energy turnover was to offset the accelerated rate of evaporative cooling of the burn wound. Undoubtedly, this concept has validity and within limits explains some of the differences between the metabolic response of burn patients and that of other severely injured individuals. But, the significance of evaporative cooling as a metabolic stimulus was originally challenged by Zawachi *et al.* (1970) when they were unable to find any consistent change in metabolic activity of burn patients after blocking their evaporative water loss by wrapping the wound in a water-impermeable membrane. More recently, efforts were made to satisfy the central reference temperatures of a group of burn patients by heating them with radiant heat lamps (Aulick *et al.*, 1979). The average rectal temperature for the group increased from 38.6 to 39.4°C without having a consistent effect on resting metabolic rate. Many of these subjects began to sweat from unburned skin, and a few awoke stating that they 'felt hot', indicating that this external heat load had exceeded their central thermostatic requirements. Additional studies have shown that while increased room temperature may reduce the hypermetabolic response of some patients and small animals, energy turnover never returns to normal resting levels (Aulick *et al.*, 1979; Wilmore *et al.*, 1974*a*; Davis, Lamke and Liljedahl, 1977; Herndon, Wilmore and Mason, 1978). In fact, the Q_{10} effect of increased body temperature explained no more than about 10–20% of the hypermetabolism. Resting metabolism in excess of that predicted by increases in body temperature has been observed not only in burn patients but in a wide variety of other injured and postoperative patients (Kinney and Roe, 1962; Roe and Kinney, 1963). This suggests that unlike the infected, non-injured patients of DuBois (1921) who developed a consistent linear relationship between body temperature and resting energy expenditure (approximately a 13% increase in metabolic rate per degree centigrade rise in rectal temperature above normal), injured patients do not simply become hypermetabolic because they are hot. Taken together, all these studies suggest that, while there are obvious thermoregulatory influences on the metabolic response to injury, the increased rate of energy turnover is primarily determined by other metabolic requirements, that is, *injury hypermetabolism is temperature sensitive but not temperature dependent.*

If the patient is unable to satisfy the elevated central reference temperature by heat conservation mechanisms, then he or she will make an effort to increase heat production above that set by injury alone. It is this patient who profits from an increase in ambient temperature, for it will not only lower overall energy turnover but it has also been shown to reduce the rate of protein catabolism and minimize disturbances in plasma proteins (Cuthbertson *et al.*, 1972; Davies, Liljedahl and Birke, 1971; Fleck, 1976). In addition to these metabolic effects, higher environmental temperatures increase blood flow and healing rate of superficial wounds (Aulick *et al.*, 1977*b*; Gimbal and Farris, 1965; Cuthbertson and Tilstone, 1967). So, while body temperature demands may not be a major determinant of metabolism, they remain important consideration in the treatment of any seriously injured patients.

An all too common complication of injury is infection, the metabolic and circulatory consequences of which have already been described. The associated changes in body temperature are not only affected by the type and severity of infection but also the thermoregulatory status of the patient at the time of infection. Patients with minimal injuries and low-grade fevers will frequently respond dramatically when they become infected, while little or no change in body temperature may develop when more severely injured, febrile patients initially became bacteraemic (Wilmore *et al.*, 1980). If the infection cannot be brought under control, metabolic and circulatory failures are frequently signalled by a fall in body temperature back toward and even below normal.

In summary, the thermoregulatory adjustments of the injured patient are determined by the nature and extent of trauma. During the initial shock, or ebb phase, thermoregulation appears to be severely impaired, leaving the patient vulnerable to changes in the thermal environment. In the subsequent flow or catabolic phase, there is good evidence that the patient is truly thermoregulating around an elevated set point temperature. The generation and maintenance of traumatic fever is promoted by a non-temperature-dependent increase in metabolic heat production, and in the case of burn injury, it is complicated by the insulative deficit created by the presence of a large surface wound. The physician must then insure that the patient is cared for in an ambient environment which will meet the thermoregulatory requirements of injury and minimize the metabolic cost of fever.

9.7 TREATMENT

The care and treatment of injured patients is dictated by many factors, not the least of which is the accelerated rate of energy turnover. A detailed description of the metabolic management of the critically ill has been provided elsewhere (Wilmore, 1977) and is beyond the scope of this review, but the basic components can be summarized:

1. Minimize external stimuli which would further increase the metabolic cost of injury. Pain, anxiety, cool, drafty environments and inadequate rest are all factors which contribute to inefficient energy utilization.

2. Vigorously feed the patient. Severe erosion of the body fuel stores will occur unless the increased energy and substrate requirements of injury are provided exogenously.

3. Maintain a planned progressive exercise programme to prevent the atrophy of disuse and encourage amino acid deposition and incorporation into skeletal muscle protein.

4. Protect against and aggressively treat infection.

5. Close the wound in a timely fashion. The metabolic response to injury is largely in support of the wound and will not be eliminated until this stimulus is removed.

9.8 SUMMARY

Metabolic, circulatory and thermoregulatory adjustments to injury vary with the nature and extent of trauma. The initial response is a generalized depression of all physiological activity which begins at the moment of insult and continues until blood volume is restored. The patient then enters a phase of accelerated function characterized by increases in metabolic heat production, total body circulation and body temperatures. The catabolic or 'flow' phase reaches a peak in early recovery and then slowly returns to normal with wound healing and restoration of function.

Baroreceptor and chemoreceptor reflexes play a dominant role in the initiation of the acute responses to injury, but the bulk of the data suggests that humoral mediators rather than neurogenic signals determine the magnitude and duration of the subsequent hypermetabolic phase. These substances presumably originate in the wound and exert their primary influence on hypothalamic centres.

The catabolic phase of injury is characterized by peripheral substrate mobilization and hepatic glucose production. The primary oxidative fuel is fat, while the wound is predominately a glycolytic organ. The increase in total body blood flow occurs in response to two factors. First, the major portion of the extra circulation occurs in the wound where it acts to support the metabolic needs of this new highly vascularized tissue. Secondly, it answers the needs imposed by the generalized increase in total body energy requirements following injury. The increase in metabolic heat production contributes to a controlled rise in body temperature, and the injured patient maintains the febrile state in an appropriate and predictable thermoregulatory manner.

Many factors effect the character of the metabolic response to injury. The physician will attempt to minimize those which increase the metabolic burden (cold stress, infection, pain, anxiety, etc.) while insuring that the patient receives sufficient substrate and calories to meet the accelerated energy cost of injury.

ACKNOWLEDGEMENT

This work was supported by NIH Trauma Center grant no. NIGM P50 GM 29327–01 – US Army Contract no. DADM 17–80–C–0–148.

The opinions or assertions contained herein are the private views of the authors and are not to be construed as official or as reflecting the views of the Department of the Army or the Department of Defense.

Human subjects participated in these studies after giving their free and informed voluntary consent. Investigators adhered to AR 70–25 and USAMRDC Reg 70–25 on Use of Volunteers in Research.

In conducting the research described in this report the investigators adhered to the *Guide for Laboratory Animal Facilities and Care,* as promulgated by the Committee on the Guide for Laboratory Animal Facilities and Care of the Institute of Laboratory Animal Resources, National Academy of Sciences National Research Council.

REFERENCES

Adams. P.B. and Rowan. K.S. (1970) Glycolytic control of respiration during aging of carrot root tissue. *Plant Physiol.*, **45**, 490.

Anggård, E. and Jonsson. C.-E. (1971) Efflux of prostaglandins in lymph from scalded tissue. *Acta Physiol. Scand.*, **81**, 440.

Arturson, G. (1978) Prostaglandins in human burn wound secretion. *Burns*, **3**, 112.

Asch, M.J.. Meserol, P.M.. Mason, A.D., jr. and Pruitt, B.A., jr. (1971) Regional blood flow in the burned unanesthetized dog. *Surg. Forum*, **22**, 55.

Aulick, L.H.. Baze, W.B.. McLeod, C.G. and Wilmore, D.W. (1980) Control of blood flow in a large surface wound. *Ann. Surg.*, **191**, 249.

Aulick, L.H.. Goodwin, C.W.. Becker, R.A. and Wilmore, D.W. (1981) Visceral blood flow following thermal injury. *Ann. Surg.*, **193**, 112.

Aulick, L.H.. Hander, E.W.. Wilmore, D.W.. Mason, A.D., jr. and Pruitt, B.A., jr. (1979) The relative significance of thermal and metabolic demands on burn hypermetabolism. *J. Trauma*, **19**, 559.

Aulick, L.H. and Wilmore, D.W. (1979) Increased peripheral amino acids release following burn injury. *Surgery*, **85**, 560.

Aulick, L.H.. Wilmore, D.W., Mason, A.D., jr. and Pruitt, B.A., jr. (1978) Muscle blood flow following thermal injury. *Ann. Surg.*, **188**, 778.

Aulick, L.H.. Wilmore, D.W., Mason, A.D., jr. and Pruitt, B.A., jr. (1977a) Peripheral blood flow in thermally injured patients. *Fed. Proc.*, **36**, 417.

Aulick. L.H.. Wilmore, D.W.. Mason, A.D., jr. and Pruitt, B.A., jr. (1977b) Influence of the burn wound on peripheral circulation in thermally injured patients. *Am. J. Physiol.*, **233**, H520.

Aulick, L.H., Wilmore, D.W., Mason, A.D., jr. and Pruitt, B.A., jr. (1982) Depressed reflex vasomotor control of the burn wound. *Cardiovascular Res.*, **16**, 113.

Barr. P.-O.. Birke. G.. Liljedahl, S.-O. and Plantin, L.-O. (1968) Oxygen consumption and water loss during treatment of burns with warm dry air. *Lancet*, **i**, 164.

Barrett, E.. Ferrannini, E.. Bevilacqua, S. and DeFronzo, R. (1981) Effects of free fatty acids on glucose metabolism in man. *ESPEN Annual Meeting Abstract*, FC36.

Becker, R.A.. Johnson, D.W.. Woeber, K.A. and Wilmore, D.W. (1976) Depressed serum triiodothyronine (T_3) levels following thermal injury. *Fed. Proc.*, **35**, 316.

Beisel, W.R. (1975) Metabolic response to infection. *Ann. Rev. Med.*, **26**, 9.

Beisel, W.R. and Wannemacher, R.W., jr. (1980) Gluconeogenesis, ureagenesis, and ketogenesis during sepsis. *J. Parent. Ent. Nutrition*, **4**, 277.

Birkhahn, R.H., Long, C.L., Fitkin, D. and Jeevanandam, M. and Blakemore, W.S. (1981) Whole-body protein metabolism due to trauma in man as estimated by L-[^{15}N]alanine. *Am. J. Physiol.*, **241**, E64.

Black, P.R., Brooks, D.C., Bessey, P.Q., Wolfe, R.R. and Wilmore, D.W. (1982) Mechanisms of insulin resistance following injury. *Ann. Surg.*, **196**, 42.

Brandt. M.R.. Kehlet, H.. Skovsted, L. and Hansen, J.M. (1976) Rapid decrease in plasma-triiodothyronine during surgery and epidural analgesia independent of afferent neurogenic stimuli and of cortisol. *Lancet*, **ii**, 1333.

Bromage, P.R., Shibata, H.R. and Willoughby, H.W. (1971) Influence of prolonged epidural blockade on blood sugar and cortisol responses to operations upon the upper part of the abdomen and the thorax. *Surg. Gyn. Obst.*, **132**, 1051.

Caldwell, F.T., Casali, R.E., Boswer, B., Smith, V., Enloe, J. and Rose, D. (1971) On the failure of heat production in the immediate postburn period. *J. Trauma*, **11**, 936.

Caldwell, F.T., jr., Osterholm, J.L., Sower, N.D. and Meyer, C.A. (1959) Metabolic response to thermal trauma of normal and thyroprivic rats at three environmental temperatures. *Ann. Surg.*, **150**, 976.

Cerra, F.B., Siegel, J.H., Border, J.R., Peters, D.M. and McMenamy, R.P. (1979) Correlations between metabolic and cardiopulmonary measurements in patients after trauma, general surgery and sepsis. *J. Trauma*, **19**, 621.

Chen, R.W. and Postlethwait, R.W. (1964) The biochemistry of wound healing. Monograph, *Surg. Sci.*, **1**, 215.

Clark, E.R. and Clark, E.L. (1939) Microscopic observations on growth of blood capillaries in living mammal. *Am. J. Anat.*, **64**, 251.

Clark, E.R., Clark, E.L. and Williams, R.G. (1934) Microscopic observations in the living rabbit of the new growth of nerves and the establishment of nerve-controlled contractions of newly formed arterioles. *Am. J. Anat.*, **55**, 47.

Crane, C.W., Picou, D., Smith, R. and Waterlow, J.C. (1977) Protein turnover in patients before and after elective orthopaedic operations. *Br. J. Surg.*, **64**, 129.

Cronenwett, J.L. and Lindenauer, S.M. (1979) Direct measurement of arteriovenous anastomotic blood flow in the septic canine hindlimb. *Surgery*, **85**, 275.

Cuthbertson, D.P. (1930) The disturbance of metabolism produced by bony and non-bony injury with notes on certain abnormal conditions of bone. *Biochem. J.*, **24**, 1244.

Cuthbertson, D.P. and Tilstone, W.J. (1967) Effect of environmental temperature on the closure of full thickness skin wounds in the rat. *Quart. J. Exp. Physiol.*, **52**, 249.

Cuthbertson, D.P., Fell, G.S., Smith, C.M. and Tilstone, W.J. (1972) Metabolism after injury. I: Effects of severity, nutrition, and environmental temperature on protein, potassium, zinc and creatine. *Br. J. Surg.*, **59**, 68.

Cuthbertson, D.P. (1976) Surgical metabolism: historical and evolutionary aspects. in *Metabolism and the Response to Injury* (eds A.W. Wilkinson and D.P. Cuthbertson), Year Book Medical Publishers, Chicago.

Cuthbertson, D.P. (1980) Alterations in metabolism following injury: Part I. *Injury*, **11**, 175.

Danielsson, U., Arturson, G. and Wennberg, L. (1976) The elimination of hypermetabolism in burn patients. *Burns*, **2**, 110.

Davies, J.W.L., Lamke, L.-O. and Liljedahl, S.-O. (1977) Treatment of severe burns. *Acta Chir. Scand. (Suppl. 468)*.

Davies, J.W.L., Liljedahl, S.-O. and Birke, G. (1971) Protein and energy metabolism in burned patients treated in a warm (32°C) or cool (22°C) environment. in *Research in Burns* (eds P. Matter, T.L. Barclay and Z. Konicková), Hans Huber Publishers, Bern, p. 542.

DuBois, E.F. (1921) The basal metabolism in fever. *J. Am. Med. Assoc.*, **77**, 325.

Edery, H. and Lewis, G.P. (1963) Kinin-forming activity and histamine in lymph after tissue injury. *J. Physiol. (Lond.)*, **169**, 568.

Eichna, L.W., Berger, A.R., Rader, B. and Becker, W.H. (1951) Comparison of intracardiac and intravascular temperatures in man. *J. Clin. Invest.*, **30**, 353.

Eriksson, E. and Zarem, H.A. (1977) Growth and differentiation of blood vessels. in *Microcirculation* I (eds G. Kaley and B.M. Altura), University Park Press, Baltimore, p. 393.

Exton, J.H. (1972) Gluconeogenesis. *Metabolism*, **21**, 945.

Finley, R.J., Duff, J.H., Holliday, RL., Jones, D. and Marchuk, J.B. (1975) Capillary muscle blood flow in human sepsis. *Surgery*, **78**, 87.

Fleck, A. (1976) The influence of the nature, severity and environmental temperature on the response to injury. in *Metabolism and the Response to Injury* (eds A.W. Wilkinson and D.P. Cuthbertson), Year Book Medical Publishers, Inc., Chicago.

Folkow, B. (1962) Nervous adjustments of the vascular bed with special reference to patterns of vasoconstrictor fibre discharge. in *Shock: Pathogenesis and Therapy* (ed. K.D. Block), Springer-Verlag, Berlin, p. 61.

Garber, A.J., Karl, I.E. and Kipnis, D.M. (1976) Alanine and glutamine synthesis and release from skeletal muscle. IV. beta-Adrenergic inhibition of amino acid release. *J. Biol. Chem.*, **251**, 851.

Giacometti, T. (1979) Free and bound glutamate in natural products. in *Glutamic Acid*. (eds L.J. Filer, jr., *et al.*), Raven Press, New York, p. 25.

Gimbal, N.S. and Farris, W. (1965) Skin grafting: The influence of surface temperature on the epithelization rate of split thickness skin donor sites. *Arch. Surg.*, **92**, 554.

Goodall, McC. and Alton, H. (1969) Dopamine (3-hydroxytyramine) replacement and metabolism in sympathetic nerve and adrenal medullary depletions after prolonged thermal injury. *J. Clin. Invest.*, **48**, 1761.

Goodall, McC. and Moncrief, J.A. (1965) Sympathetic nerve depletion in severe thermal injury. *Ann. Surg.*, **162**, 893.

Goodwin, C.W., Aulick, L.H., Powanda, M.C., Wilmore, D.W. and Pruitt, B.A., jr. (1980) Glucose dynamics following severe injury. *Eur. J. Surg. Res.*, *Suppl.* *12*, 126.

Gordon, M.L. (1950) An evaluation of afferent nervous impulses in the adrenal cortical response to trauma. *Endocrinology*, **47**, 347.

Grant, M.E. and Prockop, D.J. (1972) The biosynthesis of collagen. *N. Engl. J. Med.*, **286**, 291.

Gump, F.E., Price, J.B. and Kinney, J.M. (1970) Whole body and splanchnic blood flow and oxygen consumption measurements in patients with intraperitoneal infections. *Ann. Surg.*, **171**, 321.

Gump, F.E., Price, J.B., jr. and Kinney, J.M. (1970) Blood flow and oxygen consumption in patients with severe burns. *Surg. Gynecol. Obstet.*, **130**, 23.

Haist, R.E. and Hamilton, J.K. (1944) Reversibility of carbohydrate and other changes in rats shocked by a clamping technique. *J. Physiol.*, **102**, 471.

Haist, R.E. (1960) Influence of environment on the metabolic response to injury. in *The Biochemical Response to Injury* (eds H.B. Stoner and C.J. Threlfall), C.C. Thomas, Springfield.

Harrison, T.S., Saton, J.F. and Feller, L. (1967) Relationship of increased oxygen consumption to catecholamine excretion in thermal burns. *Ann. Surg.*, **165**, 169.

Haskins, H.N. (1907) The effect of transfusion of blood on the nitrogenous metabolism of dogs. *J. Biol. Chem.*, **3**, 321.

Hawk, P.B. and Gies, W.J. (1904) The influence of external hemorrhage on chemical changes in the organism, with particular reference to protein catabolism. *Am. J. Physiol.*, **2**, 171.

Henderson, Y., Prince, A.L. and Haggard, H.W. (1971) Observations on surgical shock. A preliminary note. *J. Am. Med. Assoc.*, **69**, 965.

Hermreck, A.S. and Thal, A.P. (1969) Mechanisms for the high circulatory requirements in sepsis and septic shock. *Ann. Surg.*, **170**, 677.

Herndon, D.N., Wilmore, D.W. and Mason, A.D., jr. (1978) Development and analysis of a small animal model simulating the human postburn hypermetabolic response. *J. Surg. Res.*, **25**, 394.

Herndon, D., Wilmore, D.W., Mason, A.D., jr. and Pruitt, B.A., jr. (1977) Humoral mediators of non-temperature dependent hypermetabolism in 50% burned adult rats. *Surg. Forum*, **28**, 37.

Herrmann, V.M., Clarle, D., Wilmore, D.W. and Moore, F.D. (1980) Protein metabolism: effect of disease and altered intake on the stable ^{15}N curve. *Surg. Forum*, **31**, 92.

Hillman, J. and Lundvall, J. (1980) Beta-adrenergic dilator interaction with the constrictor response in resistance vessels of skeletal muscle during hemorrhage. *Acta Physiol. Scand.*, **108**, 77.

Hume, D.M. (1969) The endocrine and metabolic response to injury. in *Principles of Surgery* (ed. S.I. Schwartz), McGraw-Hill, New York, p. 2.

Hume, D.M. and Egdahl, R.H. (1959) The importance of the brain in the endocrine response to injury. *Ann. Surg.*, **150**, 697.

Iversen, J. (1973) Adrenergic receptors and the secretion of glucagon and insulin from the isolated, perfused canine pancreas. *J. Clin. Invest.*, **52**, 2102.

Jennings, M.A. and Florey, H.W. (1970) Healing. in *General Pathology* (ed. H.W. Florey), W.B. Saunders Co., Philadelphia, p. 480.

Jonsson, C.-E., Granström, E., Hanberg, M. (1979) Prostaglandins and thromboxanes in burn injury in man. *Scand. J. Plastic Reconstr. Surg.*, **13**, 45.

Kehlet, H., Brandt, M.R. and Rem, J. (1980) Role of neurogenic stimuli in mediating the endocrine metabolic response to surgery. *J. Parent. Ent. Nutrition*, **4**, 152.

Kinney, J.M. and Roe, C.F. (1962) Caloric equivalent of fever: I. Patterns of postoperative response. *Ann. Surg.*, **156**, 610.

Kluger, M.J. (1978) The evolution and adaptive value of fever. *Am. Sci.*, **66**, 38.

Kluger, M.J. (1979) *Fever: Its Biology, Evolution and Function.* Princeton University Press, Princeton.

Kovách, A.G.B. and Fonyó, A. (1960) Metabolic responses to injury in cerebral tissue. in *The Biochemical Response to Injury* (eds H.B. Stoner and C.J. Threlfall), C.C. Thomas, Co., Springfield, p. 129.

Kovách, A.G.B. and Sándor, P. (1976) Cerebral blood flow and brain function during hypotension and shock. *Ann. Rev. Physiol.*, **38**, 571.

Lamke, L.-O., Nilsson, G.E. and Reitner, H.L. (1977) The evaporative water loss from burns and the water-vapour permeability of grafts and artificial membranes used in the treatment of burns. *Burns*, **3**, 159.

Lerman, M.I., Abakumova, O.Y., Kucenko, N.G. and Kobrina, E.M. (1977) Stimulation of growth of connective tissue by low-molecular weight constituents from rapidly growing tissues. *Lancet*, **i**, 1225.

Lewis. D.H. (1981) The effect of injury on the microcirculation. in *Advances in Physiological Sciences* (eds Zs. Biró, A.G.B. Kovách, J.J. Spitzer and H.B. Stoner). Vol. 26: Homeostasis in Injury and Shock. Pergamon Press, New York, p. 109.

Lewis, D.H. and Lim, R.C., jr. (1970) Studies on the circulatory pathophysiology of trauma. II. Effect of acute soft tissue injury on the passage of macroaggregated albumin (^{131}I) particles through the hindleg of the dog. *Acta Ortho. Scand.*, **41**, 37.

Lieberman, Z.H. and Lansche, J.M. (1956) Effects of thermal injury on metabolic rate and insensible water loss in the rat. *Surg. Forum*, **7**, 83.

Long, C.L., Haverberg, L.N., Young, V.R. *et al.* (1975) Metabolism of 3-methylhistidine in man. *Metabolism*, **24**, 929.

Long, C.L. Jeevanandam, M., Kim, B.M. and Kinney, J.M. (1977) Whole body protein synthesis and catabolism in septic man. *Am. J. Clin. Nutr.*, **30**, 1340.

Marshall, H.W. and Stoner, H.B. (1979) The effect of dopamine in shivering in the rat. *J. Physiol.*, **288**, 393.

Miller, H.I. and Fegerson, J.L. (1977) Regional blood flow change during early burn shock in guinea pig. *Proc. Int. Union. Physiol. Sci.*, **13**, 512.

Moore, F.D. (1959) *The Metabolic Care of the Surgical Patient*, W.B. Saunders, Philadelphia.

Needham, A.E. (1955) Nitrogen excretion in *Carcinides manenas* (Pennant) during the early stages of regeneration. *J. Embryol. Exp. Morphol.*, **3**, 189.

Needham, A.E. (1958) The pattern of nitrogen excretion during regeneration. in *Oligochaetes. J. Exp. Zool.*, **138**, 369.

Newsome, H.H. and Rose, J.C. (1971) The response of human adrenocorticotrophic hormone and growth hormone to surgical stress. *J. Clin. Endocrinol. Metab.*, **33**, 481.

O'Donnell, T.F., Clowes, G.H.A., Blackburn, G.L. and Ryan, N.T. (1974) Relationship of hindlimb energy fuel metabolism to the circulatory responses in severe sepsis. *J. Surg. Res.*, **16**, 112.

O'Keefe, S.J.D., Sender, P.M. and James, W.P.T. (1974) 'Catabolic' loss of body nitrogen in response to surgery. *Lancet*, **ii**, 1035.

Picou, D. and Taylor-Roberts, T. (1969) The measurement of total protein synthesis and catabolism and nitrogen turnover in infants in different nutritional states and receiving different amounts of dietary protein. *Clin. Sci.*, **36**, 283.

Porte, D. jr. and Robertson, R.P. (1973) Control of insulin secretion by catecholamines, stress, and the sympathetic nervous system. *Fed. Proc.*, **32**, 1792.

Radcliffe, A., Wolfe, R.R., Muhlbacher, F. and Wilmore, D.W. (1981) Failure of exogenous ketones to reduce post-septic glucogenesis. *Surg. Forum*, **32**, 96.

Rector, F., Suresh, G., Rosenberg, I.K. and Lucas, C.E. Sepsis: A mechanism for vasodilatation in the kidney. *Ann. Surg.*, **178**, 222.

Roe, C.F. and Kinney, J.M. (1963) The caloric equivalent of fever: II. Influence of major trauma. *Ann. Surg.*, **161**, 140.

Rowell, L.B. (1974) Human cardiovascular adjustments to exercise and thermal stress. *Physiol. Rev.*, **54**, 75.

Rybeck, B. (1974) Missile wounding and hemodynamic effects of energy absorption. *Acta Chir. Scand. (Suppl)*, **450**, 1.

Rybeck, B., Lewis, D.H., Sandegård, J. and Seeman, T. (1974) Cardiovascular effects of venous blood from missile wounds. *Eur. Surg. Res.*, **7**, 193.

Schønheyder, F., Heilskov, N.S.C. and Olesen, K. (1954) Isotopic studies on the mechanism of negative nitrogen balance produced by immobilization. *Scand. J. Clin. Lab. Invest*, **6**, 178.

Shamoon, H., Hendler, R. and Sherwin, R.S. (1980) Altered responsiveness to cortisol, epinephrine, and glucagon in insulin-infused, juvenile-onset diabetics. *Diabetes*, **29**, 284.

Sherwin, R.S., Hendler, R.G. and Felig, P. (1975) Effect of ketone infusions in amino acid and nitrogen metabolism in man. *J. Clin. Invest.*, **55**, 1382.

Slater, G., Vladeck, B.C., Bassin, R., Kark, A.E. and Shoemaker, W.C. (1973) Sequential changes in distribution of cardiac output in hemorrhagic shock. *Surgery*, **73**, 714.

Stitt, J.T. (1976) Inhibition of thermoregulatory outflow in conscious rabbits during periods of sustained arousal. *J. Physiol.*, **260**, 31P.

Stolwijk, J.A.J. (1970) Mathematical model of thermoregulation. in *Physiological and Behavioral Temperature Regulation.* (eds J.D. Hardy, A.P. Gagge and J.A.J. Stolwijk), Charles C. Thomas, Springfield, p. 703.

Stoner, H.B. (1954) Studies on the mechanism of shock. The effect of limb ischemia on tissue temperature and blood flow. *Br. J. Exp. Path.*, **35**, 487.

Stoner, H.B. (1958) Studies on the mechanism of shock. The influence of environment on the changes in oxygen consumption, tissue temperature and blood flow produced by limb ischemia. *Br. J. Exp. Path.*, **39**, 251.

Stoner, H.B. (1969*a*) The effect of environment on the response to injury in the rat. *Postgrad. Med. J.*, **45**, 555.

Stoner, H.B. (1969*b*) Studies on the mechanism of shock. The impairment of thermoregulation by trauma. *Br. J. Exp. Path.*, **50**, 125.

Stoner, H.B. (1970) The acute effects of trauma on heat production. in *Energy Metabolism in Trauma* (eds R. Porter and J. Knight), J.A. Churchill, London, p. 1.

Stoner, H.B. (1971) Effects of injury on shivering thermogenesis in the rat. *J. Physiol.*, **214**, 599.

Stoner, H.B. (1972) Effect of injury on the responses to thermal stimulation of the hypothalamus. *J. Appl. Physiol.*, **33**, 665.

Stoner, H.B. (1976) Causative factors and afferent stimuli involved in the metabolic response to injury. in *Metabolism and the Response to Injury* (eds A.W. Wilkinson and D.P. Cuthbertson), Year Book Medical Publishers, Inc., Chicago, p. 202.

Stoner, H.B. (1977) The role of catecholamine in the effects of trauma on thermoregulation, studied in rats treated with 6-hydroxydopamine. *Br. J. Exp. Path.*, **58**, 42.

Stoner, H.B. (1978) The effect of ascending noradrenergic nerve fibers on shivering in the rat. in *New Trends in Thermal Physiology* (eds Y. Houdas and J.D. Guieu), Masson, Paris, p. 133.

Stoner, H.B. and Marshall, H.W. (1977) Localization of the brain regions concerned in the inhibition of shivering by trauma. *Br. J. Exp. Path.*, **58**, 50.

Taylor, J.W., Hander, E.W., Skreen, R. and Wilmore, D.W. (1976) The effect of central nervous system narcosis on the sympathetic response to stress. *J. Surg. Res.*, **20**, 313.

Threlfall, C.J., Stoner, H.B. and Galasko, C.S.B. (1981) Patterns in the excretion of muscle markers after trauma and orthopedic surgery. *J. Trauma*, **21**, 140.

Tyssebotn, I. and Kirdebø, A. (1977) The effect of indomethacin on renal blood flow distribution during hemorrhagic hypotension in dogs. *Acta Physiol. Scand.*, **101**, 15.

Ukai, M., Moran, W.H., jr. and Zimmerman, B. (1968) The role of visceral afferent pathways on vasopressin secretion and urinary excretory patterns during surgical stress. *Ann. Surg.*, **168**, 16.

Unger, R.H. (1971) Glucagon and the insulin:glucagon ratio in diabetes and other catabolic illnesses. *Diabetes*, **20**, 834.

Waterlow, J.P., Garlick, P.J. and Millward, D.J. (1978) *Protein Turnover in Mammalian Tissues and in the Whole Body*, Elsevier, North Holland, Amsterdam–New York.

Wilmore, D.W. (1975–1976) Studies of the effect of variations of temperature and humidity on energy demands of the burned soldier in a controlled metabolic room. in *U.S. Army Institute of Surgical Research Annual Research Progress Report*. Ft. Detrich, MD: U.S. Army Research and Development Command, 1 July 1975–30 June 1976, p. 251.

Wilmore, D.W. (1976) Hormonal responses and their effect on metabolism. in *Surg. Clin. North Am.* (ed. G.A. Clowes, jr.), **56**, 999.

Wilmore, D.W. (1977) *The Metabolic Management of the Critically Ill.* Plenum Co., New York.

Wilmore, D.W., Aulick, L.H., Mason, A.D., jr. and Pruitt, B.A., jr. (1977) The influence of the burn wound on local and systemic responses to injury. *Ann. Surg.*, **186**, 444.

Wilmore, D.W., Aulick, L.H. and Pruitt, B.A., jr. (1978) Metabolism during the hypermetabolic phase of thermal injury. in *Advances in Surgery* (ed. C. Rob), Year Book Medical Publishers, Inc., Chicago, p. 193.

Wilmore, D.W., Goodwin, C.W., Aulick, L.H., Powanda, M.C., Mason, A.D., jr. and Pruitt, B.A., jr. (1980) Effect of injury and infection on visceral metabolism and circulation. *Ann. Surg.*, **192**, 491.

Wilmore, D.W., Lindsey, C.A., Moylan, J.A., Faloona, G.R., Pruitt, B.A., jr. and Unger, R.H. (1974*b*) Hyperglucagonaemia after burns. *Lancet*, **i**, 73.

Wilmore, D.W., Long, J.M., Mason, A.D., jr., Skreen, R.W. and Pruitt, B.A., jr. (1974*a*) Catecholamines: mediator of the hypermetabolic response to thermal injury. *Ann. Surg.*, **180**, 653.

Wilmore, D.W., Mason, A.D., jr., Johnson, D.W. and Pruitt, B.A., jr. (1975*a*) Effect of ambient temperature on heat production and heat loss in burn patients. *J. Appl. Physiol.*, **38**, 593.

Wilmore, D.W., Orcutt, T.W., Mason, A.D., jr. and Pruitt, B.A., jr. (1975*b*) Alterations in hypothalamic function following thermal injury. *J. Trauma*, **15**, 697.

Wolfe, R.R., Durkot, M.J., Allsop, J.R. and Burke, J.F. (1979) Glucose metabolism in severely burned patients. *Metabolism*, **28**, 1031.

Wright, C.J., Duff, J.H., McLean, A.P.H. and MacLean, L.D. (1971) Regional capillary blood flow and oxygen uptake in severe sepsis. *Surg. Gynecol. Obst.*, **132**, 637.

Zawacki, B.E., Spitzer, K.W., Mason, A.D., jr. and Johns, L.A. (1970) Does increased evaporative water loss cause hypermetabolism in burned patients? *Ann. Surg.*, **171**, 236.

Chapter Ten

Thermogenesis and Fever

Steven M. Eiger and Matthew J. Kluger

10.1 TEMPERATURE REGULATION AS A REFLEX

The regulation of body temperature is often conceptualized as a negative feedback control system complete with afferent limb, integrating area and effector limb. Most vertebrates regulate body temperature and these animals have been classified into two groups – ectotherms and endotherms. Ectotherms rely largely on behavioural responses to raise or lower their body temperature. Among the vertebrates, most fishes, amphibians and reptiles are ectotherms. Endotherms rely to a large extent on physiological responses to raise or lower their body temperature. Often this includes increasing or decreasing metabolic heat production. Birds and mammals are endotherms. The distinction between endotherms and ectotherms, however, is not absolute. For example, a large percentage of the thermoregulatory effector responses of endotherms is behavioural modifications designed to avoid extremes in temperature. This is why most desert animals are nocturnal, or in cold climates many animals either migrate to warmer climates, avoid the intense cold by burrowing under the snow or hibernate. The result of these numerous thermoregulatory effector responses is that the deep-body temperature of both endotherms and ectotherms is regulated around a hypothetical thermoregulatory set-point.

Deep-body temperature can nevertheless change for a variety of reasons. For example, exposure to a warm environment may prevent an organism from dissipating heat, since heat transfer is dependent largely on a difference between its skin temperature and the environmental temperature. Exposure to a warm environment generally triggers a variety of thermoregulatory responses designed to prevent deep-body temperature from rising; however, if the heat-stressed organism is unsuccessful in losing a sufficient amount of heat, then body temperature can rise above the set-point temperature and hyperthermia develops. Hyperthermia may also develop if heat generation becomes so great (as in certain diseases) that the heat dissipating mechanisms are overwhelmed. An often tragic example of an explosive rise in body temperature due to both the excess production of heat and an impaired ability to dissipate heat occurs during the syndrome 'malignant hyperthermia', to be discussed later in this review.

In contrast to the inability to dissipate a heat load, fever occurs when

305

stimuli affect either the afferent and/or integrating limbs of the thermo-regulatory reflex resulting in the elevation of the thermoregulatory set-point. As a result, the febrile organism 'feels' cold and uses virtually all its thermoregulatory effector responses to raise body temperature. Lieber-meister (1887) was probably the first to suggest that a fever results in the regulation of body temperature around a higher level. This was based on his observations that the body temperature of a febrile human subject returned to its previously raised level after warming or cooling of that patient. Liebermeister differentiated between passive rises in body temperature that might occur during exposure to a hot environment, and fever. In contrast to exposure to a hot environment, during a fever, the person's heat production increased while at the same time the heat losing responses decreased. Much data have been accumulated during the past one hundred years to support Liebermeister's definition of fever. For example, Cooper, Cranston and Snell (1964) found that during fever in man thermoregulatory responses were still functional. The primary difference between febrile and afebrile subjects was that during the rising phase of fever the person's response to an external heat load was diminished or absent. In their particular study they immersed the arms of subjects in warm water. This led to vasodilation in all subjects when the deep body temperature equalled set-point temperature, regardless of whether they were afebrile or febrile; but during the rising fever, when deep-body temperature was less than set-point temperature, this heat load had little effect on skin blood flow.

In recent years there have been numerous studies that suggest fever has a long evolutionary history (see Kluger, 1979, for review). Studies on such non-mammalian vertebrates as reptiles, amphibians and fishes support the notion that the observed elevation in their body temperature during infection is the result of an elevated thermoregulatory set-point. Since these animals are ectotherms, and must raise or lower their body temperature by behavioural means, the selection of a warmer ambient temperature, resulting in the elevation of their deep-body temperature, must be attribut-able to a raised thermoregulatory set-point. Figure 10.1 illustrates this point. Lizards were placed in a simulated desert environment. To regulate their body temperature at their preferred temperature of about 38°C, the lizards had to shuttle between cool and warm areas on the sand. Resting in one area for too long led to a body temperature considerably above or below 38°C. During infection the lizards developed a fever and this resulted in the selection of a warmer portion of the chamber. As a result, body temperature rose to approximately 42°C, a clear demonstration of the regulation of body temperature at a raised thermoregulatory set-point during fever.

10.2 THE BIOLOGY OF FEVER

Fever is initiated by many substances. Regardless of whether these inducers of fever are bacteria, viruses, fungi or some other antigen, they all appear to result in the production of a small-molecular-weight protein, endogenous pyrogen (see Dinarello and Wolff, 1978, for review). Endogenous pyrogen is thought to be produced and released by many different types of leukocytes. During bacterial infections, it is thought that the production and release of endogenous pyrogen is initiated by the direct contact of these phagocytes with the bacteria. In the case of tumour-induced and hypersensitivity-induced fevers, it is thought that endogenous pyrogen is produced and released in response to an intermediary substance, lymphokine, secreted by lymphocytes. Endogenous pyrogen then circulates in

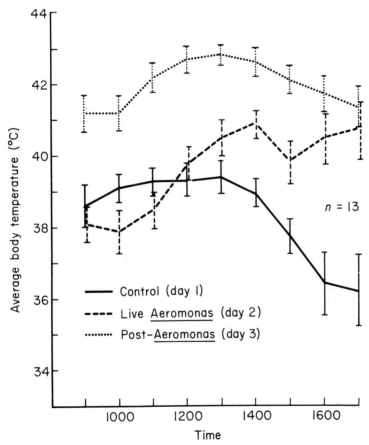

Fig. 10.1 Changes in the body temperatures of 13 desert iguanas after injection with live *Aeromonas hydrophila*. On control days (day 1), lizards had a mean body temperture of about 38°C. Following injection with bacteria (day 2) the lizards selected a warmer environment and as a result their body temperature rose to ca. 40°C within about five hours. Body temperature was further elevated the following day (day 3). (From Bernheim and Kluger, 1976).

blood and acts on the brain's central thermostat to raise the thermoregulatory set-point via some secondary agent. The brain region that
appears to be most sensitive to endogenous pyrogen is the anterior region of
the hypothalamus, although there is no direct evidence that endogenous
pyrogen actually crosses the blood–brain barrier and acts directly on this
region of the brain.

Many substances have been suggested as possible links between endogenous pyrogen and the raised thermoregulatory set-point. There are considerable data that suggest that prostaglandins, particularly of the E series,
directly affect thermally sensitive neurons in the region of the hypothalamus. For example, injection of nanogram amounts of prostaglandin E_1
into the anterior hypothalamus of many vertebrates results in fever after a
very short latency. In addition, most drugs which are known to reduce fevers
by acting at the level of the anterior hypothalamus such as aspirin, indomethacin, etc. are potent inhibitors of prostaglandin synthesis. However,
within the past several years data have appeared that argue against prostaglandins being an integral part of pathogen-induced fevers. For example in
the rabbit, a species commonly used in studies investigating various aspects
of fever, one can block the action or even the release of prostaglandins and
fevers will still develop (Cranston, Hellon and Mitchell, 1975). In addition,
in some newborn mammals an intrahypothalamic injection of prostaglandin
does not induce a fever whereas injection with heat-killed bacteria produces
large increases in body temperature (Pittman, Veale and Cooper, 1977).
Clearly one is able to dissociate prostaglandin-induced fevers from bacteria-
induced fevers under certain conditions. The brain may have evolved two
pathways enabling the development of fever with one involving prostaglandins and the other bypassing prostaglandins and working through some
as yet unidentified neurotransmitter.

The febrile organism raises its body temperature using many effector
mechanisms (for a review see Kluger, 1979). Some are designed to decrease
heat loss and others to increase heat production or heat gain. Heat loss is
decreased in endotherms by reducing skin blood flow (peripheral vasoconstriction), by reducing sweat rate and by subtle and gross changes in
posture. These postural changes can result in both a decrease in heat loss and
often an increase in energy absorption from the environment.

Additional heat can be generated by shivering or by non-shivering
thermogenesis. Shivering is essentially the repeated synchronous contractions of antagonistic muscle groups. As a result of this large increase
in internal work, without any external work, a large amount of heat is
liberated. Heat is also generated by the catabolism of fat and other
metabolic processes. This latter process occurs without any visible body
movement and is termed non-shivering thermogenesis (see Chapter 3).
There is ample evidence that pyrogens induce both shivering and non-
shivering thermogenesis (Horwitz and Hanes, 1976; Blatteis, 1976; Szekely
and Szelenvi, 1979).

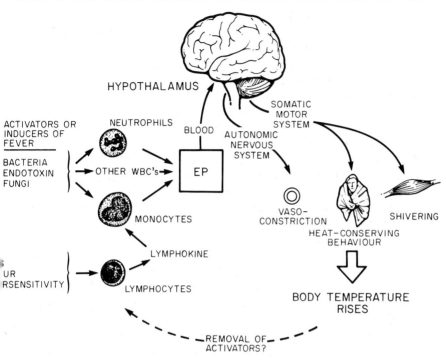

Fig. 10.2 The biology of fever (see text for explanation)

The basic biology of fever is summarized in Fig. 10.2. Many agents induce or trigger the formation of endogenous pyrogens. These protein(s?) either directly or indirectly result in the elevation of the thermoregulatory set-point, and as a result the febrile organism actively raises its body temperature by behavioural and physiological mechanisms.

There is considerable evidence in the literature supporting the idea that fever is beneficial to the infected animal (see Kluger, 1979; Roberts, 1979 for reviews). That is, the elevation in body temperature, as the result of the elevated thermoregulatory set-point, either directly or indirectly results in the removal of the activator of fever (dashed line in Fig. 10.2). As a result, the reflex resulting in an elevation in body temperature during an infection might actually constitute a homeostatic control system; the stimulus being some pathogen, the response being, among other things, an elevation in body temperature, and the negative feedback component of the control system being the harmful effects of elevated temperature on the pathogenic organisms.

10.3 THE BIOLOGY OF MALIGNANT HYPERTHERMIA

In 1962 a paper by Denborough *et al.* described the unusual case of a young man showing more concern with undergoing general anaesthesia than for his fractured leg. He barely survived the anaesthesia, his novel reaction to the anaesthetic providing the first clinical description of malignant hyperthermia (MH). Reconstruction of his family tree, revealed ten anaesthetic-related deaths within his family, thus alerting anaesthesiologists to the pharmacogenetic nature of the disease (Denborough *et al.*, 1962). Recent evidence suggests the disease is multifactorial, polygenetic and inherited in an autosomal dominant manner (Ellis, Cain and Harriman, 1978). The nature and severity of the disease differs among families (variable expressivity) (Britt, Locher and Kalow, 1969). In addition, unknown environmental qualifying factors determine whether a susceptible individual will succumb to an attack (Kalow, Britt and Chan, 1979). Despite the low incidence (1 in 50 000 humans (Kalow *et al.*, 1979)) much is now understood concerning the aetiology and biochemical derangements of MH. Predictive screening, prophylactic regimens, and effective therapeutic drugs are now available (for review see Gronert, 1980).

These rapid advances would not have been possible without the fortuitous discovery of a strikingly parallel model of human MH – porcine (pig) stress syndrome (PSS) (Hall, Trimm and Woolf, 1972). Certain pigs, when stressed, develop metabolic and respiratory acidosis, muscle rigidity, elevated body temperature and ultimately die, the same clinical symptoms that are seen in human MH. In stricken pigs the combination of low pH and high temperature denatures muscle protein, thereby 'stewing' the meat *in*

Table 10.1 Models of malignant hyperthermia

Species	Reference
Reports of MH or apparent MH episodes	
Man	(Denborough *et al.*, 1962)
Swine	(Ludvigsen, 1953)
Dogs	(Bagshaw *et al.*, 1978)
Cats	(de Jong *et al.*, 1974)
Deer	(Pertz and Sundberg, 1978)
Birds	(Henschel and Louw, 1978)
Horses	(Klein, 1975)
Wild animals during capture (zebra, wildebeest, tsessebe, blesbok)	(Harthoorn *et al.*, 1974)
In vitro skeletal muscle studies relating to MH	
Human	(Moulds and Denborough, 1972)
Porcine	(Nelson, 1978)
Frog	(Strobel and Bianchi, 1971)
Rat	(Harrison, 1973)
MH-like episodes in drug pretreated, not genetically susceptible animals	
Ryanodine-pretreated mice	(Fairhurst *et al.*, 1980)
Ryanodine-pretreated frogs	(Casson and Downes, 1978)
Ryanodine-pretreated rabbit	(Durbin and Rosenberg, 1979)

vivo and producing what is known as pale soft exudative (PSE) pork (Gronert, 1980).

The triggering agents of MH in man, potent volatile anaesthetics, skeletal muscle relaxants, and local amide anaesthetics, have also been shown to trigger PSS in swine. However, stress, a common triggering factor in swine has only recently been considered a triggering agent in human beings (Wingard, 1977).

While the MH susceptible swine has proven invaluable to MH research, studies using other models have already, or may in the future, contribute to our understanding of MH (Table 10.1). Human, pig, frog, and rat muscle have been studied *in vitro* using various triggering agents and drugs. In addition to susceptible humans and swine, caffeine-treated rabbits and ryanodine treated mice, cats, and frogs have been used as *in vivo* models of MH. While these drug-pretreated animals are susceptible to MH triggering agents like halothane (Casson and Downes, 1978), inconsistencies exist between their responses, and those of genetically predisposed animals. For example, in pigs genetically susceptible to MH, pretreatment with dantrolene will block or attenuate halothane-induced symptoms (Harrison, 1975). In contrast to this, in ryanodine-pretreated mice, pretreatment with dantrolene does not enhance survival rates after halothane exposure (Fairhurst, Hamamoto and Macri, 1980). Clearly, cautious interpretation of data from ryanodine- or caffeine-pretreated animals is warranted. Although not currently used in research, MH-like episodes have been reported in dogs, cats, deer, birds, horses, and wild animals during capture.

10.3.1 Symptoms of malignant hyperthermia

The two outstanding symptoms of a full-blown MH attack are elevated body temperature and muscle rigor. Less severe or treated episodes may lack these symptoms. Elevated body temperature, the hallmark of MH, is not a primary event but a consequence of hypermetabolism associated with the attack, coupled with an impaired ability to dissipate heat (this will be discussed later). Of all symptoms, the rise in core temperature correlates closest with mortality (Britt, Kwong and Endrenyi, 1977) and has therefore been used as an indicator of the severity of an MH episode. This elevation in temperature usually occurs once an attack is well underway. Muscle rigidity appears in 80% of all MH attacks (Britt and Kalow, 1970). The time course of muscle rigidity is dependent on at least three variables, the genetic disposition of the subject, the severity of the episode, and whether or not a skeletal muscle relaxant like succinylcholine is administered at the induction of anaesthesia. If succinylcholine is given to the subject, muscle rigor usually appears soon afterward (Steward, 1979).

In MH-susceptible patients not given succinylcholine at induction of anaesthesia, cardiac arrythmias, increased heart rates and unstable blood pressures are frequently the first observed changes (Britt *et al.*, 1977). These early occurring cardiovascular changes are thought to be due to both a direct MH rigor of cardiac muscle and as a consequence of the changes in plasma electrolytes, pH, and catecholamine secretion (Britt *et al.*, 1977). Another early indicator is a mottled cyanosis of the skin, which may initially be interspersed with red and flushed patches (Britt *et al.*, 1977). The cyanosis is a consequence of the local vasoconstriction leading to anaerobic metabolism. The localized flushing probably reflects normal local control mechanisms overriding aberrant ones. The cause of this vasoconstriction is unknown. There are at least four plausible explanations. (1) It may be a homeostatic response to keep blood pressure elevated; i.e., shunting blood from the periphery to the hypermetabolizing skeletal muscle and liver. However, clinical descriptions of 'bloodless, gelatinous muscles' at the woundsite suggest muscle blood flow is also compromised (Britt, 1979). In addition, an increase in arterial blood pressure is often observed along with this cutaneous vasoconstriction (Williams *et al.*, 1978). (2) A second possibility is that the sympathetic system is responsible for this constriction. The high blood catecholamine levels support this explanation for a decreased skin blood flow (Gronert, 1980). (3) A third possible explanation for cutaneous vasoconstriction during MH is that the cutaneous smooth muscle has undergone a rigor analagous to the situation observed in skeletal muscle. There are, however, no published data to support this hypothesis. (4) Lastly, it is possible that changes in a circulating electrolyte, for example, the rise in Ca^{2+} or K^+, both known to cause smooth muscle contraction, may be responsible.

Increased rate and depth of respiration is another early sign of MH (Britt *et al.*, 1977). The rising acidosis is probably responsible for these respiratory effects. Venous pO_2 falls somewhat, indicating an increased cellular respiration, and both venous and arterial pCO_2 rise dramatically (Berman *et al.*, 1970). These later changes are thought to be due to both an increase in aerobic metabolism (Gronert and Theye, 1976) and to the buffering of the lactic acidosis by bicarbonate resulting in increased CO_2 levels and a negative base excess (Harrison, 1979).

Blood lactate rises 10–20-fold (Berman *et al.*, 1970). Serum Ca^{2+}, K^+, PO_4^{3-} all rise during MH. This movement of ions from one compartment to another may not be via normal physiological routes but across damaged cell membranes (Britt *et al.*, 1977). Further swings in the levels of these ions may also be a consequence of the diueretics often used as therapy to prevent renal complications. Another characteristic of MH in human beings is the presence of profuse sweating (Denborough *et al.*, 1962).

If remedial steps are not quickly taken then rigor, hyperthermia, increases in blood clotting time, and ventricular failure may ensue. Even if the patient survives an acute crisis, myoglobin escaping from ruptured muscle cells may precipitate renal failure (Britt and Kalow, 1970). In addition, for some unknown reason, perhaps relating to the changes in pH or temperature, brain damage may occur during the acute crisis.

10.3.2 Mechanisms of malignant hyperthermia

The primary or causal cellular events which trigger MH are unknown. The most widely accepted hypothesis is that the triggering agent disturbs excitation–contraction (E–C) coupling within skeletal muscle resulting in a rise in myoplasmic free calcium (Ca_i) (Kalow *et al.*, 1970; Britt, 1979; Gronert, 1980). The evidence for this comes from studies with agents which affect events prior, or distal to E–C coupling (Okumura, Crocker and Denborough, 1980). This elevated Ca_i level in turn mediates the primary biochemical changes responsible for the acute MH crisis.

Since Ca_i has not been measured during an MH crisis the evidence supporting this hypothesis is indirect. There are however, several lines of evidence.

(1) Drugs known to raise Ca_i such as caffeine or lidocaine will worsen the prognosis of a patient during an MH crisis (Britt and Kalow, 1970). Drugs known to lower Ca_i such as procaine or dantrolene help the patient during an MH crisis (Harrison, 1975).

(2) Many of the perturbations seen *in vivo* can be reproduced *in vitro* by altering the bath calcium concentrations directly or by adding an agent which is known to raise or lower the Ca_i within intact cells (Britt, 1979).

(i) The production of lactate is greatly increased during an MH crisis. *In vitro*, the enzyme which breaks down glycogen to glucose-1-PO_4 (glycogen phosphorylase) is activated by increases in Ca_i (Krebs *et al.*, 1966). Since blood lactate rises before tissue ATP levels fall appreciably, this enhanced rate of glycolytic flux is probably not due to falling ATP levels initially, although later in an attack this may happen (Gronert and Theye, 1976). In addition, all glycolytic intermediates increase in concentration during an MH crisis – to biochemists, this suggests a flash activation of glycogen phosphorylase and not another enzyme within the glycolytic pathway (Berman, Conradie and Kench, 1972).

(ii) Cellular respiration is increased *in vivo* (Berman *et al.*, 1970). The rise in body temperature will account for about one half the increase in aerobic metabolism if one uses Q_{10} values for normal subjects across the appropriate temperature range (Berman *et al.*, 1970). The balance of the aerobic metabolism may be, in part, explained by the mitochondrial uptake of calcium. This active uptake of calcium dissipates the energy gradient across the mitochondrial membrane (Chance, 1965). Further rises in myoplasmic calcium may cause mitochondrial uncoupling. This uncoupling is thought to be a result of membrane damage (Chance, 1965). Histological studies on muscle taken from patients after a recent MH attack have shown swollen mitochondria with disrupted cristae (Isaacs, 1978).

(iii) Contraction of skeletal muscle is triggered by a rise in Ca_i. In turn relaxation is brought about by a fall in Ca_i. The hypothesized sustained rise in Ca_i would result in a prolonged contraction progressing to contracture as all the individual sarcomeres become maximally shortened. Muscle contraction is an active process generating heat and using ATP stores. While rigor or contracture is usually associated with a fall in ATP levels as in rigor mortis, this is not the likely cause in MH since ATP levels, while falling, are above the critical level necessary for rigor mortis to develop (Berman and Kench, 1973).

(iv) There has been speculation that the lesion in MH-susceptible patients is not confined to skeletal muscle but may affect other organ systems as well. A final step in the exocytotic theory of hormone secretion is thought to be mediated by intracellular Ca^{2+} (Malaisse, Brisson and Bourd, 1973). During an MH crisis catecholamine levels rise 100-fold. Insulin levels also rise. These findings are difficult to interpret. The rise in catecholamines is probably a secondary response since it occurs after the fall in blood pH and the metabolic changes (Gronert and Theye, 1976). Likewise the increased insulin secretion may be triggered by hyperglycaemia. This rise in blood glucose is in part due to the rising epinephrine levels and the conversion of lactate into sugar taking place in the liver. This latter reaction is thought to explain the high liver temperatures recorded during MH (Berman *et al.*, 1970). However, it is worth noting that one must be careful in ascribing mechanisms to events which are preceded by drastic changes in normal physiology.

In this light, perhaps, the studies which have been done on MH-susceptible subjects while they are not having acute MH attacks might be the most informative. MH-susceptible humans were found to respond to a glucose load with an exaggerated release of insulin relative to control subjects (Denborough *et al.*, 1974). MH-susceptible swine also have higher than normal circulating levels of catecholamines (Lister *et al.*, 1974), an observation used by some to implicate the sympathetic nervous system as being the major defect in MH swine (Williams, 1977). The cutaneous vasoconstriction, the rise in blood catecholamines, and the triggering of an attack by stress all lend support to this hypothesis. However, following sympathetic denervation, halothane still triggers an MH attack in swine (Gronert *et al.*, 1977). Clearly a functioning sympathetic system is not needed, casting much doubt on this hypothesis.

If most researchers working with MH subjects are correct, and the acute crisis is triggered by a rise in myoplasmic calcium, where is the specific lesion? The most likely candidates, those structures which actively pump calcium out of the myoplasm, are the mitochondria, the sarcoplasmic reticulum, and the sarcolemma. Halothane depolarizes the resting sarcolemmal membrane potential by 5–10 mV from MH susceptible but not from control subjects (Kendig and Bunker, 1972). Dantrolene will return this membrane potential towards a more normal resting value (Gallant, Godt and Gronert, 1979). In addition, the presence of muscle enzymes in the circulation during an MH episode suggest that the sarcolemma becomes leaky (Britt, 1975). This sarcolemmal leakiness may contribute to the irreversibility of MH. Mitochondria sequester calcium. Mitochondria from normal swine accumulate about 8–10 times as much calcium as mitochondria isolated from MH-susceptible swine (Britt *et al.*, 1975). In addition halothane will uncouple (thereby inhibiting Ca^{2+} uptake) mitochondria in low pH environments (Mitchelson and Hird, 1973). Low pH also inactivates Ca^{2+} transport into the sarcoplasmic reticulum (Berman *et al.*, 1977). Clearly once an MH episode has been initiated and cellular pH is falling the control of Ca_i levels will be compromised. The protection from, or reversal of, MH attacks by the drug dantrolene, which affects excitation–contraction coupling by attenuating Ca^{2+} release independent of Ca^{2+} uptake (Morgan and Bryant, 1977), has prompted researchers to suspect a primary lesion within the excitation–contraction coupling mechanism (Nelson, 1978).

10.3.3 Positive feedback nature of malignant hyperthermia

In swine, gaseous halothane may be administered for up to 20 min (rigor will appear) and if then turned off, the swine will recover (Williams, 1977). However, if the anaesthesia is continued, the attack will become fulminant or irreversible, which suggests the syndrome is progressing to a positive-feedback system. In addition to membrane damage, the occurrence of acidosis and elevated temperature, by denaturing proteins and other sensitive molecules, will lead to a loss of homeostatic control. The fall in pH resulting from initial metabolic and respiratory derangements may be exacerbated by the rising catecholamine levels during an MH attack. One of the results of the cAMP-triggered cascade responsible for the β-adrenergic actions of catecholamines is the activation of glycogen phosphorylase by protein kinase (Krebs *et al.*, 1966). This may increase the already elevated glycolytic flux rate. Whether this actually happens in skeletal muscle is dependent on the existence of appropriate receptors and effector mechanisms. Lactate produced during anaerobic metabolism is converted by the liver into glucose. MH-susceptible swine have a defect within this gluconeogenic pathway, resulting in a compromised ability to counteract metabolic acidosis (DiMarco *et al.*, 1976).

In addition to the heat produced from neutralization of acid, muscle contraction, aerobic and anaerobic metabolism, and ion transport mechanisms substrate cycling has been observed to be 10- to 50-fold higher in MH-susceptible swine (Clark *et al.*, 1973). The phosphorylation of fructose-6-phosphate to fructose-1,6-diphosphate and the hydrolysis back again results in a net loss of one high-energy phosphate bond for each cycle turn. While the heat generated by this cycling cannot account for all of the heat produced in an attack, it adds significantly to the total heat production.

The rise in muscle temperature may also play a direct role in the rigor which is often observed. Studies with actomyosin preparations have shown a loss in calcium sensitivity as bath temperatures rise from a normal physiological range to temperatures seen in MH (Fuchs, 1975). In addition, since in these studies, ATP and magnesium appear to be protective, their hypothesized falling levels within the cell would contribute to the contraction. This rigor would in turn lower blood flow to the muscle, rendering muscle area hypoxic, thereby contributing to the anaerobic state.

10.3.4 Is malignant hyperthermia a fever?

As described earlier, fever implies a raised thermoregulatory set-point. Accordingly, the febrile organism actively raises its body temperature to the elevated level using physiological and behavioural mechanisms. Body temperature is then regulated around a new elevated temperature, perhaps 38 or 39°C. Few fevers are ever above 41°C (DuBois, 1949). The *regulation* of body temperature whether during health or febrile episodes, involves negative feedback control. For example, if the thermoregulatory set-point is raised from 37 to 39°C, once the resulting elevation in body temperature is sensed, this information integrated, and the net result is a diminution in the 'drive' to elevate body temperature further – the negative feedback component of the regulatory system.

On the other hand, during an MH attack, body temperature is not effectively regulated, often rising to as high as 44–46°C (Britt, 1979). This rise in body temperature may be as rapid as $0.2°C\ min^{-1}$, while the subject is trying to lose heat as evidenced by the high sweat rate. During an MH attack, heat production may become transiently explosive in nature, suggesting a positive-feedback system. As discussed in this review, there is evidence that the initial rise in Ca_i triggers events such as the fall in pH, rise in temperature, and fall in cellular ATP levels which appear to feedback in such a manner as to cause further rises in Ca_i, falls in pH, and rises in temperature.

While unsubstantiated, body temperature itself may be caught in a positive feedback system. An anaesthetized MH-susceptible subject clearly cannot use behavioural temperature regulation, and the presence of cutaneous vasoconstriction has changed a potential heat dissipating surface to a thermal insulating layer. Because of the inherent effects of temperature on virtually all biochemical reactions, each increase in temperature of 1°C will result in approximately a 7–10% increase in metabolic heat production. Given a compromised ability to dissipate this added heat, an upwardly spiralling body temperature might occur, until a point is reached where the low pH and high temperature combine to denature metabolic enzymes and ultimately slow down this process. There are other conditions which lead to large increases in internal heat production, for example exercise. However, providing environmental temperature is moderate, heat dissipating mechanisms are generally adequate to stabilize body temperature at biologically tolerable limits.

REFERENCES

Bagshaw, R.J., Cox, R.H., Knight, D.H., *et al.* (1978) Malignant hyperthermia in a greyhound. *J. Am. Vet. Med. Assoc.*, **172**, 61–2.

Berman, M.C., Conradie, P.J. and Kench, J.E. (1972) The mechanism of accelerated skeletal muscle glycogenolysis during malignant hyperthermia in swine. *S. Afr. Med. J.*, **46**, 1810.

Berman, M.C., Harrison, G.G., Bull, A.B., *et al.* (1970) Changes underlying halothane-induced malignant hyperpyrexia in Landrace pigs. *Nature*, **225**, 653–5.

Berman, M.C. and Kench, J.E. (1973) Biochemical features of malignant hyperthermia in Landrace pigs. in *International Symposium on Malignant Hyperthermia* (eds R.A. Gordon, B.A. Britt and W. Kalow), Charles C. Thomas, Springfield, IL.

Berman, M.C., McIntosh, D.B. and Kench, J.E. (1977) Protein inactivation of Ca^{2+} transport by sarcoplasma reticulum. *J. Biol. Chem.*, **256**, 994.

Bernheim, H.A. and Kluger, M.J. (1976) Fever and antipyresis in the lizard *Dipsosaurus dorsalis*. *Am. J. Physiol.*, **213**, 198–203.

Blatteis, C.M. (1976) Fever: exchange of shivering by nonshivering pyrogenesis in cold-acclimated guinea pigs. *J. Appl. Physiol.*, **40**, 29–34.

Britt, B.A. (1975) Malignant hyperthermia in clinical anesthesia. in *Metabolic Aspects of Anesthesia*, vol. 11 (ed. P.J. Cohen), Davis, Philadelphia.

Britt, B.A. (1979) Etiology and pathophysiology of malignant hyperthermia. *Fed. Proc.*, **38**, 44–8.

Britt, B.A., Endrenyi, L., Cadman, D.L., *et al.* (1975) Porcine malignant hyperthermia: effects of halothane on mitochondrial respiration and calcium accumulation. *Anesthesiology*, **42**, 292–300.

Britt, B.A. and Kalow, W. (1970) Malignant hyperthermia: a statistical review. *Can. Anaesth. Soc. J.*, **17**, 293–315.

Britt, B.A., Kwong, FH.-F. and Endrenyi, L. (1977) Management of malignant-hyperthermia susceptible patients – a review. in *Malignant Hyperthermia, Current Concepts* (ed. E.O. Hanschel), Appleton-Century-Crofts, New York.

Britt, B.A., Locher, W.G. and Kalow, W. (1969) Hereditary aspects of malignant hyperthermia. *Can. Anaesth. Soc. J.*, **16**, 89–98.

Casson, H. and Downes, H. (1978) Ryanodine toxicity as a model of malignant hyperthermia. in *Second International Symposium on Malignant Hyperthermia* (eds J.A. Aldrete and B.A. Britt), Grune and Stratton, New York, pp. 3–10.

Chance, B. (1965) The energy-linked reaction of calcium with mitochondria. *J. Biol. Chem.*, **240**, (6), 2729–48.

Clark, M.G., Williams, C.H., Pfeifer, W.F., *et al.* (1973) Accelerated substrate cycling of fructose-6-phosphate in the muscle of malignant hyperthermic pigs. *Nature*, **245**, 99–101.

Cooper, K.E., Cranston, W.I. and Snell, E.S. (1964) Temperature regulation during fever in man. *Clin. Sci.*, **27**, 345–56.

Cranston, W.I., Hellon, R.F. and Mitchell, D. (1975) A dissociation between fever and prostaglandin concentration in cerebrospinal fluid. *J. Physiol. (Lond.)*, **253**, 583–92.

de Jong, R.H., Heavner, J.E. and Amory, D.W. (1974) Malignant hyperpyrexia in the cat. *Anesthesiology*, **41**, 608–9.

Denborough, M.A., Forster, J.F.A., Lovell. R.R.H., *et al.* (1962) Anaesthetic deaths in a family. *Br. J. Anaesth.*, **34**, 395–6.

Denborough, M.A., Warne, G.L., Moulds, R.F.W., *et al.* (1974) Insulin secretion in malignant hyperpyrexia. *Br. Med. J.*, **3**, 493–5.

DiMarco, N.W., Beitz, D.C., Young, J.W., *et al.* (1976) Gluconeogenesis from lactate in liver of stress-susceptible and stress-resistant pigs. *J. Nutr.*, **106**, 710–16.

Dinarello, C.A. and Wolff, S.M. (1978) Pathogenesis of fever in man. *New Engl. J. Med.*, **298**, 607–12.

DuBois, E.F. (1949) Why are fevers over 106°F rare? *Am. J. Med. Sci.*, **217**, 361.

Durbin, C.G., Jr. and Rosenburg. H. (1979) A laboratory animal model for malignant hyperpyrexia. *J. Pharmacol. Exp. Ther.*, **210**, 70–4.

Ellis, F.R., Cain, P.A. and Harriman, D.G.F. (1978) Multifactorial inheritance of malignant hyperthermia susceptibility. in *Second International Symposium on Malignant Hyperthermia* (eds J.A. Aldrete and B.A. Britt), Grune and Stratton, New York, pp. 329–38.

Endo, M. (1977) Calcium release from the sarcoplasmic reticulum. *Physiol. Rev.*, **57**, 71–108.

Fairhurst, A.S., Hamamoto, V. and Macri, J. (1980) Modification of ryanodine toxicity by dantrolene and halothane in a model of malignant hyperthermia. *Anesthesiology*, **53**, 199–204.

Fuchs, F. (1975) Thermal inactivation of the calcium regulatory mechanism of human skeletal muscle actomyosin: a possible contributing factor in the rigidity of malignant hyperthermia. *Anesthesiology*, **42**, 584–9.

Gallant, E.M., Godt, R.E. and Gronert, G.A. (1979) Role of plasma membrane defect of skeletal muscle in malignant hyperthermia. *Muscle Nerve*, **2**, 491–4.

Gronert, G.A. (1980) Malignant hyperthermia. *Anesthesiology*, **53**, 395–423.

Gronert, G.A. and Theye, R.A. (1976) Halothane-induced porcine malignant hyperthermia: metabolic and hemodynamic changes. *Anesthesiology*, **44**, 36–43.

Gronert, G.A., Milde, J.H. and Theye, R.A. (1977) Role of sympathetic activity in porcine malignant hyperthermia. *Anesthesiology*, **47**, 411–5.

Hall, L.W., Trimm, N. and Woolf, N. (1972) Further studies in porcine malignant hyperthermia. *Br. Med. J.*, **2**, 145–50.

Harrison, G.G. (1973) A pharmacological *in vitro* model of malignant hyperpyrexia. *S. Afr. Med. J.*, **47**, 774–6.

Harrison, G.G. (1975) Control of the malignant hyperpyrexic syndrome in MHS swine by dantrolene sodium. *Br. J. Anaesth.*, **47**, 62–5.

Harrison, G.G. (1979) Porcine malignant hyperthermia. in *Malignant Hyperthermia*, vol. 17 (ed. B.A. Britt), International Anesthesia Clinic, pp. 25–62.

Harthoorn, A.M., van der Walt, K. and Young, E. (1974) Possible therapy for capture myopathy in captured wild animals. *Nature*, **247**, 577.

Henschell, J.R. and Louw, G.N. (1978) Capture stress, metabolic acidosis and hyperthermia in birds. *S. Afr. J. Sci.*, **74**, 305–6.

Horwitz, B.A. and Hanes G.E. (1976) Propranolol and pyrogen effects on shivering and nonshivering thermogenesis in rats. *Am. J. Physiol.*, **230**, 637–42.

Isaacs, H. (1978) Myopathy and malignant hyperthermia. in *Second International Symposium on Malignant Hyperthermia* (eds J.A. Aldrete and B.A. Britt), Grune and Stratton, New York, pp. 89–120.

Jones, E.W., Nelson, T.E., Anderson, I.L., *et al.* (1972) Malignant hyperthermia of swine. *Anesthesiology*, **36**, 42–51.

Kalow, W., Britt, B.A. and Chan, F-Y. (1979) Epidemiology and inheritance of malignant hyperthermia. in *Malignant Hyperthermia* vol. 17, (ed. B.A. Britt), International Anesthesia Clinic, 119–39.

Kalow, W., Britt, B.A., Terreau, M.E., *et al.* (1970) Metabolic error of muscle metabolism after recovery from malignant hyperthermia. *Lancet*, **ii**, 895–8.

Kendig, J.J. and Bunker, J.P. (1972) Alterations in muscle resting potentials and electrolytes during halothane and cyclopropane anesthesia. *Anesthesiology*, **36**, 128–31.

Klein, L.V. (1975) Case report: a hot horse. *Vet. Anesth.*, **2**, 41–2.

Kluger, M.J. (1979) *Fever: Its Biology, Evolution and Function.* Princeton Univ. Press, Princeton, NJ.

Krebs, E.G., DeLange, R.J., Kemp, R.G. and Riley, W.D. (1966) Activation of skeletal muscle phosphorylase. *Pharmacol. Rev.*, **18**, 163–71.

Liebermeister, C. (1887) *Vorlesungen uber specielle Pathologie und Therapie.* Verlag von F.C.W. Vogel, Leipzig.

Lister, D., Hall, G.M. and Lucke, J.N. (1974) Catecholamines in suxamethonium-induced hyperthermia in Pietrain pigs. *Br. J. Anaesth.*, **46**, 803–4.

Ludvigsen, J. (1953) Muscular degeneration in hogs. *International Veterinary Congress*, 15th Congress, Stockholm, **1**, 602–6.

Malaisse, W.J., Brisson, G.R. and Baird, L.E. (1973) Stimulus-secretion coupling of glucose-induced insulin release. X. Effect of glucose on ^{45}Ca efflux from perifused islets. *Am. J. Physiol.*, **224**, 389–94.

Mitchelson, K.R. and Hird F.J.R. (1973) Effect of pH and halothane on muscle and liver mitochondria. *Am. J. Physiol.*, **225**, 1393–8.

Morgan, K.G. and Bryant S.H. (1977) The mechanism of action of dantrolene sodium. *J. Pharmacol. Exp. Ther.*, **201**, 138–47.

Moulds, R.F.W. and Denborough, M.A. (1972) Procaine in malignant hyperthermia. *Br. Med. J.*, **4**, 526–8.

Nelson, T.E. (1978) Excitation-contraction coupling: a common etiologic pathway for malignant hyperthermic susceptible muscle. in *Second International Symposium on Malignant Hyperthermia* (eds J.A. Aldrete and B.A. Britt), Grune and Stratton, New York, pp. 23–36.

Okumura, F., Crocker, B.D. and Denborough M.A. (1980) Site of the muscle cell abnormality in swine susceptible to malignant hyperpyrexia. *Br. J. Anaesth.*, **52**, 377–83.

Pertz, C. and Sundberg, J.P. (1978) Malignant hyperthermia induced by etorphine and xylazine in a fallow deer. *J. Am. Vet. Med. Assoc.*, **173**, 1243.

Pittman, Q.J., Veale, W.L. and Cooper, K.E. (1977) Absence of fever following intrahypothalamic injections of prostaglandins in sheep. *Neuropharm.*, **16**, 743.

Roberts, N.J., jr., (1979) Temperature and host defense. *Microbiol. Rev.*, **43**, 241–59.

Rosenberg, H. (1979) Sites and mechanisms of action of halothane on skeletal muscle function *in vitro*. *Anesthesiology*, **50**, 331–5.

Steward, D.J. (1979) Malignant hyperthermia: the acute crisis in malignant hyperthermia. *Int. Anesthesiol. Clin.*, **17**, 1–9.

Strobel, G.E. and Bianchi, C.P. (1971) An *in vitro* model of anesthetic hypertonic hyperpyrexia, halothane-caffeine-induced muscle contractures: prevention of contracture by procainamide. *Anesthesiology*, **35**, 465–73.

Szekely, M. and Szelenyi, Z. (1979) Age-related differences in thermoregulatory responses to endotoxin in rabbits. *Acta Physiol. Acad. Sci. Hung.*, **54**, 389–99.

Williams, C.H. (1977) The development of an animal model for the fulminant hyperthermia porcine stress syndrome. in *Malignant Hyperthermia – Current Concepts* (ed. E.O. Hanschel), Appleton-Century-Crofts, New York.

Williams, C.H., Shanklin, M.D., Hedrick, H.B., *et al.* (1978) The fulminant hyperthermia-stress syndrome: genetic aspects, hemodynamic and metabolic measurements in susceptible and normal pigs. in *Second International Symposium on Malignant Hyperthermia* (eds J.A Aldrete and B.A. Britt), Grune and Stratton, New York, pp. 113–41.

Wingard, D.W. (1977) Malignant hyperthermia – acute stress syndrome of man? in *Malignant Hyperthermia: Current Concepts* (ed. E.O. Henschel). Appleton-Century-Crofts, New York, pp. 79–95.

Chapter Eleven

Pharmacology of Thermogenesis

Donald Stribling

11.1 INTRODUCTION

Mammalian thermogenesis, or acceleration of metabolic rate in excess of essential metabolism, cannot be considered in isolation since it is co-ordinated with functions controlling heat loss to defend the body temperature. Further complexity is introduced when one considers that heat loss is effected by three discrete processes (vasoconstriction/vasodilation, sweating and changes in behaviour) and thermogenesis is also derived from three independent processes (shivering, voluntary physical activity and non-shivering thermogenesis). After recognition of a deviation in body temperature from the defended norm, heat loss is modified in a reciprocal manner to thermogenesis and, depending on age or previous training, the various sources of heat are used to achieve the necessary thermogenesis. For example, adaptation of rodents to a cold environment enhances capacity for non-shivering thermogenesis (NST) such that shivering plays a less important role in the response to cold exposure (Jansky, Bartunkova and Ziesberger, 1967).

Thermogenesis can also be affected by nutritional status. In addition to the acute rise in metabolic rate associated with feeding (thermic effect or specific dynamic action of food), it was noted some years ago that the effects of feeding a low-protein diet to rats (Stirling and Stock, 1968) had a similar effect on thermogenesis to cold adaptation. The more recent use of a cafeteria diet to enhance voluntary food intake in rats (Rothwell and Stock, 1979) has highlighted the role of nutrient intake in modifying thermogenesis.

On reviewing the pharmacological control of thermogenesis, only a small proportion of authors have determined metabolic rate as such; most measure body temperature. However, since it is necessary to consider both the control of individual processes and their integration in a total regulatory system this does not necessarily prevent the discussion of some of these experiments.

11.2 CENTRAL CONTROL OF THERMOGENESIS

11.2.1 Sites responsible for initiating responses to temperature

Thermosensitive neurons have been identified in the pre-optic anterior region of the hypothalamus, in the spinal cord and at other locations in the

CNS (especially cerebral cortex, thalamus, midbrain and medulla-pons). For a review see Reaves and Hayward (1979).

Two types of thermosensitive neurons can be distinguished; warm type which increase firing rates in response to a rise in temperature and cold type which show a similar response to a fall in local temperature. Although neurons with the appropriate neurophysiological responses can be identified, their organization into a control system is still relatively obscure.

The hypothalamus was first thought to be the primary site of temperature control. In addition to thermodetector neurons sensitive to change in temperature, it receives afferent inputs from the peripheral and spinal thermoreceptors and monoamine pathways from the brain stem (Fuxe and Hokfelt, 1969). However, thermoregulation still persists after destruction of the pre-optic and other hypothalamic areas provided that the spinal cord is intact (Chai and Lin, 1973, Carlisle, 1969). Warming and cooling the spinal cord rapidly activates the same effector responses as similar treatment of the hypothalamus (Jessen and Simon, 1971).

The inter-relationships between the major thermosensitive hypothalamic and extra-hypothalamic areas have been explored by transection (Chambers, Siegel and Lin, 1974), which indicates a series of inhibitory and facilitatory neuronal circuits involved in the integration of temperature control. For example by independent manipulation of hypothalamic and spinal temperatures it is possible to obtain additive or subtractive effects (Jessen and Ludwig, 1971), but spinal neurons appear to prevail over hypothalamic in some species, e.g. the guinea pig (Wunnenberg and Bruck 1970).

There is some evidence that the cooling of the preoptic areas in the rat has a selective effect on NST whilst shivering is selectively activated by cooling the spinal cord (Banet, Hansel and Lieberman, 1978; Fuller, Horowitz and Horwitz, 1977). However, heating the spinal cord inhibits the activation of NST elicited by cooling the preoptic anterior hypothalamus indicating that the two areas are still mutually interdependent.

11.2.2 Sites sensitive to nutrient

Hypoglycaemia is associated with a fall in body temperature (Freinkel *et al.*, 1963). Administration of 2-deoxyglucose which causes cellular glucopaenia (Brown 1962) causes a similar hypothermic response which appears to be centrally mediated (Freinkel *et al.*, 1972). Stereotactic injection of 2-deoxyglucose into the hypothalamus causes a reduction in core temperature of the rat (Shiraishi and Mager, 1980). Whilst one area, the ventral premammillary nucleus showed the clearest, fastest, dose-related response, hypothermic effects were also obtained from injections into other hypothalamic sites.

The effects of 2-deoxyglucose are not necessarily indicative of a hypothalamic glucoreceptor linked to thermogenesis. Inhibition of energy metabolism in neurons responsible for thermoregulation could give disturbances in core temperature.

There are no reports of an increase in metabolic rate or core temperature in response to central administration of glucose or other nutrients which would be clearer indices of a nutrient receptor. Although hypothalamic neurons have been shown to be sensitive to insulin (Anand *et al.*, 1964), insulin-dependent neurons with an impact on thermogenesis have not been recognized as such. It would appear that insulin may be the primary signal for activation of the sympathetic nervous system since in the absence of changes in blood glucose, insulin infusion increases plasma noradrenaline levels in man (Rowe *et al.*, 1981). By contrast, a hyperglycaemic glucose clamp had no effect.

The hypothalamus plays a major role in the regulation of food intake and body weight. The ventromedial hypothalamus exerts an inhibitory influence on food intake and removal of this control would have an inevitable impact on thermoregulation through the activation of diet induced thermogenesis (DIT). Whilst most lesions of the ventromedial hypothalamus (VMH) cause hyperphagia, which has been considered as the mechanism of obesity (Brobeck, Tepperman and Long, 1943), it is possible to generate lesions of the VMH which do not cause hyperphagia but still cause obesity (Rabin, 1974). Non-hyperphagic, VMH lesioned young rats are not intolerant to cold exposure and their maintenance energy requirement is unaffected whilst the efficiency of energy retention beyond maintenance is markedly increased (Vander Tuig *et al.*, 1980). Similarly, gold thioglucose (GTG) treatment of mice which deposits gold in the VMH, disrupts the dietary regulation of sympathetic activity without affecting sympathetic activation by cold exposure (Young and Landsberg, 1979). GTG-lesioned mice are neither able to increase noradrenaline turnover in response to overfeeding nor are they able to reduce turnover in under-nutrition. This implies that the VMH plays a role in the control of DIT rather than being the common descending pathway controlling NST.

Stimulation of the VMH increases brown adipose tissue (BAT) temperature in the rat (Perkins *et al.*, 1981) with a characteristic pattern similar to that seen after nerve stimulation (Flaim, Horwitz and Horowitz, 1977). The neural connection between BAT and the VMH is further supported by the observation that local administration of tetracaine in the interscapular region prevents the rise in BAT temperature after VMH stimulation (Perkins *et al.*, 1981). It would appear that the VMH also plays an important role in the trophic control of BAT since the thermogenic response to nerve stimulation or exogenous noradrenaline is diminished within three days of lesioning the VMH (Seydoux *et al.*, 1981).

The VMH could either contain a nutrient receptor or receive an afferent signal from a remote sensor. No receptors have been identified as such but the region of the VMH which affects DIT does have a high rate of insulin-dependent glucose utilization since parenteral injection of gold thioglucose deposits gold in the appropriate region in normal but not diabetic mice (Debous *et al.*, 1968, 1970). However, some part of the protection of diabetic mice could be due to the relative hyperglycaemia.

Insulin has been shown to affect turnover of catecholamines in the hypothalamus (Sauter *et al.*, 1981). However, there have been no detailed studies of hypothalamic insulin/nutrient administration on thermogenesis comparable with those on satiety which showed that the effects of [^{14}C]glucose injected into the VMH are dependent on incorporation of the label into tissue components (Panksepp, 1974). There is no evidence that nutritional status affects shivering thermogenesis, other than by an indirect potentiation of a capacity for NST and it would be easy to postulate a simplistic theory of hypothalamic control of brown adipose tissue as the primary regulator of DIT. However, the innervation of BAT is complex (Flaim, Horowitz and Horwitz, 1976) and there is ample evidence for the existence of other control systems. Just as thermoregulation is dependent on ascending and descending facilitatory and inhibitory pathways within the CNS, involving the hypothalamus and the brain stem, it is likely that central regulation of DIT will prove to be a complex integration of afferent and efferent signals with appetite, behaviour and peripheral metabolism as inter-dependent variables. (See Chapters 3 and 4 for further discussion of the role of VMH.)

11.2.3 Neurotransmitters affecting central thermoregulation

The measurement of body temperature offers a simple way to assess the effects of neurotransmitters or pharmacological agents on thermoregulation. Unfortunately it fails to discriminate between a primary effect on the putative set point as opposed to a direct effect on either heat loss or thermogenesis. The effects of intervention are dependent on ambient temperature which modulates any response by feedback regulation. Similarly the time course of any changes is critical since the system is geared to nullify any perturbation by compensatory changes in the other elements of thermoregulation.

After allowing for these possible causes of variability, there is still a wide species variation in the response to centrally injected neurotransmitters. Some discrepancies could be due to the differing relative importance of inhibitory/stimulatory pathways and pre- and post-synaptic actions of the same transmitters.

Various models have been proposed to encompass the observations of effects of centrally administered neurotransmitters and antagonists on thermoregulation (Bligh, Colhe and Maskery, 1971; Frens, 1980). Over the years, the range of neurotransmitters thought to be involved in central thermoregulation has extended and now includes acetylcholine, serotonin, dopamine, noradrenaline and histamine (Lomax and Schonbaum, 1979). γ-Aminobutyric acid has also been implicated as the neurotransmitter involved in cross-inhibition between cold sensor and warm sensor pathways (Bligh, Smith and Baumann, 1980). To add to the complexity, there is now an accumulating literature on the effects of neuropeptides on thermo-regulation (Metcalf, Dettmar and Watson, 1980).

11.3 NEURAL CONTROL OF SHIVERING AND NON-SHIVERING THERMOGENESIS

The primary motor centre for shivering is in the caudal hypothalamus with efferent fibres passing down the ventrolateral columns of the spinal cord distinct from the fibres carrying the voluntary input from higher brain areas (Hemingway, 1963; Jung, Doupe and Carmichael, 1937). Local control is via proprioreceptors and the motor nerve afferent arc system (Stuart *et al.*, 1963) and shivering can be effectively prevented by blockade of the neuro-muscular junction with D-tubocurarine (Davis and Mayer, 1955).

The sympathetic nervous system may have an impact on shivering since sympathetic ganglion blocking agents have been reported to prevent shivering (Anderson *et al.*, 1964) and noradrenaline has been shown to affect the post-synaptic neuromuscular junction (Teskey, Horwitz and Horowitz, 1975; Moravec and Vyskocil, 1976).

In contrast to shivering, non-shivering thermogenesis is primarily effected by the sympathetic nervous system. Whilst the target tissue responsible for non-shivering thermogenesis was still obscure, precise definition of the neural control of NST was impossible. However, with the recognition that BAT is the primary tissue, at least in rodents, progress has been facilitated.

The ventromedial nucleus is an important element of the organization of the sympathetic neural output from the hypothalamus (Ban, 1975). The efferent sympathetic nerves give rise to preganglionic fibres from nerve cells in the thoracic and upper lumbar segments which leave the spinal cord by the ventral root ganglia. The preganglionic fibres of the sympathetic system are cholinergic but the post-ganglionic are noradrenergic. Stimulation of the VMH activates BAT (Perkins *et al.*, 1981) but also affects many other processes dependent on the sympathetic nervous system, raising glucose and glucagon, decreasing insulin and increasing lipolysis (Ban, 1975; Shimazu and Amakawa, 1968). Similarly, lesions of the VMH which induce obesity reduce many peripheral responses mediated via the sympathetic nervous system (Nishizawa and Bray, 1978).

The hypothalamus also controls parasympathetic outflow (Ban, 1975) and obesity induced by VMH lesions can be prevented by subdiaphragmatic vagotomy (Inoue and Bray, 1977). Although this primarily disrupts the parasympathetic system, some sympathetic fibres do join the vagus. The effects of vagotomy on obesity may be mediated by the prevention of the hyperinsulinaemia (Inoue and Bray, 1977) which is an early change in the induction of VMH obesity (Hustveldt and Loro, 1972). This conflicts with the suggestion that VMH obesity is due to a defect in DIT in BAT. However, DIT in BAT is dependent on insulin secretion (Rothwell and Stock, 1981) and there may be a more direct effect of the parasympathetic on thermogenesis (Rothwell, Saville and Stock, 1981a).

The innervation of BAT is complex with variations in diameter and myelination of fibres (Flaim *et al.*, 1976). Whilst some of the nerves may be cutaneous afferent or efferent fibres there is the possibility of separate innervation of BAT cells and the vasculature. On the basis of a heterogeneous response to immunosympathectomy, Derry, Schönbaum and Steiner (1969) proposed that a system of long adrenergic neurones supplied the blood vessels whilst short adrenergic neurones, derived from ganglia within the BAT fat pad, supplied the adipocytes.

Only a small proportion of the sympathetic nerve supply needs to be intact to achieve a full response to electrical stimulation. This could be explained by the observation that clusters of brown adipocytes may be electrically coupled through gap-junctions (Linck *et al.*, 1973; see also Chapter 3), thus facilitating an integrated response to neural stimulation.

11.4 RECEPTORS MEDIATING THERMOGENIC RESPONSES IN BROWN ADIPOSE TISSUE

The process of NST in brown adipose tissue is complex. For example, in addition to the various biochemical processes which are activated or inhibited within the adipocyte in the switch from inactivity to maximal thermogenesis, a sharp change in blood flow is required to supply oxygen and substrates and carry away the heat produced. The complex integration of the contributory processes could depend on the interplay of more than one agonist/hormone, a mixed population of membrane receptors, and differing intracellular mediators.

Stimulation of the VMH (Perkins *et al.*, 1981) or the nerve supplying BAT (Flaim *et al.*, 1976) causes an initial decrease in tissue temperature followed by a prolonged rise. The initial fall in temperature which may be due to a change in blood flow (Hull and Segall, 1965) is prevented by the α-adrenergic antagonist, phentolamine (Flaim *et al.*, 1977). This is consistent with the lack of inhibitory effects of α-agonists on isolated brown adipocytes. The subsequent rise in temperature is prevented by pre-treatment with the β-adrenergic antagonist, propranolol (Flaim *et al.*, 1977).

Cold exposure and overfeeding both cause an increase in urinary excretion of catecholamines (Gale, 1973; Landsberg and Young, 1981) which could reflect either increased sympathetic activity or adrenal secretion. However, it would appear that sympathetic nerves are more important than the adrenal medulla since adrenalectomy impairs thermoregulation less than immunosympathectomy (Himms-Hagen, 1975). This is consistent with the observation that the increase in interscapular BAT temperature following stimulation of the VMH can be blocked by local administration of tetracaine whereas the response to infused noradrenaline is unaffected (Perkins *et al.*, 1981). Furthermore, the blood levels of adrenaline and noradrenaline required to achieve maximal stimulation of thermogenesis or BAT function are well above the range of normal circulating levels (Rothwell and Stock, 1979; Nicholls, 1979) presumably because the thermogenic response depends on the extravascular influx of infused catecholamine (Mejsnar and Jirak, 1981).

Noradrenaline depolarizes brown adipocytes but a similar effect can be attained with phenylephrine and isoproterenol. The effects of phenyl-ephrine are prevented by the α-antagonist phentolamine whereas isoproter-enol is blocked by propranolol indicating that both α- and β-adrenergic receptors could be involved in the response to sympathetic stimulation (Fink and Williams, 1976).

Addition of L-noradrenaline to isolated brown adipocytes rapidly stimu-lates oxygen consumption. β-Adrenergic agonists stimulate respiration in a rank order or potency of isoproterenol >noradrenaline >adrenaline sug-gesting the involvement of the β_1-receptor (Bukowiecki *et al.*, 1980). The effects of noradrenaline are completely reversed by addition of propranolol and reinstated by further addition of excess noradrenaline confirming competitive inhibition at the β-receptor. However, the α_1-antagonists, phentolamine and phenoxybenzamine have some partial effect on the response to the mixed α/β-agonist noradrenaline but not the β-agonist isoproterenol.

The α_1-agonist phenylephrine stimulates BAT cell respiration to the same maximum achieved by noradrenaline but at a 100 times higher con-centration. The effects of maximal levels of phenylephrine and isoproterenol are not additive suggesting that they are affecting the same rate limiting step (Bukowiecki *et al.*, 1980). Phenylephrine does have a weak β-agonist activity which could account for this action.

These experiments *in vitro* are consistent with the observation that stimulation of non-shivering thermogenesis and DIT is dependent on the β-receptor *in vivo* (Flaim *et al.*, 1977; Rothwell and Stock, 1979). However, the α-blocker phentolamine does have a small effect on oxygen consumption (Rothwell, Stock and Wyllie, 1981*b*) and phenylephrine causes a significant increase in oxygen consumption of both normal and cafeteria fed rats (Rothwell, Stock and Stribling, 1982*c*) as predicted from its effects *in vitro*.

Hamster brown fat-cells have a high density of specific, high affinity [³H]dihydrolalprenolol-binding sites which, from the relative capacity of ligands to displace the dihydrolalprenolol, are of the β_1-subtype (Svoboda *et al.*, 1979). Although ligand binding and effects on oxygen consumption of hamster BAT cells indicate the predominance of β_1-receptors, the effects of noradrenaline on oxygen consumption of BAT cells from the rat cannot be completely blocked by a β_1-selective antagonist atenolol (Table 11.1). A greater effect is attained with the selective β_2-antagonist ICI 118551 (Holloway and Venters, unpublished observations, 1981).

Table 11.1 The relative potency of selective β_1- and β_1-antagonists on noradrenaline stimulated respiration of isolated rat brown adipocytes
Cells were prepared and incubated according to methods of Bukowiecki *et al.* (1980). Basal respiration = 61.5 ± 9.5 ($n = 20$) ng mol O_2 min^{-1} per 10^6 cells. Noradrenaline respiration = 377.0 ± 48.5 ($n = 17$). The residual rate of respiration in the presence of antagonist is expressed as percentage of the increment in oxygen consumption caused by addition of noradrenaline (10^{-6}M) $\pm$ SEM.
Superscripts denote the numbers of independent observations of the mean.

Concentration of antagonist (M)	Propranolol (β_1/β_2)	Atenolol (β_1)	ICI 118551 (β_2)
10^{-7}	100[3]		
10^{-6}	62.7 ± 10.6[7]	95.2 ± 5.8[5]	91.8 ± 8.2[6]
10^{-5}	16.4 ± 2.8[6]	101.0 ± 1.0[3]	63.0 ± 8.6[6]
10^{-4}	—	70[1]	0[1]

The β_1- and β_2-selective agonists, prenalterol and clenbuterol respectively, stimulate oxygen consumption in isolated BAT cells indicating that β_1- and β_2-receptors are functional and the effects are antagonized by the appropriate antagonists (Table 11.2). A similar mixed β_1/β_2-receptor

Table 11.2 The effect of selective β agonists on the rate of respiration of isolated rat brown adipocytes
Details are described in the legend to Table 11.1. The rate of respiration in the presence of the selective antagonists is expressed as a percentage of the increment in respiration caused by the appropriate agonist in each experiment.

Agonist	Stimulated respiration (ng mol O_2 g min^{-1} 10^{-6} cells)	Atenolol (β_1 antagonist) (10^{-5}M)	ICI 118551 (β_2 antagonist) (10^{-6}M)	(10^{-5}M)
Prenalterol (β_1) 10^{-5}M	400.4[2]	51.0[2]	—	—
Clenbuterol (β_2) 10^{-5}M	379[2]	—	88.6[2]	11.9
Noradrenaline (β_1/β_2) 10^{-6}M	377.0[17]			

population can be demonstrated *in vivo* using selective antagonists to block the increment in interscapular BAT temperature during noradrenaline infusion (Wheeler and Stribling, 1983).

Membranes prepared from hamster BAT also bind the non-selective α-antagonist, [^{3}H]dihydroergocryptine and a high proportion of this can be displaced by the α_1-specific ligand, phentolamine (Svartengren, Mohell and Cannon, 1980). β- and α-receptors are present in the ratio of about 5:1. Stimulation of isolated BAT cells from rats with adrenaline causes an activation of phosphatidyl inositol labelling which is inhibited by the selective, α_1-antagonist prazosin (Garcia Sainz, Hasler and Fain, 1980) indicating that the α-receptor response may be mediated by a different cellular mechanism than that of the β-receptor.

In addition to α- and β-agonists, glucagon, ACTH and TSH which stimulate lipolysis in white adipose tissue, also stimulate oxygen consumption by BAT although the effects of the various stimuli differ in magnitude (Joel, 1966). Noradrenaline stimulates release of free fatty acids from BAT cells (Nedergaard and Lindberg, 1979) and provision of free fatty acids to isolated BAT cells causes a marked increase in oxygen consumption (Prusiner, Cannon and Lindberg, 1968). Hence it is possible that most if not all of the effects are mediated via lipolysis which could explain the wide range of receptor types apparently involved.

Brown adipocytes differ from white adipocytes in having an inducible 32 000 dalton (32k) peptide in the inner mitochondrial membrane which operates a proton conductance pathway dissipating the proton gradient generated by respiration. The 32 000 dalton peptide binds GDP and ADP with high affinity and the proton conductance of the mitochondrial membrane is proportional to the number of GDP binding sites (Nicholls, 1979). Acute cold exposure causes an approximate doubling in the number of GDP binding sites per unit of mitochondrial protein with no apparent change in affinity for GDP (Desautels, Zaror-Behrens and Himms-Hagen, 1978; De Sautels and Himms-Hagen, 1981). Part of this increase in sites possibly reflects an increased accessibility of the 32k peptide to GDP rather than an increase in the amount per mitochondrion such as that seen during longer cold adaptation (Desautels and Himms Hagen, 1981). A similar increase in GDP-binding sites can be achieved within 30 min of administering noradrenaline although this is not apparent in the immediate rise in interscapular BAT temperature during a 10 min noradrenaline infusion (Wheeler, unpublished observations, 1981).

The rapid increase in GDP binding in response to noradrenaline (0.5 mg kg^{-1} s.c.) is dependent on the β-receptor, with both β_1 and β_2-subtypes involved since a total inhibition can only be achieved by a combination of β_1- and β_2-receptor antagonists. This is consistent with the observation that phentolamine pretreatment does not affect the response to noradrenaline administration (Rothwell *et al.*, 1982).

A similar mechanism underlies the physiological activation of BAT since the acute responses in GDP-binding to cold exposure or drinking 20% glucose are similarly prevented by β-blockade (P. Winter, unpublished observations). The chronic adaptive responses to cold exposure include both an increase in [^{3}H]thymidine incorporation into the stem cells of BAT and an enhanced synthesis of mitochondrial components within existing cells. Both elements can be inhibited by administration of β-blockers (Rothwell *et al.*, 1982). It is not easy to explore the pharmacological control of the chronic adaptive responses to cafeteria feeding since the stimulus is dependent on voluntary selected and ingested nutrient. Any diminution of the response could be dependent on a trivial change in the stimulus. However, addition of propranolol to a 20% glucose solution does not markedly reduce the amount of glucose ingested whilst the trophic responses of BAT are inhibited (Table 11.3).

11.5 INTRACELLULAR MEDIATORS OF THE ACUTE, SUBACUTE AND CHRONIC EFFECTS OF SYMPATHETIC STIMULATION

Sympathetic stimulation of brown fat increases intracellular cAMP concentrations (Petterson and Vallin, 1976), a response which is potentiated by phosphodiesterase inhibitors such as theophylline (Hittelman, Bertin and Butcher, 1974). The increase in cAMP leads to activation of a number of discrete protein kinases which in turn correlates with the activity of lipase in fragments of BAT (Skala and Knight, 1977).

This results in a correlation between cAMP levels in brown fat and the rate of glycerol release with a concomitant increase in O_2 consumption. The increase in tissue cAMP levels in BAT of cold-exposed rabbits is prevented by sectioning the cervical sympathetic nerve. This indicates a physiological role for cAMP but it is possible to achieve maximal stimulation of lipolysis *in vitro* with only marginally increased cAMP levels (Knight, 1974).

The impact of cAMP on tissue oxygen consumption may be through provision of increased free fatty acids since administration of exogenous fatty acids has an effect similar to that of noradrenaline on O_2 consumption (Prusiner *et al.*, 1968). However, isolated cells contain high levels of free fatty acids which remain either constant or even decrease after the stimulation of lipolysis and oxygen consumption by noradrenaline (Bieber, Petterson and Lindberg, 1975). Release of fatty acids from BAT cells is O_2 dependent and is claimed to be independent of the energy status of the cell (Bieber *et al.*, 1975); an observation so far lacking a clear explanation.

The precise mechanism determining the fate of liberated fatty acids is unknown. The capacity for activation, transport and oxidation of fatty acids by mitochondria from BAT is high but the rate of mitochondrial fatty acid oxidation *in vitro* is dependent on the ATP/ADP ratio in the incubation

Table 11.3 The effects of chronic (16 days) ingestion of a 20% glucose solution in place of drinking water on interscapular BAT of rats – inhibition by simultaneous administration of propranolol. Groups of six Alderley Park rats were supplied a 20% (w/v) glucose solution in one case containing propranolol (0.005% w/v) in place of drinking water for 16 days. Daily intake of propranolol averaged 9 mg kg⁻¹ day⁻¹. Results expressed as mean ± SEM ($n = 6$)

Treatment	Total (16 day) consumption of glucose solution (g/two rats)	Interscapular BAT as % body weight	Total IBAT mitochondrial protein (mg rat⁻¹)	GDP-binding (nmol mg⁻¹ mitochondrial protein)	IBAT cytochrome oxidase (μmol O₂ min⁻¹ rat⁻¹)
Control	0	0.136 ± 0.006	18.39 ± 1.66	0.285 ± 0.021	41.3 ± 4.5
20% glucose *ad lib*	1674 ± 61	0.194 ± 0.01	31.0 ± 2.39	0.676 ± 0.031	62.2 ± 4.7
		< 0.001*	< 0.01*	< 0.001*	< 0.02*
20% glucose + 0.005% Propranolol	1460 ± 27	0.145 ± 0.009	23.34 ± 2.77	0.442 ± 0.019	43.5 ± 5.5
	NS†	< 0.01†	< 0.05†	< 0.001†	< 0.05†

Significance (by Students t): *vs. water control group; †vs. 20% glucose group.

medium suggesting a regulation of acyl-CoA synthetase by cytoplasmic energy status (Pettersson and Vallin, 1976). Fatty acids will uncouple mitochondria (Flatmark and Pedersen, 1975) and their effects are more pronounced in mitochondria from BAT than from liver (Heaton and Nicholls, 1976) and their action may be to some extent independent of GDP binding to the 32k peptide (for review see Chapter 2).

Proton conductance of the mitochondrial membrane is inhibited by the binding of nucleotides to the 32k peptide. However, cytosolic levels of nucleotides do not change sufficiently to give the necessary range of control and indicators of both high and low energy status (ATP and GTP or ADP and GDP) bind with high affinity and inhibit proton conductance. This implies the presence of a competitive antagonist of nucleotide binding or an independent uncoupler of respiration, such as fatty acid. Fatty acyl-CoA has been suggested as an antagonist of nucleotide binding and does decrease the affinity of GDP binding (Cannon, Sundin and Ranert, 1977) and increases ion permeability, albeit at a level which also uncouples liver mitochondria.

The activation of BAT by noradrenaline involves both uncoupling via existing GDP-binding peptide and an increase in the number of sites available for binding or ion translocation. These two processes could be controlled separately. Mitochondria prepared from BAT of cold-adapted rats which have been returned to thermoneutrality, have high levels of

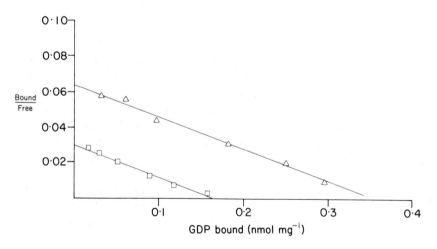

Fig. 11.1 Scatchard analysis of GDP-binding to mitochondria preincubated at either pH 7.1 or pH 7.9.

Mitochondria were prepared from the interscapular brown adipose tissue of rats housed at 22°C by the method of Cannon *et al.* (1977) and preincubated at 24°C in sucrose—TES buffer at either pH 7.1 (□—□) or 7.9 (△—△) for 15 min before washing and resuspending in pH 7.1 buffer for measurement of [³H]GDP-binding at various concentrations of ligand (Nicholls, 1979).

32k peptide most of which is masked or unavailable for binding. Incubation of these mitochondria *in vitro* with various levels of cAMP fails to increase the levels of GDP binding.

However, preincubating the mitochondria in media with pH greater than 7.1, leads to an increase in specific GDP-binding. This is due to an increase in the number of binding sites. No change in affinity could be demonstrated by Scatchard analysis (Fig. 11.1) (Davidson and Stribling unpublished data). The effects are comparable to those achieved by overnight exposure to 4°C and suggest that cytosolic pH could play a role in the control of the unmasking of GDP-binding sites. Stimulation of BAT activates the membrane Na^+, K^+-ATPase which would lead to an increase in cytosolic pH. However, the effects of incubation at alkaline pH may reflect a change in mitochondrial energy state which could be achieved via a different process *in vivo*.

In addition to the activation of proton conductance and unmasking of additional sites, chronic cold exposure and overfeeding leads to a trophic increase in BAT mitochondria and differentiation of stem cells to form functional adipocytes. The intracellular control processes governing these actions remain obscure, although they are important since they determine the maximal capacity for thermogenesis.

11.6 IMPACT OF VARIOUS PHARMACOLOGICAL AGENTS ON THERMOGENESIS

The first part of this chapter has concentrated on a description of the control systems known to affect thermogenesis in so far as this might facilitate either interpretation or prediction of the effects of pharmacological intervention. Whilst there is extensive literature on the effects of pharmocological probes on body temperature this is rarely clarified as effects on thermogenesis or heat loss. Measurements of total oxygen consumption similarly do not distinguish between effects on physical activity, increases in thermogenesis or secondary responses to cooling. With the identification of brown fat as the tissue largely responsible for non-shivering thermogenesis, at least in rodents, there is greater opportunity for exploring pharmacological control. In the subsequent sections, agents will be discussed according to type with some attempt to fit their effects to the proposed control cascade, but with the emphasis on total metabolic rate.

11.6.1 Sympathomimetic agents

(a) Direct agonists

Non-shivering thermogenesis is activated by administration of noradrenaline. Both α- and β-receptor types have been demonstrated in BAT and, in acute experiments, the effects of noradrenaline can to some extent be reproduced using direct agonists of these receptors (e.g. isoprenaline (β_1/β_2) phenylephrine (α_1)). However, in the rat, it is reported that increased oxygen uptake in response to exogenous adrenergic agents is primarily mediated by the β_1-receptor with little impact from either β_2- or α_1-receptor activation (Arnold and McAuliff, 1969; Fregly *et al.*, 1977).

There is no comparable data regarding the receptor subtypes involved in human NST. However, it would appear that both β_1- and β_2-agonists are lipolytic as in the rat since salbutamol, a selective β_2-stimulant, is lipolytic even in the presence of β_1-blockade with practolol (Goldberg *et al.*, 1975). Stimulation of lipolysis by a mixed α/β-agonist gives a lower maximal lipolytic response than a pure β-agonist because of the inhibitory effect of the α_1-receptor (Burns and Langley, 1970, 1971). Chronic administration of a β_1-stimulant does not form part of normal medical practice. However, there are no observations of useful weight loss from chronic treatment of asthmatics with β_2-stimulants such as salbutamol, and no evidence for obesity resulting from chronic administration of β-adrenergic antagonists. β-Stimulants have been claimed to reduce deposition of fat in growing farm animals and a recent report suggested that chronic administration of such agents would lead to a reduction in adiposity in obese rodents and dogs independent of effects on food intake (Shaw *et al.*, 1981). Nonetheless, there are several disadvantages to this approach. Adrenergic agonists have powerful effects on the cardiovascular system which would be contra-indicated in the treatment of the obese. Use of a β_2-stimulant may, to some extent, reduce these effects but still results in a marked rise in circulating free fatty acid levels (Golberg *et al.*, 1975; Shaw *et al.*, 1981), with possible deleterious effects on cardiac metabolism (Oliver, 1972; Jung, Shetty and James, 1980).

Complications in the use of adrenergic agonists could also arise from possible central effects on either thermoregulation or behaviour. Central injections of noradrenaline into the hypothalamus cause marked hypo-thermia (Metcalf and Myers, 1978). However, this central action is dependent on the α-receptor since pretreatment with phentolamine blocks the response to noradrenaline (Saxena, 1973) whereas β-receptor antagonists are without effect (Burks, 1972). This is consistent with the observation that central injections of phenylephrine mimic the action of noradrenaline (Rudy and Wolf, 1971; Lin *et al.*, 1980). Appetite control is similarly dependent on central catecholaminergic pathways. Intrahypo-thalamic injection of noradrenaline induces eating in satiated rats (Grossman, 1962) although the opposite effect is achieved by selective injection into the lateral hypothalamus. A balance has been proposed between an α-adrenergic hunger system and a β-adrenergic satiety system (Liebowitz, 1970). Since circulating metabolic levels have a major impact on appetite, it is hard to predict the balance of effects of a peripherally administered β-stimulant on food intake but ostensibly the effect should tend to favour appetite suppression in accord with increased thermogenesis.

In addition to the use of agonists of the post-synaptic receptor, the activity of noradrenergic terminals can be increased by preventing the destruction or reuptake of noradrenaline or through actions at presynaptic receptors which modulate noradrenaline release.

(b) Amine releasers

As well as having weak direct agonist effects, amphetamine and related compounds stimulate the release and competitively inhibit the reuptake of both noradrenaline and dopamine.

Amphetamine readily crosses the blood–brain barrier and causes increased wakefulness, tremor, agitation and motor activity as well as a reduction in food intake. In addition, amphetamine modifies thermoregulation, high doses being associated with hyperthermia (Jordan and Hampson, 1960). This action could be either centrally or peripherally mediated. Intracerebroventricular amphetamine causes hypothermia possibly through activation of heat dissipation whilst its peripheral action is to increase thermogenesis (Jellinek, 1971). In common with many other agents acting on central thermoregulation, the effects of amphetamine are dependent on housing temperature. In cold-housed rats parenterally administered amphetamine is hypothermic, the effects being inhibited by the dopamine-receptor antagonists haloperidol and pimozide (Yehuda and Wurtman, 1972). In rats housed at ambient temperature, amphetamine is hyperthermic. This effect can also be antagonized by the dopamine blocker, pimozide (Matsumoto and Griffin, 1971) but is also impaired by depletion of stores of noradrenaline (Caldwell, Sever and Trelinski, 1974) and by pretreatment with the β-blocker, propranolol (Mantegazza et al., 1970). Cyproheptadine, a 5-HT antagonist, also prevents amphetamine hyperthermia (Frey, 1975) which implies that amphetamine is affecting a complex control cascade involving serotinergic pathways, possibly in the preoptic nucleus.

Peripheral sympathectomy blocks amphetamine hyperthermia whilst depletion of catecholamine stores and β-blockade could also be reducing the response by a peripheral action. The ganglion blocker, chloroisondamine does not antagonize the hyperthermic effect of amphetamine, further supporting a direct effect on catecholamine release from peripheral sympathetic nerve endings (Caldwell et al., 1974). Administration of amphetamine is associated with an increase in circulating free fatty acid levels and although inhibition of this rise by nicotinic acid prevents the hyperthermic response, the two effects are not essentially coupled since the amine uptake inhibitor, desimipramine potentiates the hyperthermic response whilst antagonizing the lipolysis (Matsumoto and Shaw, 1971). Similarly, whilst both (+) and (−) isomers of amphetamine are lipolytic, hyperthermic effects are restricted to the (+) isomer (Hajors and Garattini, 1973).

The central stimulant effects of amphetamine which give rise to problems of dependence and abuse are minimized in ephedrine which does not readily penetrate in the the CNS. In addition to losing the stimulant effects of behaviour, food intake is not affected yet and chronic administration of ephedrine to rodents is associated with increased oxygen consumption and decreased body weight (Massoudi and Miller, 1977). Similar effects are evident in man (Evans and Miller, 1977), although all agents which depend on non-specific stimulation of release of noradrenaline are subject to pharmacological tolerance through depletion of amine stores (Harrison, Ambrus and Ambrus, 1952).

Non-shivering thermogenesis in the obese is reported to be insensitive to administration of noradrenaline (Jung, Shetty and James, 1979). Although it is hard to separate changes due to obesity from those due to reduced food intake, it is probable that sympathomimetic agents acting as either direct post-receptor agonists or as enhancers of release would be subject to the same insensitivity.

(c) Amine uptake inhibitors

A potentiation of sympathetic stimulation can be achieved by blocking reuptake of catecholamines. This is less likely to result in tachyphylaxis than an amine release mechanism since the effect is restricted to enhancing endogenous signals rather than a non-specific release. Amine uptake inhibitors are widely used in the treatment of depression but one of the conventional screening methods depends on reversal of reserpine-induced hypothermia.

In normal animals, tricyclic antidepressants such as imipramine decrease body temperature (Garrattini and Jori, 1979). This is related to a sedative action similar to that of the phenothiazines such as chlorpromazine and may be dependent on a peripheral vasodilation enhancing heat loss (Borbely and Loepfe-Hinnkannen, 1979). However, the tricyclic antidepressants interact with both hypothermic and hyperthermic stimuli; inhibiting the effects of agents such as reserpine and oxotremorine (Costa, Garattini and Valzelli, 1960; Rathburn and Slater, 1963) whilst potentiating the effects of amphetamine and noradrenaline (Mopurgo and Theobald, 1965; Jori and Garattini, 1965). Administered before reserpine, the tricyclics potentiate the initial hyperthermic response and inhibit the subsequent fall in temperature. When administered during the hypothermic phase after reserpinization, body temperature is returned toward normal (Garattini and Jori, 1967). The hyperthermic effect of reserpine is due to an initial surge of noradrenaline release from existing stores. The magnitude of the response is therefore enhanced by an amine uptake inhibitor. When, as a result of reserpine preventing protective storage of noradrenaline in granules, the rate of release had fallen to the level of *de novo* synthesis, inhibition of reuptake would enhance the levels of noradrenaline available to the post-synaptic receptors by protecting against mitochondrial inactivation (Sulser *et al.*, 1969; Costa *et al.*, 1966). These effects can be blocked by α-antagonists such as phentolamine (possibly centrally) or β-antagonists such as propranolol (peripherally) and are mediated by sympathetic innervation since ganglion blockade with chloroisondamine also inhibits the effect (Jori, Carrara and Garattini, 1966; Garattini and Jori, 1967).

The reversal of reserpine-hypothermia by tricyclic antidepressants is prevented by pretreatment with α-methyl-*p*-tyrosine which inhibits the synthesis of catecholamines (Jori *et al.*, 1966). Sensitivity can be restored by administration of L-dopa which replenishes the precursor for noradrenaline and dopamine synthesis (Jori and Bernardi, 1970). The effect is probably specific to noradrenaline since L-dopa cannot alleviate the block induced by diethyl dithiocarbamate which inhibits dopamine β-hydroxylase and selectively depletes noradrenaline (Jori *et al.*, 1966). A dopaminergic element may be involved in mice, however, since apomorphine, a dopamine agonist, alleviates reserpine hypothermia in this species (Cox and Tha, 1975).

Amphetamine hyperthermia is potentiated by inhibitors of amine re-uptake but this is possibly due to interference with metabolic clearance of amphetamine (Sulser, Owens and Dingle, 1966). However, the hyperthermic response to exogenous noradrenaline is potentiated (Jori and Garattini, 1965). A dramatic hyperthermia can also be induced by administering the monoamine oxidase inhibitor, pargyline before the amine uptake inhibitor, imipramine (Gory and Rodgers, 1971).

The effects of the inhibitors of amine uptake on reserpine hypothermia are prevented by adrenalectomy and thyroidectomy (Cowan and Whittle, 1972; Garattini and Jori, 1979), whilst tri-iodothyronine pretreatment renders desmethylimipramine hyperthermic in normal rats presumably as a result of sensitization of thermogenesis to sympathetic activation (Breese, Taylor and Prange, 1972). BAT would appear as the likely site of increased thermogenesis since desmethylimipramine elevates free fatty acid levels in brown fat of reserpinized rats.

The hypothermic effects of amine uptake inhibitors in normal animals make their use in treatment of obesity unlikely. However, sympathetic drive is increased by feeding and decreased during dietary restriction (Landsberg and Young, 1981; James *et al.*, 1981). One might predict, therefore, that the tricyclic antidepressants would be valuable in sustaining the sympathetic drive on thermogenesis in the obese, especially during dieting. However, their use is associated with increases in body weight especially when used in combination with monoamine oxidase inhibitors (Arenillas, 1964; Winston and McCann, 1972). This appears to relate to a stimulation of appetite (Gander, 1965), which would clearly offset any slimming action through thermogenesis.

It may be possible to shift the emphasis of central effects on the appetite centre to peripheral effects on thermogenesis. This may have been, to some extent, achieved in ciclazindol which was found to reduce weight in depressed patients (Ghose *et al.*, 1978). It has subsequently been shown to assist weight loss in the treatment of obesity (Greenbaum and Harry, 1980). This effect could relate either to a reduction in appetite, since ciclazindol is related to the appetite suppressant mazindol, or to a stimulation of metabolic rate. Ciclazindol has been shown to increase resting metabolic rate independent of effects of arousal in rats (Fletcher *et al.*, 1981). This may be due to an inhibition of noradrenaline uptake in brown adipose tissue (Latham *et al.*, 1981) but could also be due to a more selective action on central noradrenaline and dopamine uptake since ciclazindol is devoid of the central anticholinergic and antihistaminic effects of amitriptyline (Sugden, 1974) and has no significant α-antagonist activity (Ghose *et al.*, 1978).

11.6.2 Autoregulation of noradrenaline release

The release of noradrenaline from adrenergic nerves is affected by a number of other controls beyond firing rate and reuptake/degradation (Westfall, 1977). Continued stimulation leads to a decrease in the amount of noradrenaline secreted per nerve impulse, an effect which may be related to a presynaptic α_2-receptor (Langer *et al.*, 1971; Langer, 1977; Starke, 1971). The presence of this receptor system has been identified in the hypothalamus (Bryant *et al.*, 1975) and yohimbine (an α_2-antagonist) has been shown to potentiate the effects of submaximal stimulation of GDP-binding in BAT mitochondria (Rothwell *et al.*, 1982).

Prostaglandins, especially E_1 and E_2 have been shown to decrease noradrenaline release in response to nerve stimulation (Hedqvist, 1969) but this is not a pronounced effect in adipose tissue in the dog (Fredholm and Hedqvist, 1973). Prostaglandins may, however, mediate the increase in blood flow through adipose tissue during infusion of noradrenaline, and endogenous prostaglandins also inhibit lipolysis at a post receptor site (Iliano and Cuatrecasas, 1971). These effects could be apparent in BAT but administration of the cyclo-oxygenase inhibitor, flurbiprofen to cafeteria-fed rats selectively inhibits rather than potentiates the DIT (Rothwell *et al.*, 981*b*). These is clearly further scope for research in this area with the recent identification of thromboxane, prostacyclin and other arachidonic acid products.

An autoregulatory control which may be relevant to the sympathetic activation of BAT is the purinergic system demonstrated in canine adipose tissue. Adenosine is released from the adipocytes in response to sympathetic activation. Addition of adenosine inhibits the accumulation of cAMP and has a lesser effect on lipolysis. (Fredholm, 1974, 1976). The relative potency of purinergic agonists is in the order phenyl isopropyl adenosine >

adenosine $>$ ADP $>$ ATP characteristic of the P_1 subtype of high affinity purinergic receptors (Fain, 1973). Adenosine release from adipocytes is blocked by the α_1-antagonist, phenoxybenzamine suggesting the possible involvement of adenosine in the inhibitory action of α_1-agonists in adipose tissue. Theophylline, in addition to being a phosphodiesterase inhibitor, is a purinergic receptor antagonist. Addition of theophylline to isolated brown adipocytes increases O_2 consumption although similar effects can be achieved by simple phosphodiesterase inhibitors. *In vivo* theophylline raises metabolic rate but more detailed dose–response studies or a more selective agent would be needed to clarify the role of the P_1 receptor in this action.

A role in the modulation of sympathetic innervation has also been proposed for muscarinic receptors in the rabbit heart (Loffenholz and Muscholl, 1969). The muscarinic antagonist atropine has dual effects on body temperature. Systemically administered, atropine causes hypothermia in rats, an effect which is also produced by the quaternary analogue, atropine methyl nitrate which does not cross the blood–brain barrier (Kirkpatrick and Lomax, 1967). This effect of high doses is likely to be due to increased heat loss through vasodilation. Centrally administered atropine has differing effects between species and according to environmental temperatures, reflecting an action on thermoregulatory control.

In normal rats, subcutaneous injection of atropine potentiates the thermic response to a liquid meal. A similar, but greater effect is evident in genetically obese Zucker rats such that the previously depressed thermic effect approaches the level of the atropine-treated lean rats (Rothwell *et al.*, 1981*a*). This implies an enhanced parasympathetic inhibition of thermogenesis in the obese rodent. There is some evidence to suggest that human obesity is associated with increased parasympathetic drive in that salivary responses to food are enhanced in the obese and during dieting, both situations of depressed thermogenesis (Klajner *et al.*, 1981). The level of this muscarinic control on dietary-induced thermogenesis is not defined. It could reflect a central inhibitory input, an effect on the autonomic ganglia or a direct parasympathetic influence on the brown adipocyte.

Nicotinic receptors are associated with release of catecholamines and sympathetic activation both by stimulation of the adrenal medulla and autonomic ganglia and through direct action on noradrenergic nerve terminals (Kiser, Bocher and Watts, 1955; Su and Bevan, 1970; Lee and Shiderman, 1959). These actions may explain the observation of weight gain after cessation of smoking (Khosla and Lowe, 1971). In view of the plethora of pharmacological actions of nicotine, there seems little opportunity to exploit this action.

11.6.3 Tissue level potentiation of thermogenesis

At least part of the thermogenic response within BAT depends on cAMP-mediated mechanisms. In accordance with this, caffeine, the most commonly ingested phosphodiesterase inhibitor, stimulates oxygen consumption in man (Miller, Stock and Stuart, 1974). This effect may be indirect through releasing fatty acids (Oberman *et al.*, 1975) which in turn stimulate oxygen consumption in man (Jung *et al.*, 1980) or via release of adrenaline from the adrenals (Robertson *et al.*, 1978) rather than a direct effect on thermogenesis. The effects of caffeine are not reduced by β-blockade with propranolol nor by sympathectomy in animals (Jung *et al.*, 1981: Strubelt, 1969) suggesting an independent effect on thermogenesis.

As discussed above, the intracellular mediation of thermogenesis is not completely clear. A distinction between white and brown fat lies in the fate of lipids taken up from plasma or synthesized *in situ*. Whereas in white fat the proportion oxidized is minor with a majority being stored, brown fat in the neonate or cold-exposed animal is committed to oxidation. In the adult animal, brown fat can assume a storage role although it retains sensitivity to catecholamines and the basic machinery for thermogenesis. This transition is particularly marked in the harp seal, where, beyond three days of age the dorsal subcutaneous adipose tissue which previously has defended body temperature against extreme conditions in the neonate, rapidly accumulates lipid and becomes the lipid-rich, white blubber (Blix, Grav and Ronald, 1979). This transition is particularly interesting since, within the same animals, the internal depots surrounding the venous plexuses and the kidney still retain their function as thermogenic BAT. Although this discrimination could be due to a difference in receptors for an extracellular signal, it reflects a marked change in the fate of lipid within different depots. The detailed mechanism for this discrepancy is not known.

There is some potential for the modulation of thermogenic response to sympathetic stimulation by affecting the capacity for oxidation of lipid. For example, carnitine, which is essential for the transport of activated fatty acids into the mitochondria, potentiates the thermogenic response to noradrenaline (Hahn, Skala and Davies, 1971) and other lipolytic stimuli. There is no evidence, however, to link obesity with carnitine deficiency but there may well be additional ways of facilitating thermogenesis within BAT as opposed to release of fatty acids following a sympathetic stimulus. Further exploration of the mechanism underlying the expression of the 32K GDP binding peptide may reveal a critical control which can be exploited and, because of the unique nature of the mechanism, engender tissue selectivity.

11.7 HORMONAL CONTROL OF THERMOGENESIS

11.7.1 Thyroid hormones

Hyperthyroidism and hypothyroidism are respectively associated with elevated and depressed metabolic rates. Changes in thyroid hormones occur in response to feeding (Hesse, Spahn and Plenert, 1981), overfeeding (Rothwell and Stock, 1979) and cold exposure (Reichlin *et al.*, 1973). Chronic treatment with high doses of thyroid hormones causes an enhanced response to noradrenaline (Leblanc and Villemaire, 1970) and increases the number of β-receptors (Kempson, Mannetti and Shaw, 1978) whilst hypo-thyroidism is associated with a decrease in β-receptor number and affinity (Seydoux, Giacobino and Girardier, 1980). Bilateral lesions of the ventro-medial hypothalamus reduce thyroid secretion rate and it has been sug-gested that this contributes to the VMH-obese syndrome (Hinman and Griffith, 1973). There is no evidence, however, that simple obesity is associated with a lack of circulating thyroid hormones, although in estab-lished obesity there may be differences in kinetic parameters (Hofmann *et al.*, 1975; Riviere *et al.*, 1975). Administration of physiological doses of thyroxine T_3 suppresses the activity of the thyroid gland resulting in little overall change in metabolic rate. At doses which induce thyrotoxicosis, metabolic rate is elevated and a marked potentiation of the effects of sympathomimetic agents can be seen. This does not offer a safe, general treatment for obesity. However, some part of the fall in metabolic rate during dieting may reflect a change in thyroid function which could limit thermogenesis (Shetty, Jung and James, 1979).

Effects on metabolic rate are not restricted to T_3/T_4. TSH has been suggested as having a direct effect on BAT development independent of effects on the thyroid (Doniach, 1975). Similarly TRH has been reported to cause marked hypothermia after intra-cerebroventricular injection in the cat (Metcalf, 1974). In line with this observation, the ob/ob mouse which has a defective thermoregulatory response to cold and over-eating has elevated hypothalamic levels of TRH (Donaldson, 1980).

11.7.2 Other hormones

Administration of glucagon to rats raises metabolic rate (Davidson, Salter and Best, 1960) and on chronic administration reduces fat deposition (Salter, 1960). These effects may be dependent on an increase in NST since glucagon stimulates heat production by BAT *in vivo* (Heim and Hull, 1966; Cockburn, Hull and Walton, 1968) and chronic administration of glucagon improves cold tolerance and increases BAT depots and functional adapta-tion of BAT mitochondria (Yahata, Ohno and Karoshima, 1981). *In vitro* glucagon stimulates oxygen consumption by isolated BAT cells (Kuroshima and Yahata, 1979) but maximal responses are only achieved at glucagon concentrations outside the physiological range.

Cold exposure causes a rise in plasma glucagon levels in both man and rodents (Seitz *et al.*, 1981, Kuroshima *et al.*, 1981). However, there is little evidence to suggest that obesity is associated with a defect in glucagon secretion. Glucagon, in addition to being a stress hormone, forms with insulin an integrated part of the glucose homeostatic mechanism. It has been proposed that insulin and glucagon are also involved in the regulation of a 'set point' for body weight (de Castro, Raullin and Dehugas, 1975). Glucagon secretion is inhibited by elevated plasma glucose levels but is stimulated by most amino acids. It could therefore contribute to the regulation of the thermic effect of food, accounting for the greater response to a protein stimulus (Pittet, Gygax and Jecquier, 1974) It remains to be seen whether glucagon secretion could be modulated pharmacologically to achieve a useful contribution to the treatment of obesity without deleterious effects on metabolic homeostasis.

Of the other hormones which affect metabolic rate, growth hormone perhaps merits consideration. Growth hormone increases metabolic rate in obese patients and plasma levels of growth hormone are generally lowered in the basal state and in response to various stimuli in the obese (Hansen, 1973; Londono, Gallacher and Bray, 1969; Kopelman *et al.*, 1979). It may be that the age-dependent decrease in capacity to accomodate an excessive dietary intake and to cold adapt is related to a reduction in growth hormone secretion with age.

11.8 A RATIONALE FOR SELECTING THERMOGENIC AGENTS

It is apparent from this review of the pharmacological control of thermogenesis that, whilst a great deal is known about certain elements of the control system, many confusions and uncertainties persist. With the recognition of a target tissue for cold- and diet-induced thermogenesis in rodents, a more direct measure of NST is possible. Whereas the response to cold is complicated by an interplay between shivering, NST and attempts to reduce heat loss, DIT contributes only a fraction of metabolic rate and any interference with the nutrient stimulus could as easily affect the response.

There is still disagreement about the elements of metabolic rate which may be defective in obesity but the elegant experiments of Sims *et al.* (1973) emphasized the discrepancy in energy consumption required to maintain an elevated body weight in people who tended to be lean as opposed to those who tended to be fat. In treating obesity one is not merely seeking to correct the underlying abnormality. Any treatment must also dispose of the accumulated adipose depots from the previous imbalance between intake and expenditure. In a 100 kg individual who is about 150% of ideal body weight, the additional 30 kg of adipose tissue has an energy value of about 750 MJ. If an isocaloric diet originally corresponded to 10 MJ day^{-1} a reasonable reducing diet of 4.2 MJ would give a deficit of intake from expenditure of 5.8 MJ day^{-1}. If the patient could sustain metabolic rate during this diet a steady return to normal body weight would be achieved in six months. However, metabolic rate declines on dieting more rapidly than the decline in body weight. This is associated with a decrease in thermogenesis as indicated by a lesser inhibition of metabolic rate by propranolol. With the increase in metabolic efficiency, after an initial period of weight loss, a new equilibrium can be reached still well above the target weight. This is associated with a fall in levels of both circulating T_3 and catecholamines (Jung *et al.*, 1979). Administration of T_3 sustains metabolic rate and weight loss but causes further decreases in T_3-receptor population (Moore *et al.*, 1981), and is associated with a damaging loss of lean body mass (Bray, Melvin and Chopra, 1973). However, sustaining the sympathetic drive with L-dopa administration prevents the fall in resting metabolic rate without affecting the fall in T_3 levels (Shetty *et al.*, 1979). This implies an effect on NST but no longer-term studies to evaluate the effect on weight loss have been reported.

It is unrealistic to seek a treatment for obesity which is not first preceded by dieting, and use of a thermogenic agent in conjunction with unrestrained food intake could well lead to an uncomfortable sensation of warmth if not dangerous hyperthermia. However, the L-dopa experiments possibly indicate a novel approach to a therapy which would assist dieting. The fall in sympathetic drive could result from some autoregulatory feedback such as those described above, or to an additional inhibitory input e.g. parasympathetic as in the Zucker rat. This could be exploited in a thermogenic agent which disinhibited the depressed thermogenesis rather than elicited an inappropriate enhancement of sympathetic activity with the attendant problems of elevated fatty acids, hypertension and cardiac effects.

It is unclear whether the decrease in sympathetic drive results from an inhibitory input peripherally or centrally. Clearly, the effectiveness of agents which potentiated the effects of noradrenaline at peripheral synapses would be ineffective if there were to be a pronounced reduction in the central outflow during dieting. In this situation, effective treatment would demand an attenuation of the central inhibitory input. Brown fat is the only tissue so far reported which does not show receptor supersensitivity when the endogenous drive is depleted. Whereas one would predict an enhanced sensitivity to exogenous noradrenaline, the converse is true. This could, conceivably, reflect the atrophy of the post-receptor mechanism required for thermogenesis masking the predicted rise in β-receptor population.

Agents which stimulate metabolic rate will inevitably provide an indirect stimulus to cardiac function. However, many of the agents which have been identified as likely to potentiate the sympathetic drive on brown fat would have comparable, direct effects on the heart. One primary medical justification for the treatment of obesity is to reduce the attendant complications of hypertension and cardiac disorders. Any agent which reduced body weight but which intensified the long-term risks would not achieve wide acceptance. This factor proved to be the downfall of work directed at thyroxine analogues and could also prove difficult to resolve with agents which stimulate both thermogenesis and lipolysis.

The homeostatic mechanisms of thermogenesis are finely geared and it would appear that the peripheral receptor systems which are pro-thermogenic act centrally to inhibit thermogenesis and stimulate food intake. It is likely that the ideal combination will be derived empirically but in this current surge of interest in thermogenic agents there is the added advantage that, in rodents at least, the thermogenic tissue has been identified.

Little attention has been paid in this review to the trophic effects of sustained stimulation of BAT. Current evidence would suggest that to a large extent, those factors which can elicit an acute thermogenic response in BAT, can also mediate a trophic response in BAT both by increasing mitochondrial capacity within the existing BAT cells and by accelerating division and differentiation of stem cells (Rothwell *et al.*, 1982). However, care in the choice of models for depressed thermogenesis is warranted since the acute response attained could be severely reduced by the duration of a previous defective drive on BAT. In this respect, whilst genetically obese rodents are invaluable for demonstrating the contribution of BAT to NST and the control cascade leading to weight control, there are dangers implicit in selecting a model with an inherited defective drive on BAT.

Provided that it can be shown that a BAT-like mechanism exists in higher mammals and that an adequate thermogenic capacity exists to have an effect on energy balance, one could predict the discovery of an agent which both aided dieting by sustaining weight loss and perhaps thereafter removed from those who tended to be fat, the never ceasing chore of counting calories.

REFERENCES

Anand, B.K., China, G.S., Sharma, K.N., Dua, S. and Singh, B. (1964) Activity of a single neurone in the hypothalamic feeding centres: effects of glucose. *Am. J. Physiol.*, **207**, 1146–54.

Anderson, B., Brook, A.H., Gale, C.C. and Hokfelt, B. (1964) The effects of a ganglion blocking agent on the thermoregulatory response to preoptic cooling. *Acta Physiol. Scand.*, **61**, 393–9.

Arenillas, L. (1964) Amititriptyline and weight gain. *Lancet*, **i**, 432.

Arnold, A. and McAuliff, J.D. (1969) Correlation of calorigenesis with other β-mediated responses to catecholamines. *Arch. Int. Pharmacodyn. Ther.*, **179**, 381–7.

Ban, T. (1975) Fibre connections in the hypothalamus and some autonomic functions. *Pharmacol. Biochem. Behav.*, **3**, 3–13.

Banet, M., Hensel, H. and Lieberman, H. (1978) Central control of shivering and non-shivering thermogenesis in the rat. *J. Physiol.*, **283**, 569–84.

Bieber, L.L., Pettersson, B. and Lindberg, O. (1975) Studies of norepinephrine induced effects of free fatty acid from hamster brown adipose tissue cells. *Eur. J. Biochem.*, **58**, 375–81.

Bligh, J., Colhe, W.H. and Maskery, M. (1971) Influence of ambient temperature on the thermoregulatory responses to 5-HT, noradrenaline and acetylcholine injected into the lateral cerebral ventricles of sheep, goats and rabbits. *J. Physiol. (Lond.)*, **212**, 377–92.

Bligh, J., Smith, C.A. and Bauman, I. (1980) Central GABA and thermoregulation in sheep. in *4th Int. Symp. on Pharmacology of Thermogenesis* (eds B. Cox *et al.*), Karger, Basel, pp. 9–11.

Blix, A.S., Grav, H.J. and Ronald, K. (1979) Some aspects of temperature regulation in newborn harp seal pups. *Am. J. Physiol.*, **236**, R188–R197.

Borbely, A. and Leopfe-Hinnkannen, M. (1979) Phenothiazines. in *Body Temperature Regulation, Drug Effects and Therapeutic Implications.* (eds P. Lomax and E. Schonbaum), Marcel Dekker, New York and Basel, pp. 403–26.

Bray, G.A., Melvin, K.E. and Chopra, I. (1973) Effect of triiodothyronine on some metabolic responses of obese patients. *Am. J. Clin. Nutr.*, **26**, 715–21.

Breese, C.R., Taylor, T.D. and Prange, A.J. (1972) The effect of triiodothyronine on the disposition and actions of imipramine. *Psychopharmacologia*, **25**, 101–11.

Brobeck, J.R., Tepperman, J. and Long, C.N.H. (1943) Experimental hypothalamic hyperphagia in the albino rat. *Yale J. Biol. Med.*, **15**, 831–53.

Brown, J. (1962) Effects of 2-deoxyglucose on carbohydrate metabolism. *Metabolism*, **11**, 1098–112.

Bryant, B.J., McCalloch, M.W., Rand, M.J. and Story, D.F. (1975) Release of ³H-noradrenaline from guinea pig hypothalamic slices – effects of adrenceptor agonists and antagonists. *Br. J. Pharmacol.*, **53**, 454.

Bukowiecki, L., Folles, N., Paradis, A. and Collet, A. (1980) Stereospecific stimulation of brown adipocytes respiration by catecholamines via the β_1-adrenoreceptor. *Am. J. Phsyiol.*, **238**, E552–63.

Burks, T.F. (1972) Central alpha-adrenergic receptors in thermoregulation. *Neuropharmacology*, **11**, 615–24.

Burns, T.W. and Langley, P.E. (1970) Lipolysis by human adipose tissue. The role of cAMP and adrenergic receptor sites. *J. Lab. Clin. Med.*, **75**, 983–97.

Burns, J.W. and Langley, P.G. (1971) Differential effects of α- and β-adrenergic blocking agents on basal and stimulated lipolysis of human and rat isolated adipose tissue cells. *Pharmacol. Res. Commun.*, **3**, 271–7.

Caldwell, J., Sever, P.S. and Trelinski, M. (1974) On the mechanism of the hyperthermia induced by amphetamine in the rat. *J. Pharm. Pharmacol.*, **26**, 821–3.

Cannon, B., Sundin, U. and Ranert, L. (1977) Palmitoyl-CoA: a possible physiological regulator of nucleotide binding to brown adipose tissue mitochondria. *FEBS Lett.*, **74**, 43–6.

Carlisle, H.J. (1969) Effect of preoptic and anterior hypothalamic lesions on behavioural thermoregulation in the cold. *J. Comp. Physiol. Psychol.*, **69**, 391–402.

Chai, C.Y. and Lin, M.T. (1973) Effects of thermal stimulation of medulla oblongata and spinal cord of decerebrated rabbits. *J. Physiol. (Lond.)*, **234**, 409–19.

Chambers, B.N., Siegel, M.S. and Liu, C.N. (1974) Thermoregulatory responses of decerebrate and spinal cats. *Exp. Neurol.*, **42**, 282–99.

Cockburn, F., Hull, D. and Walton, I. (1968) The effect of lipolytic hormones and theophylline on heat production in brown adipose tissue *in vivo*. *Br. J. Pharmacol.*, **31**, 568–71.

Costa, E., Boullin, D.J., Hammer, W., Vogel, W. and Brodie, B.B. (1966) Interactions of drugs with adrenergic neurons. *Pharmacol. Rev.* **18**, 577–97.

Costa, E., Garattini, S. and Valzelli, L. (1960) Interactions between reserpine, chlorpromazine and imipramine. *Experientia*, **16**, 461–3.

Cowan, A. and Whittle, B.A. (1972) Inability of antidepressants and morphine to reverse reserpine-induced hypothermia in adrenalectomized mice. *Br. J. Pharmacol.*, **27**, 242–7.

Cox, B. and Tha, S.J. (1975) The role of dopamine and noradrenaline in temperature control of normal and reserpine pretreated mice. *J. Pharm. Pharmacol.*, **27**, 242–7.

Davidson, I.W.F., Salter, J.M. and Best, C.H. (1960) The effect of glucagon on the metabolic rate of rats. *Am. J. Clin. Nutr.*, **8**, 540–6.

Davis, T.R.A. and Mayer, J. (1955) Nature of the physiological stimulus for shivering. *Am. J. Physiol.*, **181**, 669–74.

Debous, A.F., Krimsky, I., From, A. and Cloutier, R.J. (1970) Site of action of gold thioglucose in the hypothalamic satiety centre. *Am. J. Physiol.*, **219**, 1397–402.

Debous, A.J., Krimsky, I., Likuski, H.J. From, A. and Cloutier, R.J. (1968) Gold thioglucose damage to the satiety centre: inhibition in diabetes. *Am. J. Physiol.*, **214**, 652–8.

de Castro, J.M., Raullin, S.K. and Dehugas, G.M. (1978) Insulin and glucagon as determinants of body weight set point and microregulation in rats. *J. Comp. Physiol.*, **92**, 571–9.

Derry, D.M., Schönbaum, E. and Steiner, G. (1969) Two sympathetic nerve supplies to brown adipose tissue of the rat. *Can. J. Physiol. Pharmacol.*, **47**, 57–63.

Desautels, M. and Himms-Hagen, J. (1981) Brown adipose tissue mitochondria of cold acclimated rats: Change in characteristics of purine nucleotide control of the protom electrochemical gradient. *Can. J. Biochem.*, **59**, 619–25.

Desautels, M., Zaror-Behrens, G. and Himms-Hagen, J. (1978) Increase purine nucleotide binding, altered polypeptide composition and thermogenesis in BAT mitochondria of cold acclimated rats. *Can. J. Biochem.*, **56**, 378–83.

Donaldson, A. (1980) Neuroendocrine mechanisms in obesity. Thesis submitted to Univeristy of London, p. 110.

Doniach, D. (1975) Possible stimulation of thermogenesis in brown adipose tissue by thyroid-stimulating hormone. *Lancet*, **ii**, 160–1.

Ellis, S. (1980) Effects on the metabolism in adrenergic activators and inhibitors part I. *Handbook Exp. Pharm.* Vol. 54. (ed. L. Szekeres), Springer Verlag.

Evans, E. and Miller, D.S. (1977) The effects of ephedrine on the oxygen consumption of fed and fasted subjects. *Proc. Nut. Soc.*, **36**, 136A.

Fain, J.N. (1973) Inhibition of adenosine cyclic 3′-5′ monophosphate accumulation in fat cells by adenosine, N^6-Phenyl isopropyl adenosine and related compounds. *Mol. Pharmacol.*, **9**, 595–604.

Fink, S.A. and Williams, J.A. (1976) Adrenergic receptors mediating depolarisation in brown adipose tissue. *Am. J. Physiol.*, **231**, 700–6.

Flaim, K.E., Horowitz, J.M. and Horowitz, B.A. (1976) Functional and anatomical characteristics of the nerve-brown adipose tissue interaction in rats. *Pflugers Arch.*, **365**, 9–14.

Flaim, K.E., Horwitz, B.A. and Horowitz, J.M. (1977) Coupling of signals to brown fat and β-adrenergic responses in intact rats. *Am. J. Physiol.*, **232**, R101–9.

Flatmark, T. and Pedersen, J.I. (1973) Brown adipose tissue mitochondria. *Biochim. Biophys. Acta*, **416**, 53–103.

Fletcher, A., Green, D., Rothwell, N.J., Stephens R.J., Stock, M.J. and Wyllie M.G. (1981) Ciclazindol increases metabolic rate in rats in the absence of CNS stimulation. *Br. J. Pharmac.*, **74**, 770P.

Fredholm, B.B. (1974) Vascular and metabolic effects of theophylline, dibutyryl cAMP and dibutyryl cGMP in canine subcutaneous adipose tissue *in situ*. *Acta Physiol. Scand.*, **90**, 226–36.

Fredholm, B.B. (1976) Release of adenosine-like material from isolated perfused dog adipose tissue following sympathetic nerve stimulation and its inhibition by adrenergic α-receptor blockade. *Acta Physiol. Scand.*, **96**, 422–30.

Fredholm, B.B. and Hedqvist, P. (1973) Role of pre- and post-junctional inhibition by prostaglandin E$_2$ of lipolysis induced by smpathetic nerve stimulation in dog, subcutaneous adipose tissue perfused *in situ*. *Br. J. Pharmacol.*, **47**, 711–8.

Fregly, M.J., Field, F.P., Nelson, E.L., Tyler, P.E. and Dasler, R. (1977) Effect of chronic exposure to cold on some responses to catecholamines. *J. Appl. Physiol.*, **42**, 349–354.

Freinkel, N., Metzger, B.E., Harris E., Robinson, S. and Mager, M. (1972) The hypothermia of hypoglycaemia. *New Engl. J. Med.*, **287**, 841–5.

Freinkel, N., Singer, D.L., Arky, R.A. *et al.* (1963) Carbohydrate metabolism of patients with clinical alcohol hypoglycaemia and the experimental reproduction of the syndrome with pure ethanol. *J. Clin. Invest.*, **42**, 1112–33.

Frens, J. (1980) Neurotransmitter mapping in central thermoregulation. in *4th Int. Symp. on Pharmacol. of Thermoregulation* (eds B. Cox *et al.*), Karger, Basel, pp. 1–5.

Frey, H.H. (1975) Hyperthermia induced by amphetamine, *p*-chloroamphetamine, and fenfluramine in the rat. *Pharmacology*, **13**, 163–76.

Fuller, K.E., Horowitz, J.M. and Horwitz, B.A. (1977) Spinal cord thermo-sensitivity and sorting of neural signals in cold exposed rats. *J. Appl. Physiol.*, **42**, 154–8.

Fuxe, K. and Hokfelt, T. (1969) Catecholamines in the hypothalamus and the pituitary gland. in *Frontiers of Neuroendocrinology* (eds W.F. Ganong and L. Martini), Oxford, New York, pp. 47–96.

Gale, C.C. (1973) Neuroendocrine aspects of thermoregulation. *Ann. Rev. Physiol.*, **35**, 391–430.

Gander, D.R. (1965) Treatment of depressive illnesses with combined anti-depressants. *Lancet*, **ii**, 107.

Garcia Sainz, J.A., Hasler, A.K. and Fain, J.N. (1980) α$_1$-adrenergic activation of phosphatidyl inositol labelling in isolated brown fat cells. *Biochem. Pharmacol.*, **29**, 3330–3.

Garattini, S. and Jori, A. (1967) Interactions between imipramine-like drugs and reserpine on body temperature. in *Antidepressant drugs* (eds S. Garattini and M.N.G. Dukes), Excepta Medica, Amsterdam, pp. 179–93.

Garattini, S. and Jori, A. (1979) Tricyclic antidepressant drugs. in *Body Temperature regulation, Drug Effects and Therapeutic Implications*. (eds P. Lomax and E. Schonbaum) Marcel Dekker, New York and Basal, pp. 439–59.

Ghose, K., Rama Rae, V.A., Bailey, J. and Coppen, A. (1978) Anti-depressant activity and pharmacological interactions of ciclazindol. *Psychopharmacol.*, **57**, 109–14.

Goldberg, R., Joffe, B.I., Bersohn, I., Vau, A.S.M., Kout, L. and Seflet, H.C. (1975) Metabolic responses to selective β-adrenergic stimulation in man. *Postgrad. Med. J.*, **51**, 53–8.

Gory, S.N.C. and Rogers, K.J. (1971) Role of brain monoamines in the fatal hyperthermia induced by peltidine or imipramine in rabbits pretreated with pargyline. *Br. J. Pharmacol.*, **42**, 646P.

Greenbaum, R. and Harry, T.V.A. (1980) Ciclazindol, an adjunct to weight control. *Pharmacotherapy*, **3**, 82–3.

Grossman, S.P. (1962) Direct adrenergic and cholinergic stimulation of hypothalamic mechanisms. *Am. J. Physiol.*, **202**, 872–82.

Hahn, P., Skala, J. and Davies, P. (1971) Carnitine enhances the effect of norepinephrine on oxygen consumption in rats and mice. *Can. J. Physiol. Pharmac.*, **49**, 853–5.

Hajors, G.T. and Garattini, S. (1973) A note on the effect of (+)- and (−)-amphetamine on lipid metabolism. *J. Pharm. Pharmacol.*, **25**, 418–9.

Hansen, A.P. (1973) Serum growth hormone response to exercise in non-obese subjects. *Scand. J. Clin. Lab. Invest.*, **31**, 175–8.

Harrison, J.W.E., Ambrus, C.M. and Ambrus, J.L. (1952) Tolerance of rats towards amphetamine and metamphetamine. *J. Am. Pharm. Assoc.*, **41**, 539–41.

Heaton, G.M. and Nicholls, D.G. (1976) Hamster brown adipose tissue mitochondria. *Eur. J. Biochem.*, **67**, 511–7.

Hedqvist, P. (1969) Modulating effect of prostaglandin E₂ on noradrenaline release from the isolated cat spleen. *Acta Physiol. Scand.*, **25**, 511–2.

Heim, T. and Hull, D. (1966) The effects of propranolol on the calorigenic response in brown adipose tissue of newborn rabbits to catecholamines, glucagon, corticotrophin and cold exposure. *J. Physiol.*, **187**, 271–83.

Hemingway, A. (1963) Shivering. *Physiol. Rev.*, **43**, 397–422.

Hesse, V., Spahn, U. and Plenert, W. (1981) T_4-T_3 shift during the post prandial period in obese children before and after a hypocaloric diet, one factor for post-prandial thermogenesis. *Horm. Met. Res.*, **13**, 28–33.

Himms-Hagen, J. (1970) Adrenergic receptors for metabolic responses in adipose tissue. *Fed. Proc.*, **29**, 1388–401.

Himms-Hagen, J. (1975) Role of the adrenal medulla in adaptation to cold. in *Handbook of Physiology. Endocrinology* VI, 637–65.

Hinman, D.J. and Griffith, D.R. (1973) Effects of ventromedial hypothalamic lesions on thyroid secretion in rats. *Horm. Met. Res.*, **5**, 48–50.

Hittelman, K.J., Bertin, R, and Butcher, R.S. (1974) Cyclic AMP metabolism in brown adipocytes of hamsters exposed to different temperatures. *Biochim. Biophys. Acta*, **338**, 348–407.

Hofman, G.G. Schneider, G., Strohmeier, E., Pickardt, C.R. and Krick, L. (1975). Thyroid hormones in obesity: thyroid function, effects on isolated fat cells of man and therapeutic application. in *Recent Advances in Obesity*, Research I, (ed. A. Howard), Newman Publishing, London, pp. 383–6.

Horwitz, B.A. and Hanes, G.E. (1976) Propranolol and pyrogen effects on shivering and non-shivering thermogenesis in rats. *Am. J. Physiol.*, **230**, 637–42.

Hull, D. and Segall, M.M. (1965) Sympathetic nervous control of brown adipose tissue and heat production in the newborn rabbit. *J. Physiol. (Lond.)*, **181**, 458–67.

Hustveldt, B.E. and Loro, A. (1972) Correlation between hyperinsulinaemia and hyperphagia in rats with VMH lesions. *Acta Physiol. Scand.*, **84**, 29–33.

Iliano, G. and Cuatrecasas, P. (1971) Exogenous prostaglandins modulate lipolytic processes in adipose tissue. *Nature*, **234**, 72–4.

Inoue, S.E. and Bray, G.A. (1977) The effect of subdiaphragmatic vagotomy in rats with VMH obesity. *Endocrinology*, **100**, 108–14.

Inoue, S., Bray, G.A. and Mullen, Y.S. (1977) Effect of transplantation of pancreas on the development of hypothalamic obesity. *Nature*, **266**, 742–4.

James, W.P.T., Haralds dottir, J. Liddell, F., Jung, R.T. and Shetty, P.S. (1981) Autonomic responsiveness in obesity with and without hypertension. *Int. J. Obesity.*, **5**, Suppl. 1, 73–8.

Jansky, L., Bartunkova, R. and Ziesberger, E. (1967) Acclimation of the white rat to cold. *Physiol. Bohemoslov.*, **16**, 366–72.

Jellinek, P. (1971) Dual effect of dexamphetamine on body temperature in the rat. *Eur. J. Pharmacol.*, **15**, 389–92.

Jessen, C. and Ludwig, O. (1971) Spinal cord and hypothalamus as core sensors of temperature in the conscious dog. II Addition of signals. *Pflugers Arch.*, **324**, 205–16.

Jessen, C. and Simon, E. (1971) Spinal cord and hypothalamus as core sensors of temperature in the conscious dog. II Identity of functions. *Pflugers Arch.*, **324**, 217–26.

Joel, C.D. (1966) Stimulation of metabolism of rat brown adipose tissue by addition of lipolytic hormones *in vitro. J. Biol. Chem.*, **241**, 814–21.

Jordan, S.C. and Hampson, F. (1960) Amphetamine poisoning associated with hyperpyrexia. *Br. Med. J.*, **ii**, 844.

Jori, A. and Bernardi, D. (1970). Importance of catecholamines for the interaction between reserpine and desipramine on body temperature in rats. *Pharmacology*, **4**, 235–41.

Jori, A., Carrara, M.C. and Garattini, S. (1966) Importance of noradrenaline synthesis for the interaction between desipramine and reserpine. *J. Pharm. Pharmacol.*, **18**, 619–20.

Jori, A. and Garattini, S. (1965) Interaction between imipramine-like agents and catecholamine induced hyperthermia. *J. Pharm. Pharmacol.*, **17**, 480–8.

Jori, A., Paglialunga, S. and Garattini, S. (1966) Adrenergic mediation in the antagonism between desipramine and reserpine. *J. Pharm. Pharmacol.*, **18**, 326–7.

Jung, R., Doupe, J. and Carmichael, E.A. (1937) Shivering, a clinical study of the influence of sensation. *Brain*, **60**, 20–8.

Jung, R.T., Shetty, P.S. and James, W.P.T. (1979) Reduced thermogenesis in obesity. *Nature*, **279**, 322–3.

Jung, R.T., Shetty, P.S. and James, W.P.T. (1980) Heparin, free fatty acids and an increased metabolic demand for oxygen. *Postgrad. Med. J.*, **56**, 330–2.

Jung, R.T., Shetty, P.S. and James, W.P.T. (1981) The effect of β-adrenergic blockade on metabolic rate and peripheral thyroid metabolism in obesity. *Eur. J. Clin. Invest.*, **10**, 179–82.

Jung, R.T., Shetty, P.S., James, W.P.T., Barrand, M.A. and Callingham, B.A. (1981) Caffeine: its effects on catecholamines and metabolism in lean and obese humans. *Clin. Sci.*, **60**, 527–35.

Kempson. S., Mannetti. G.V. and Shaw. A. (1978) Hormone action at the membrane level VII Stimulation of dihydroalprenolol binding to β-adrenergic receptors in the rat heart by triiodothyronine and thyroxine. *Biochim. Biophys. Acta*, **540**, 320–9.

Khosla. T. and Lowe. C.R. (1971) Obesity and smoking habits. *Br. Med. J.*, **4**, 10–13.

Kirkpatrick. W.E. and Lomax, P. (1967) The effect of atropine on the body temperature of the rat following systemic and intracerebral injection. *Life Sci.*, **6**, 2273–8.

Kiser. J.C., Booher, W.T. and Watts, D.T. (1955) Blood epinephrine levels in dogs following the intravenous administration of nicotine sulphate. *Fed. Proc.*, **14**, 358.

Klajner, E.. Herman, F.P., Polivy, J. and Chhabra, R. (1981) Human obesity dieting and anticipatory salivation to food. *Physiol. Behav.*, **27**, 195–8.

Knight, B.L. (1974) Adenosine 3'-5'-cyclic phosphate lipolysis and oxygen consumption in BAT from newborn rabbits. *Biochim. Biophys. Acta*, **343**, 287–96.

Kopelman, P.G., White, N. Pilkington, T.R.E. and Jeffcoate, S.L. (1979) Impaired hypothalamic control of prolactin secretion in massive obesity. *Lancet*, **i**, 747–9,

Kuroshima, A, Doi, K., Yahata, T., Kurahashi, M. and Ohno, T. (1981) Glucagon and temperature acclimation. in *Satelite 28th Int. Cong. of Physiol. Sci., Pecs.* (eds Z. Szdengi and M. Szekaly), Pergamon Press, New York, pp. 305–7.

Kuroshima, A. and Yahata, T. (1979) Thermogenic responses of brown adipocytes to noradrenaline and glucagon in heat-acclimated and cold-acclimated rats. *Jpn. J. Physiol.*, **29**, 683–90.

Landsberg, L. and Young, J.B. (1981) Diet-induced changes in sympathoadrenal activity – implications for thermogenesis. *Life Sci.*, **28**, 1801–19.

Langer, S.Z. (1977) Presynaptic receptors and their role in regulation of neurotransmitter release. *Br. J. Pharmacol.*, **60**, 481–97.

Langer, S.Z., Adler, E., Energo, A. and Stegano, F.J.E. (1971) The role of the α-receptor in regulating noradrenaline overflow by nerve stimulation. *Proc. 25th Intern. Cong. Physiol. Sci.* p. 335.

Latham, A., Rothwell, N.J., Stock, M.J., White, A.C. and Wyllie, M.G. (1981) Na^+K^+ATPase activity in brown adipose tissue its relationship to resting metabolic rate and the effects of ciclazindol. *Br. J. Pharmac.*, **73**, 182P.

Leblanc, J. and Villemaire, A. (1970) Thyroxine and noradrenaline sensitivity, cold resistance and brown fat. *Am. J. Physiol.*, **218**, 1742–5.

Lee. W.C. and Shiderman, F.E. (1959) Mechanism of the positive inotropic response to certain ganglion stimulants. *J. Pharmac. Exp. Ther.*, **126**, 239–49.

Liebowitz, S.F. (1970) Hypothalamic β-adrenergic satiety system antagonises an α-adrenergic hunger system in the rat. *Nature*, **226**, 963–4.

Lin. M.T., Chandra, A., Chera. Y.F. and Tsay, B.L. (1980) Intracerebrovascular injection of sympathomimetic drugs inhibits both heat production and heat loss mechanisms in the rat. *Can. J. Physiol. Pharmacol.*, **58**, 896–902.

Linck. G., Stoeckel, M.E.. Porte, A. and Petrovic, A. (1973) An electron microscope study of the specialised cell contacts and innervation of adipocytes in the brown fat of the European hamster. *Cytobiology.*, **7**, 431–6.

Loffenholz. V.K. and Muscholl, E. (1969) A muscarinic inhibition of the noradrenaline release evoked by postganglionic nerve stimulation. *Naunyn-Schm. Arch. Pharm.*, **265**, 1–15.

Lomax, P. and Schonbaum, E. (1979) Body temperature regulation, drug effects and therapeutic implications. Marcel Dekker, New York and Basel, pp. 183–336.

Londono, H.J., Gallacher, T.F. and Bray, G.A. (1969) Effect of weight reduction, triiodothronine and diethyl-stilboestrol on growth hormone in obesity. *Metabolism*, **18**, 986–92.

Mantegazza, P., Miller, E.E., Naimzada, M.K. and Riva, M. (1970) Studies on the lack of correlation between hyperthermia, hyperactivity and anorexia induced by amphetamine. in *Amphetamines and Related Compounds* (eds E. Costa and S. Garattini), Raven Press, New York.

Massoudi, M. and Miller, D.S. (1977) Ephedrine, a thermogenic and potential slimming drug. *Proc. Nut. Soc.*, **36**, 135A.

Matsumoto, C. and Griffin, W. (1971) Antagonism of (+) -amphetamine induced hyperthermia in rats by pimozide. *J. Pharm. Pharmacol.*, **23**, 710.

Matsumoto, C. and Shaw, M.M. (1971) The involvement of plasma free fatty acids in (+) amphetamine induced hyperthermia in rats. *J. Pharm. Pharmacol.*, **23**, 387–8.

Mejsnar, J. and Jirak, E. (1981) Thermogenic response as a function of extravascular influx of infused noradrenaline. *Experientia*, **37**, 482.

Metcalf, G. (1974) TRH: a possible mediator of thermoregulation. *Nature*, **252**, 310–1.

Metcalf, M., Dettmar, P. and Watson, T. (1980) The role of neuropeptides in thermoregulation. in *4th Int. Symp. on Pharmacology of Thermoregulation* (ed. B. Cox *et al.*), Karger, Basel, pp. 175–9.

Metcalf, G. and Myers, R.D. (1978) Precise location within the preoptic area where noradrenaline produces hyperthermia. *Eur. J. Pharmacol.*, **51**, 47–53.

Miller, D.S., Stock, M.J. and Stuart, J.A. (1974) Effects of caffeine and carnitine on the oxygen consumption of fed and fasted subjects. *Proc. Nutr. Soc.*, **33**, 28A–29A.

Mishizawa, Y. and Bray, G.A. (1978) Ventromedial hypothalamic lesions and the mobilisation of fatty acids. *J. Clin. Invest.*, **61**, 714–21.

Moore, R. Mekrishi, J.N., Verdoorn, C. and Mills, I.H. (1981) The role of T_3 and its receptor in efficient metabolisers receiving very low caloric diets. *Int. J. Obesity*, **5**, 283–6.

Mopurgo, C. and Theobald, W. (1965) Influence of imipramine-like compounds and chlorpromazine on the reserpine-hypothermia in mice and the amphetamine hyperthermia in rats. *Med. Pharmacol. Exp.*, **13**, 226–32.

Moravec, J. and Vyskocil, F. (1976) Neuromuscular transmission in a hibernator. in *Regulation of Depressed Metabolism and Thermogenesis* (eds L. Jansky and X.J. Musacchia), Thomas, Springfield, pp. 81–92.

Nedergaard, J. and Lindberg, O. (1979) Norepinephrine-stimulated fatty acid release and oxygen consumption in isolated hamster brown fat cells. *Eur. J. Biochem.*, **95**, 139–45.

Nicholls, D.G. (1979) Brown adipose tissue mitochondria. *Biochim. Biophys. Acta*, **549**, 1–29.

Oberman, Z., Harell, A., Herzberg, M., Hoerer, E., Jakolska, H. and Laurian, L. (1975) Changes in plasma glucose and free fatty acids after caffeine ingestion in obese women. *Israeli J. Med. Sci.*, **11**, 33–6.

Oliver, M.F. (1972) Metabolic response during impending myocardial infarction – clinical implications. *Circulation*, **45**, 491–500.

Panksepp, J. (1974) Hypothalamic regulation of energy balance and feeding behaviour. *Fed. Proc.*, **33**, 1150–65.

Perkins, M.N., Rothwell, N.J., Stock, M.J. and Stone, T.W. (1981) Activation of brown adipose tissue thermogenesis by the ventromedial hypothalamus. *Nature*, **289**, 401–2.

Pettersson, B. and Vallin, I. (1976) Norepinephrine shift in levels of adenosine 3'-5' monophosphate and ATP parallel to increased respiratory rates and lipolysis in isolated hamster brown fat cells. *Eur. J. Biochem.*, **62**, 383–90.

Pittet, P., Gygax, P.H. and Jecquier, E. (1974) Thermic effect of glucose and amino acids in man, studied by direct and indirect calorimetry. *Brit. J. Nutr.*, **31**, 343.

Prusiner, S.B., Cannon, B. and Lindberg, O. (1968) Oxidative metabolism in cells isolated from brown adipose tissue. *Eur. J. Biochem.*, **6**, 15–22.

Rabin, B.M. (1974) Independence of food intake and obesity following ventromedial hypothalamic lesions in the rat. *Physiol. Behav.*, **13**, 769–72.

Rathburn, R.C. and Slater, I.H. (1963) Amitriptyline and nortriptyline as antagonists of central and peripheral cholinergic activation. *Psychopharmacologia*, **4**, 114–25.

Reaves, T. and Hayward, J.N. (1979) Hypothalamic and extrahypothalamic thermoregulatory centres. in *Modern Pharmacology, Toxicology*, Vol. 16. (eds P. Lomax and E. Schonbaum), Marcel Dekker, New York and Basel pp. 39–70.

Reichlin, S., Bollinger, J., Nejad, I. and Sullivan, P. (1973) Tissue thyroid concentration of rat and man determined by radioimmunoassay – biological significance. *Sinai J. Med.*, **40**, 502–10.

Riviere, J., Roger, P., Nogue, F., Emperaire, C. and Marapre, J.L. (1975) Thyroid metabolism in obesity. in *Recent Advances in Obesity Research* I, (ed. A. Howard), Newman Publishing, London, pp. 103–6.

Robertson, D., Frolich, J.C., Carr, R.K., Watson, J.T., Hollinfield, J.W., Shand, D.G. and Oates, J.A. (1978) Effects of caffeine on plasma renin activity, catecholamines and blood pressure. *New Engl. J. Med.*, **298**, 181–6.

Rothwell, N.J., Saville, M.E. and Stock, M.J. (1982) Metabolic responses to food atropine and 2-deoxyglucose in Zucker rats. *Proc. Nutr. Soc.*, **41**, 37A.

Rothwell, N.J. and Stock, M.J. (1979) A role for brown adipose tissue in diet induced thermogenesis. *Nature*, **281**, 31–3.

Rothwell, N.J. and Stock, M.J. (1981) Role for insulin in dietary induced thermogenesis. *Metabolism*, **30**, 673–8.

Rothwell, N.J., Stock, M.J. and Stribling, D. (1982) Diet induced thermogenesis. *Pharmacol. Therap.*, **17**, 251–68.

Rothwell, N.J., Stock, M.J. and Wyllie, M.G. (1981*b*) Sympathetic mechanisms in diet induced thermogenesis. Modification by ciclazindol and anorectic drugs. *Br. J. Pharm.*, **74**, 539–46.

Rowe, J.W., Young, J.B., Minaker, K.L., Stevens, A.L., Pallotta, J. and Landsberg, L. (1981) Effect of insulin and glucose infusions on sympathetic nervous system activity in normal man. *Diabetes*, **30**, 219–25.

Rudy, J.A. and Wolf, H.H. (1971) The effect of intrahypothalamically injected sympathomimetic amines on temperature regulation in the cat. *J. Pharmacol. Exp. Ther.*, **179**, 218–35.

Salter, J.M. (1960) Metabolic effects of glucagon on the Wistar rat. *Am. J. Clin. Nutr.*, **8**, 535–9.

Sauter, A., Veta, K., Engel, J. and Goldstein, M. (1981) Effects of insulin on the release and turnover of dopamine, noradrenaline and adrenaline in the rat brain. *Experientia*, **37**, 631.

Saxena, P.N. (1973) Mechanism of hypothermic action of catecholamines in the cat. *Ind. J. Pharmacol.*, **5**, 3–6.

Seitz, H.J., Krone, W., Wilke, H. and Tarnowski, W. (1981) Rapid rise in plasma glucagon induced in acute cold exposure in man and rat. *Pflugers Arch.*, **389**, 115–210.

Seydoux, J., Giacobino, J.P. and Girardier, L. (1980) Influence of thyroid status on β-receptor affinity of brown adipose tissue. *Experientia*, **36**, 700.

Seydoux, J., Rohner-Jeanrenaud, F., Assimacopoulos-Jeannet, F., Jeanrenaud, B. and Girardier, L. (1981) Functional disconnection of brown adipose tissue on hypothalamic obesity in rats. *Pflugers Arch.*, **390**, 1–4.

Shaw, W.N., Schmiegel, K.K., Yen, T.T., Toomey, R.E., Meyers, D.B. and Mills, J. (1981) LY79771. A novel compound for weight control. *Life Sci.*, **29**, 2091–101.

Shetty, P.J., Jung, R.T. and James, W.P.T. (1979) Effect of catecholamine replacement with levo-dopa on the metabolic response to semistarvation. *Lancet*, **i**, 77–9.

Shimazu, T. and Amakawa, A. (1968) Regulation of glycogen metabolism in liver by the autonomic nervous system II. Neural control of glycogenolytic enzymes. *Biochim. Biophys. Acta*, **165**, 335–48.

Shimazu, T. and Takahashi, A. (1980) Stimulation of hypothalamic nuclei has differential effects on lipid synthesis in brown and white adipose tissue. *Nature*, **284**, 62–3.

Shiraishi, T. and Mager, M. (1980) Hypothermia following injection of 2-deoxy-D-glucose into selected hypothalamic sites. *Am. J. Physiol.*, **239**, R265–9.

Sims, E.A.H., Darnforth, E., Horton, E.S., Bray, G.A., Glennon, J.A. and Salams, L.B. (1973) Endocrine and metabolic effects of experimental obesity in man. *Rec. Prog. Horm. Res.*, **29**, 457–96.

Skala, J.P. and Knight, B.L. (1977) Protein kinases in brown adipose tissue of developing rats. *J. Biol. Chem.*, **252**, 1064–70.

Stanton, H.C. (1972) Selective metabolic and cardiovascular β-receptor antagonism in the rat. *Arch. Int. Pharmcodyn. Ther.*, **196**, 246–58.

Starke, K. (1971) Influence of α-receptor stimulants on noradrenaline release. *Naturwissenschaften*, **58**, 420.

Stirling, J.L. and Stock, M.J. (1968) Metabolic origins of thermogenesis induced by diet. *Nature*, **220**, 801–2.

Strubelt, D. (1969) The influence of reserpine, propranolol and adrenal medullectomy on the hyperglycaemic actions of theophylline and caffeine. *Arch. Int. Pharmacodynam. Ther.*, **179**, 215–24.

Stuart, D.G., Eldred, E., Hemingway, A. and Kawamura, Y. (1963) Neural regulation of the rhythm of shivering. in *Temperature – its Measurement and Control in Science and Industry*, Vol. 3. pp. 545–57.

Su, C. and Bevan, J.A. (1970) Blockade of the nicotine induced norepinephrine release by cocaine, phenoxybenzamine and desipramine. *J. Pharmacol. Exp. Therap.*, **175**, 533–40.

Sugden, R.F. (1974) Action of indoramin and other compounds on the uptake of neurotransmitters into rat cortical slices. *Br. J. Pharmacol.*, **51**, 467–72.

Sulser. F.. Owens. M.L. and Dingle. J.V. (1966) On the mechanisms of amphetamine potentiation by desimipramine (DMI). *Life Sci.*, **5**, 2005–10.

Sulser. F.. Owens. M.L. Strada. S.J. and Dingle J.V. (1969) Modification by desipramine of the availability of norepinephrine released by reserpine in the hypothalamus of the rat *in vivo. J. Pharmacol. Exp. Ther.*, **158**, 272–82.

Svartengren. J.. Mohell, N. and Cannon. B. (1980) Characterisation of ^{3}H-dihydroergocrytine binding sites in brown adipose tissue. Evidence for the presence of α-receptors. *Acta Chem. Scand.*, B.. **34**, 231–2.

Svoboda. P.. Svartengren. J.. Snochowski. M.. Houstek. J. and Cannon. B. (1979) High number of affinity sites for ^{3}H-dihydrolalprenolol in isolated hamster brown fat cells. *Eur. J. Biochem.*, **102**, 203–10.

Teskey. N. Horwitz, B.A. and Horowitz. J.M. (1975) Norepinephrine induced depolarisation of skeletal muscle cells. *Eur. J. Pharmacol.*, **30**, 252–5.

Vander Tuig, J.G.. Flynn. A.M.. Rosmos, D.R. and Leveille. G.A. (1980) Energy retention and maintenance energy requirements of young rats with VMH lesions. *Fed. Proc.*, **39**, 887.

Westfall. T.C. (1977) Local regulation of adrenergic neurotransmission. *Physiol. Rev.*, **57**, 659–728.

Wheeler, H. and Stribling, D. (1983) Pharmacology of thermogenesis in the rat, assessment by means of interscapular brown adipose tissue temperature. *Br. J. Pharmacol.*, (in press).

Winston. F. and McCann. M.H. (1972) Antidepressants drugs and excessive weight gain. *Br. J. Psychiat.*, **120**, 693–7.

Wunnenberg. W. and Bruck. K. (1970) Studies on the ascending pathways from the thermosensitive region of the spinal cord. *Pflugers Arch.*, **321**, 233–41.

Yahata. Y.. Ohno,T. and Kuroshima. A. (1981) Improved cold tolerance in glucagon treated rats. *Life Sci.*, **28**, 2603–10.

Yehuda. S. and Wurtman. R.J. (1972) Release of brain dopamine as the probable mechanism for the hypothermic effect of D-amphetamine. *Life Sci.*, **11**, 851–9.

Young. J.B. and Landsberg. L. (1979) Gold thioglucose treatment selectively disrupts dietary regulation of sympathetic activity in the mouse. *Clin. Res.*, **27**, A380.

Index